CHROMOSOMES: ORGANIZATION AND FUNCTION

D1477031

Chromosomes

Organization and Function

Adrian T. Sumner

North Berwick, United Kingdom

Blackwell
Publishing

© 2003 by Blackwell Science Ltd
a Blackwell Publishing company

350 Main Street, Malden, MA 02148-5018, USA
108 Cowley Road, Oxford OX4 1JF, UK
550 Swanston Street, Carlton South, Melbourne, Victoria 3053, Australia
Kurfürstendamm 57, 10707 Berlin, Germany

First published 2003

Library of Congress Cataloging-in-Publication Data

Sumner, A.T. (Adrian Thomas), 1940–
    Chromosomes: organization and function/Adrian T. Sumner.
            p.   cm.
    Includes bibliographical references and index.
    ISBN 0-632-05407-7 (pbk.: alk. paper)
        1. Chromosomes.   I. Title.
QH600.S863 2003
572.8'7 – dc21

                                                                        2002066646

A catalogue record for this title is available from the British Library.

Set in 9.5/12 pt Bembo
by SNP Best-set Typesetter Ltd., Hong Kong
Printed and bound in the United Kingdom
by MPG Books Ltd, Bodmin, Cornwall

For further information on
Blackwell Publishing, visit our website:
http://www.blackwellpublishing.com

Contents

Preface

Several years ago, with the tidal wave of molecular biology threatening to engulf and obliterate the rest of biology, it might have seemed that the study of chromosomes was something to be left to a few old-fashioned scientists to occupy them harmlessly until they retired. In fact, nothing could be further from the truth, and recently there has been an upsurge in chromosome studies, stimulated by these advances in molecular biology but accompanied by the realization that the arrangement of biological molecules could not, on its own, explain all biological phenomena. In fact, it has long been known that the behaviour of chromosomes at mitosis and meiosis determines the nature of inheritance, and it is becoming clear that the disposition of chromosomes in interphase nuclei is also important for their functioning. Many chromosomal substructures such as heterochromatin, nucleoli, centromeres and telomeres are being studied intensively, as well as chromosomal phenomena such as imprinting. With the immense reduction in mortality from infectious disease in Western societies, genetic diseases have become much more significant, and many of these, including spontaneous abortions and cancer, are the result of chromosomal defects. These and other chromosomal topics are covered in this book, which is aimed at advanced undergraduate and postgraduate students who, it is assumed, will have a basic knowledge of chromosomes such as can be gleaned from many excellent genetics and cell biology textbooks.

Each chapter can be read in isolation, but in reality no single topic is isolated from any other, and I have cross-referenced the text quite heavily to guide the reader to further, related information. I have also included a substantial amount of tabular material, which I believe is the most satisfactory way of dealing with the vast amount of data now available on some topics. We are supposed to be living in an electronic age, and where appropriate I have referred to websites, but only when they supplement or complement the material in this book. Access to additional chromosomal websites can be obtained through www.chromosome.net/index.htm

This book could never have been written without the help of numerous scientists who not only spared the time to discuss various points and to send me numerous reprints of their work but, perhaps more importantly, offered their encouragement, and convinced me that this book would really meet a need. I am also very grateful for access to the library at the MRC Human Genetics Unit in Edinburgh. Many people have generously supplied illustrations for the book and, although they are acknowledged individually in the figure legends, I should like to thank them again here. The study of chromosomes includes strong visual and aesthetic elements as well as scientific aspects, and no book on chromosomes could be produced without being generously illustrated. I hope that the result will not merely describe the state of chromosomology at the beginning of the twenty-first century but also, by highlighting lacunae in our knowledge, stimulate further research into chromosomes.

Adrian T. Sumner
North Berwick
January 2002

Why study chromosomes?

1.1 Early studies of chromosomes

The idea of chromosomes only appeared in the last quarter of the nineteenth century. The first scientist to describe clearly the process of mitosis and the involvement of the 'chromatic nuclear figure' (i.e. chromosomes) was apparently the German zoologist Anton Schneider in 1873 (Zacharias, 2001). Before then, it was thought that cells and nuclei simply pinched in half to divide. Clear and detailed descriptions of mitotic chromosomes in plants and animals were published by Strasburger in 1875 and by Flemming in 1879–1882, respectively. Their work formed the foundation of modern studies of chromosomes. Flemming's work is readily available in English translation (Flemming, 1965) and is worth looking at for the clarity of his descriptions of mitosis (a process to which he gave the name), which can hardly be bettered today. For an account of the life and work of Walther Flemming, see Paweletz (2001). Flemming also discovered lampbrush chromosomes, so named by Rückert in 1892 (Chapter 14), and, also in the early 1880s, Balbiani discovered polytene chromosomes (Chapter 15). However, the term chromosome was not introduced until 1888 by Waldeyer, an anatomy professor in Germany (Zacharias, 2001).

Whatever their function, chromosomes inevitably became popular subjects for study, being conspicuous cellular organelles with considerable aesthetic attraction (a consideration that still draws people to them today). However, from the very earliest studies of chromosomes it became clear that chromosomes were involved in inheritance, so that as early as 1887 Weismann could put forward his chromosome theory of inheritance (Darlington, 1966). This included the following points:

1 The nuclear substance controls the form and function of every cell, and divides at mitosis to give equal products.
2 Eggs must lose half their nuclear substance in the polar body before fertilization, and this must be replaced exactly by the nuclear substance of the sperm.
3 Because sexual reproduction depends on adding together the egg and sperm nuclei in every generation, there must be a halving of the nuclear substance in both male and female germ cells. (This proposition was made before the process of meiosis had been discovered.)
4 There are no essential differences between the nuclear substance of eggs and sperm.
5 Sexual reproduction is a means of producing variability between individuals, on which natural selection can act.

Weismann's theory, which has proved to be true in all its principles, was, of course, formulated in ignorance of Mendelian genetics.

1.2 The origin of genetics, and the chromosome theory of inheritance

The story of the discovery of the principles of genetics by Gregor Mendel, their publication in

1866, their neglect for nearly 35 years and their rediscovery in 1900 is too well known to need repetition here. It is, however, worth remembering that Mendel worked out his laws in complete ignorance of the physical mechanisms that might be involved. Indeed, chromosomes had not yet been discovered when he did his famous work. However, once Mendel's Laws had been rediscovered, it was quickly realized that the behaviour of chromosomes at cell division was exactly what was required to explain the distribution of the hereditary factors (the genes) to daughter cells and organisms. The Chromosome Theory of Heredity explained many other features of inheritance that were discovered in the few years following the rediscovery of Mendel's work. Although there was no idea, a century ago, of how many genes an organism might have, it was clear that there would be many more than one per chromosome, thus providing a physical basis for the genetic phenomenon of linkage. The discovery in some organisms of an unpaired chromosome, named X because its nature was uncertain, led to the recognition of sex chromosomes and their involvement in sex determination (see Chapter 8), and to an explanation of sex linkage when that was discovered shortly afterwards. Not many years after linkage had been discovered, and associated with the presence of several genes on the same chromosome, it was found that linkage was not necessarily complete. The explanation for this was found in detailed studies of meiosis, in which physical crossing-over of homologous chromosomes could be seen, which would break up the genetic linkage previously observed. The intimate relationship between genetical phenomena and the physical behaviour of chromosomes was thus well established in the early years of the twentieth century, even though no-one had any clear idea of the nature of a gene at that time.

1.3 The chemical nature of genes and chromosomes

Back in 1868, Miescher in Basel isolated what he called nuclein, which was apparently an impure form of DNA. This was the beginning of the study of nucleic acids. Much of the work on nucleic acids during the first three-quarters of the twentieth century was concerned with their chemistry. Nevertheless, as early as the 1880s, Flemming had speculated that chromatin, 'the colourable substance of the nucleus' (and therefore of chromosomes), might be the same as the nuclein recently isolated by Miescher. The presence of DNA in chromosomes and nuclei was established unequivocally when Feulgen introduced his histochemical method for DNA (Feulgen & Rossenbeck, 1924), which later provided a means for measuring the DNA content of nuclei and chromosomes.

Miescher's work on the chemical composition of nuclei was continued by various workers, including Kossel, who discovered histones, although it became clear that the nuclear proteins were more diverse and complex than the histones alone. None of these studies, however, gave any clear indication of what substance the genes were made of. In fact, for many years it was held that genes were most likely to be made of protein, because the proteins were thought to be much more complex than the DNA. This was due to limited information about the composition of DNA, which was thought to consist of tetranucleotides, containing one of each of the four bases. Such a structure would lack the variety required for the many different genes that were known to exist by then (the 1940s).

Subsequently the work of Chargaff showed that the four nucleotides were not present in equimolar proportions. The DNA sequences could therefore be much more variable, which would be compatible with a function as genes. More important, perhaps, was the finding by Avery *et al.* (1944) that the substance responsible for bacterial transformation was in fact DNA. The model of DNA proposed by Watson and Crick in 1953, showing that it was a complementary double helix, provided the basis of the mechanisms for replication of the genetic material. The subsequent elucidation of the mechanisms of transcription of DNA into messenger RNA (mRNA) and its translation into proteins confirmed the position of DNA as the substance of the genes.

1.4 The position of chromosomes in an age of molecular biology

The Watson and Crick model for the structure of DNA might be regarded as the beginning of the era of molecular biology. In the 1970s, when writing about chromosomes it was possible to refer to the 'central position of DNA', and to suggest that 'the other chromosomal constituents are subservient to its needs' (Bostock & Sumner, 1978, p. 5).

Even then, perhaps, this was an overstatement of the position. Chromosomes remain important not simply because they carry the genes, but because their behaviour determines the mechanism of inheritance. The distribution of genes to daughter cells at mitosis and meiosis is a direct consequence of chromosome behaviour. Genetic linkage is a direct result of numerous genes being contained in the same chromosome. The crossing-over and reassortment of genes at meiotic prophase is also a chromosomal phenomenon, which has consequences at the evolutionary level by providing variation for natural selection to work on. Genetic variation is also provided by the fusion of egg and sperm nuclei at fertilization, to produce a diploid zygote containing two sets of chromosomes, each derived from a different individual.

The behaviour of DNA and genes is greatly constrained by the fact that they are incorporated into chromosomes and chromatin (which is, in effect, interphase chromosomes). The DNA can only function in replication and transcription because it is associated with proteins that control and catalyse these processes. Gene expression is controlled by modifications to histones (Sections 4.2.4 and 4.2.6) and by chromatin remodelling complexes (Section 4.2.5). Even something apparently as trivial as the position of a gene within a chromosome can greatly affect its behaviour. Heterochromatin (Chapter 7) consists of chromosomal segments that fail to decondense at the end of mitosis, and are genetically inactive.

Placing a gene next to heterochromatin may inactivate the gene, producing the effect known as position effect variegation (PEV), whereby a particular gene may be switched on in some cells and switched off in others. Such effects are turning out to be surprisingly widespread.

Heterochromatin (Chapter 7) generally consists of highly repeated short DNA sequences incapable of coding for proteins (Chapter 3). Analysis of the DNA sequences has failed to give any clue to their function, if indeed they have one. The same is true of many other DNA sequences that are not associated with genes (see Chapter 3), and which in fact make up the great bulk of the DNA in some organisms. The large quantity of such sequences, and the differences in amount between different organisms, have led to the 'C-value paradox', that the amount of DNA in a diploid nucleus of an organism (Table 3.1) is not necessarily related to the complexity of the organism and is greatly in excess of the amount required to provide all the genes needed. According to some, the extra DNA is just 'junk', while others have proposed that the extra DNA may have structural functions (Cavalier-Smith, 1978). Whatever the answer may turn out to be, it is clear that chromosomes are more than just strings of genes.

Errors in chromosome behaviour are an important cause of ill-health. In humans, foetal wastage occurs at a very high rate (Section 17.2), and a substantial proportion of this wastage is due to chromosome abnormalities, particularly trisomies and other aneuploidies. Some, such as trisomy 21 (Down's syndrome) and sex chromosome aneuploidies, give rise to individuals who grow to adulthood, but nevertheless show a variety of abnormalities (Chapter 17). The development of chromosomal abnormalities is usual in cancers, and a specific chromosome abnormality may often be one of the first events in the development of a cancer (Chapter 17). On the other hand, the possibility exists of creating artificial chromosomes and using them to treat genetic diseases (Chapter 18). Success will depend not merely on inserting genes in chromatin so that they can function properly, but on packing them so that they can be replicated and distributed properly to daughter cells.

We study chromosomes, therefore, not merely because they are interesting and aesthetically pleasing in their own right, but because their behaviour at fertilization and cell division determines the nature of inheritance, and their organization controls the activity of genes. Genes do not and cannot function properly, or be distributed regularly to daughter cells, unless they are in a chromosomal environment. Chromosomes are thus the ultimate determinants of the organization of all living organisms. In the following chapters, all aspects of eukaryotic chromosomes will be described, from their composition, structure and behaviour, and the ways in which they can control the functioning of genes, to their role in evolution and medicine, and to a future in which artificial chromosomes may be used to correct genetic abnormalities and disease.

Note

Original references have not been given for most of the historical observations mentioned in this chapter; however, summaries of historical work on chromosomes, genetics and DNA can be found in Krízenecký (1965), Schultz-Schaeffer (1976), Bostock & Sumner (1978, pp. 1–5), Adams *et al.* (1992, pp. 1–4), Blackburn & Gait (1996, pp. 1–9), Gall (1996) and Capanna (2000).

Website

The Mendel website (www.netspace.org/mendelweb) contains Mendel's classic paper in both the original German and in English translation, with commentaries. It also contains a chronology of related events in genetics and cell biology.

Mitosis, meiosis and the cell cycle

2.1 The necessity for accuracy in the cell cycle

Growth of a multicellular organism almost always requires an increase in the number of its cells. This is accomplished by the process of mitosis, at which discrete chromosomes become visible and are segregated equally to the daughter cells. Between successive mitoses the nucleus is in the interphase stage. The alternation of interphase and mitosis makes up the somatic cell cycle. A round of DNA replication in each cell cycle ensures that there is no progressive diminution in the amount of nuclear DNA. In sexual reproduction, however, it is necessary to halve the nuclear DNA content of eggs and sperm, so that the normal diploid DNA quantity is restored when sperm and egg fuse at fertilization. This halving of the DNA content occurs at meiosis, when a single round of DNA replication is followed by two rounds of chromosome division. Errors in cell division have serious consequences for the cell and the organism (Chapter 17), so it is clear that the cell cycle must be controlled very precisely.

Mitosis had been described thoroughly by the 1880s, and its essentials are well known (Fig. 2.1). In between successive mitoses, the chromosomes decondense to form the 'resting' nucleus, in which discrete chromosomes are no longer visible. Not until much later did it become clear that the interphase nucleus was anything but resting, but rather that its chromosomal DNA was very actively transcribed into the various sorts of RNA necessary for different aspects of protein synthesis and other functions, and thus for the life of the cell and the organism. Nor was it established until the 1950s that DNA synthesis, which was clearly necessary to compensate for the halving of the amount of nuclear DNA at each mitosis, occurred in the interphase nucleus. Interphase (Chapter 5) is therefore in many ways the most active part of the cell cycle, and its activity must be interrupted by cell and chromosomal division to produce more cells, which are required for the growth of the organism.

Mitosis is a very accurate process: almost every daughter cell finishes up with the correct set of chromosomes. Rates of loss of non-essential chromosomes in mammalian cell lines range from 1 in 20 000 to 1 in 250 per division (Burns *et al.*, 1999), but are likely to be lower in diploid cells *in vivo*. If the process goes wrong in a living organism, the consequences are disastrous: death, severe abnormality or cancer (Chapter 17). The need for a precise distribution of chromosomes and the genes they carry into the daughter nuclei implies that the process of chromosomal replication must be equally accurate. This is ensured by a number of checkpoints in the cell cycle, which ensure that all the processes of the cell cycle occur in the correct order (Fig. 2.2). A checkpoint is therefore a mechanism to inhibit a subsequent process while it assesses whether a preceding process has been completed (Elledge, 1996). There are checkpoints to ensure that DNA replication has been completed, that the

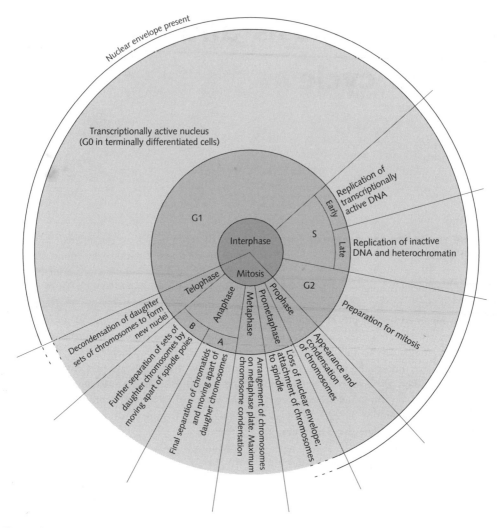

Figure 2.1 The mitotic cell cycle, showing the activities that occur at each stage.

DNA is undamaged and that all the chromosomes are properly attached to the spindle at metaphase. Failure of these checkpoints has serious consequences, and specific diseases have been identified that result from such failure (Chapter 17).

This chapter will focus on these checkpoints, on the biochemical processes that drive the cell cycle and the distribution of chromosomes into daughter cells, and on the mechanisms that ensure the accuracy of events occurring during the cell cycle.

2.2   The mitotic cycle

The mitotic cycle is controlled by a complicated pattern of protein phosphorylation, mediated by the cyclin-dependent kinases (Cdks) and reversible by protein phosphatases, and also by ubiquitin-mediated proteolysis, which is, of course, irreversible and thus provides directionality to the process. The behaviour of Cdks (reviewed by Morgan, 1997) has been worked out most clearly in yeasts, in which there is a single kinase (Table 2.1) that interacts with

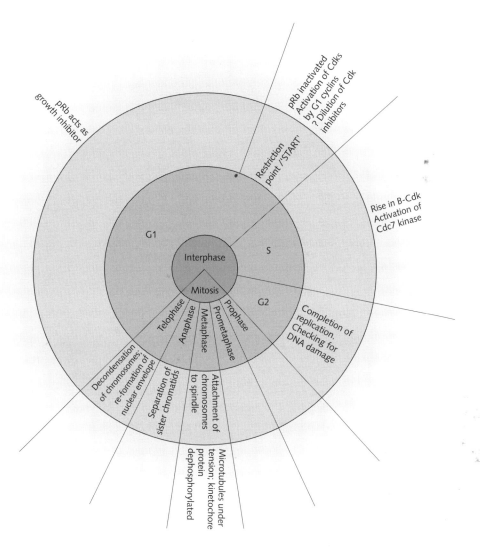

Figure 2.2 The mitotic cell cycle and its checkpoints. See text for further explanation.

different cyclins to promote different cell-cycle transitions. In mammals and plants, the situation is more complicated, with different kinases binding different cyclins to promote the different transitions (Table 2.1) (Hemerley *et al.*, 1999; Pines, 1999; den Boer & Murray, 2000). The cyclic pattern of cyclin expression to produce progression through the cell cycle is under transcriptional control, but cyclin levels are also modulated by proteolytic breakdown. Activation of Cdks requires their dephosphorylation; kinase activity can also be blocked by cyclin-dependent

kinase inhibitors (CKIs) (Sherr & Roberts, 1999). Because of the importance of Cdks in regulating the cell cycle, Cdk inhibition has been proposed as a possible approach to cancer therapy (Garrett & Fattaey, 1999). One CKI, known as p16, is in fact a tumour suppressor, deficiencies in which are associated with many cancers (Rocco & Sidransky, 2001). Phosphorylation is involved in virtually all the processes that occur during the cell cycle, while protein degradation is involved in the G1–S transition, the separation of sister chromatids at anaphase and in the

Table 2.1 Cyclin-dependent kinases.

Taxonomic group	Cell-cycle transition	Cyclin	Kinase
S. pombe	G1–S	?	Cdc2
	S–G2		Cdc2
	G2–M	Cdc13	Cdc2
S. cerevisiae	G1–S	Cln1, 2, 3	Cdc28 (= Cdk1)
	S–G2	Clb5, 6	Cdc28 (= Cdk1)
	G2–M	Clb1, 2, 3, 4	Cdc28 (= Cdk1)
Mammals	G1–S	Cyclin D	Cdk4
		Cyclin E	Cdk2
	S–G2	Cyclin A	Cdk2
	G2–M	Cyclin B	Cdk1 (= Cdc2, MPF)
Plants	G1–S	Cyclin D	?
	S–G2	Cyclin A1, A2, A3	Cdc2 (= MPF)
	G2–M	Cyclin B1, B2	Cdc2 (= MPF)

telophase–interphase transition at the end of mitosis.

2.2.1 Start

If the mitotic cell cycle can be said to begin at a specific point, it is in late G1 ('G' indicates 'gap') (Fig. 2.2). G1 is the normal state of the cycling cell, and it occupies most of the cell cycle. This is the stage during which most cellular growth occurs. Differences in the rate of cell proliferation and the cell-cycle time are correlated with the length of G1. If growth is inhibited, for example by limiting the supply of nutrients, cells are arrested in G1 (referred to as G0 if maintained for any length of time, as in terminally differentiated cells), indicating that certain G1-specific processes must be completed before the cell can proceed to mitosis. Completion of these processes allows the cells to pass a point known as START in yeasts, or the restriction point in mammalian cells, and go on to replicate their DNA and divide. After this point, the cells no longer require mitogenic stimulation, but are committed to DNA synthesis and mitosis (Planas-Silva & Weinberg, 1997).

Passage through START is regulated by G1 cyclins, which activate certain Cdks. A G1 cyclin in *Saccharomyces cerevisiae*, Cln3p, acts upstream of all other G1 cyclins, and is more active when the

cell's ribosome content is higher, and thus growing more vigorously (Polymenis & Schmidt, 1999). However, other factors such as the level of Cdk inhibitors in the cell also regulate passage through START; in mammals it may be necessary to dilute out inhibitors before DNA replication can begin (Polymenis & Schmidt, 1999). In yeasts, an independent cell-cycle oscillator coordinates events in G1 (Roussel, 2000). The restriction point involves phosphorylation of the retinoblastoma protein, pRb. Prior to reaching the restriction point, pRb acts as a growth inhibitor, by repressing transcription of genes needed for the G1–S transition (Harbour & Dean, 2000), but it becomes inactivated by extensive phosphorylation when the cell passes through the restriction point (Zhang, 1999). Phosphorylation of pRb is begun by complexes of D-type cyclins with Cdk 4 or 6, but is completed by a cyclin E–Cdk2 complex. Levels of activity of these complexes are controlled by a Cdk inhibitor, p27^{KIP1}, levels of which decline in G1 in response to mitogen stimulation.

2.2.2 DNA replication

Once a cell has satisfied all the conditions to pass through START, it can proceed to replicate its DNA and is, generally, committed to go on to mitosis and divide. The most important exception to this is endoreduplication, in which successive rounds of replication occur without cell or chromosomal division. This is the process by which polytene chromosomes are formed (Chapter 15). The normal mitotic cell-cycle controls are modified so that DNA replication is not followed by mitosis (Section 15.6).

The mechanics of DNA replication are described in Section 3.4; here we shall be concerned with the selection of the sites from which replication begins, the mechanism that ensures that all the DNA is replicated, but is only replicated once, and the temporal control of replication.

2.2.2.1 Origins of replication

In the yeast *S. cerevisiae*, replication origins have been identified as autonomously replicating sequences (ARSs) of about 150 bp that consist of an essential 11 bp ARS consensus sequence plus certain functionally conserved but structurally divergent sequences, one of which is known as a DNA-unwinding element (DUE). The DUE is an A+T-rich region and therefore its two strands are easily separated, an essential precondition for DNA replication (Gilbert, 1998). The ARS is a binding site for the origin recognition complex (ORC, a complex of six proteins, ORC1–ORC6), which is a site for binding other proteins (Cdc6, Cdc45 and the Mcm complex) required for replication. A rise in B–Cdk (cyclin B–cyclin-dependent kinase) activity, and activation of Cdc7 kinase, are then required for entry into S phase (DNA replication) (Fig. 2.2). After replication is initiated, the ORC remains at the replication origin, while the Mcm complex stays associated with the replication fork (Rowles & Blow, 1997; Gilbert, 1998; DePamphilis, 1999; Donaldson & Blow, 1999). A similar, though not identical, sequence of events is also found in the fission yeast *Schizosaccharomyces pombe*.

In multicellular organisms, in particular *Drosophila* and vertebrates, no such simple universal system of defining origins of replication has been identified, and origins of replication have been localized, in general, only to regions of DNA consisting of many kilobases (Ina *et al.*, 2001; Méchali, 2001). Indeed, in the early cleavage stages of *Xenopus* eggs, replication can start anywhere on the DNA (Gilbert, 1998). Nevertheless, initiation of replication does not seem to be random in vertebrates, and much the same proteins are involved in initiating replication in yeasts and vertebrates (DePamphilis, 1999; Pasero & Schwob, 2000). Although the precise structure of replication origins in multicellular organisms has not yet been established (if, indeed, there is a single structure), it is nevertheless possible to describe many features of such origins. They are confined to regions of 0.5–2 kb in size, and contain an ORC-binding site, and probably also A+T-rich sequences, although in mammals many are associated with CpG islands (Section 3.4). In *Drosophila* several A+T-rich initiation sites occur in a 10 kb replication origin region (Ina *et al.*, 2001). Other components of origins may be bent DNA, Alu repeats (Section 3.2.2), transcription factor binding sites and binding sites for a protein (PUR) that recognizes purine-rich stretches of single-stranded DNA.

In general, multicellular organisms have many more potential initiation sites than are normally used. Some are 'weak', and only bind ORC proteins when these are present in high concentrations, so that numerous initiation sites could be used when cells are growing and dividing rapidly, as in early embryos. When ORC proteins are less abundant, only the 'strong' sites will bind them, replication will be initiated at fewer sites and the process will be slower. Choice of initiation sites is also influenced by nuclear structure and DNA methylation. Differences in chromatin structure, such as binding of histone H1, restrict accessibility of ORC and other proteins to the DNA, thereby modulating initiation (DePamphilis, 1999).

Initiation of replication is under strict temporal control, and there is a mid-S phase checkpoint to ensure that synthesis of early replicating sequences is completed before that of late replicating sequences commences (Donaldson & Blow, 1999; Dimitrova & Gilbert, 2000; Pasero & Schwob, 2000). Differences in timing of initiation may result from differences in chromatin structure, but binding of Cdc45 to early but not late origins is another factor (Pasero & Schwob, 2000).

2.2.2.2   Licensing of replication

If the genome is to remain stable from one generation to the next, there must be mechanisms to ensure that all of it is replicated, and that it is only replicated once. Apart from the problems that might result from incorrect gene dosage, unreplicated regions of chromosomes could not separate, and would therefore cause problems at anaphase. A system of 'licensing' has therefore been postulated, so that once any sequence has been replicated during a given cell cycle, synthesis cannot be initiated again until the next cell cycle.

The licensing hypothesis postulates that a non-diffusible licensing factor marks origins of replication as competent to replicate, but that the licensing factor would be destroyed during the process of replication. Fresh licensing factor could only reach the origins of replication from the cytoplasm during the later stages of mitosis, when the nuclear envelope had broken down (Fig. 2.3). Thus, once a segment of DNA had been replicated, it would be unable to replicate again until the cell had divided, and there would automatically be only one round of DNA replication per cell cycle.

This model has proved to be substantially correct, and the nature of the replication licensing factor (RLF) has been analysed in some detail (Chong et al., 1996). In Xenopus, and probably in mammals, there are two components to the RLF: RLF-B and RLF-M. The RLF-M consists of at least three polypeptides, which are members of the Mcm (minichromosome maintenance) family of proteins, and behaves exactly as expected for a licensing factor. Rather less is known about RLF-B, but it is activated during anaphase, and in the presence of activated RLF-B the RLF-M can be assembled on to the chromatin, presumably at replication origins. The activity of RLF-B decays after anaphase, and in addition RLF-B cannot pass through the nuclear envelope, so that licensing is restricted to anaphase. Once the cell

has passed START and is committed to DNA synthesis, the S-phase promoting factor (SPF) induces initiation of replication at sites that carry the RLF, at the same time removing RLF-M from these sites. In yeast, which has a closed mitosis (i.e. the nuclear envelope does not break down), one component of the RLF, Mcm4, is actively exported from the nucleus when it is no longer needed for replication (Blow & Prokhorova, 1999). Because RLF-B is not active at this stage of the cell cycle, RLF-M cannot bind to the chromatin again, so that the DNA cannot be re-replicated during the same S phase (Fig. 2.3).

2.2.2.3   Ensuring DNA is completely replicated

Although licensing (Section 2.2.2.2) ensures that re-replication of DNA cannot occur during the same S phase, an S phase checkpoint is required to prevent the cell proceeding through G2 to mitosis if replication has not been completed (Clarke & Giménez-Abián, 2000). This depends on the detection of several proteins that reside at replication forks, and acts through proteins known as Mec1, and either Rad53 or Pds1, depending on the stage of S phase.

2.2.3   G2

As soon as DNA replication has been completed, the cell is in G2 and the DNA is assessed for damage and completeness of replication. A single DNA break is sufficient to halt the mitotic cycle at this stage. A large number of genes have been identified that are involved in the DNA damage checkpoint, and these are conserved from yeasts to higher animals (O'Connell et al., 2000). To pass on to mitosis, M-phase kinase (also known as MPF or maturation-promoting factor) has to be activated (Roberge, 1992; Ohi & Gould, 1999). The MPF is a complex of Cdc2 protein kinase with a cyclin B (Table 2.1), which regulates the activity of the protein kinase. The kinase is dephosphorylated by the protein phosphatase Cdc25 (in S. pombe; the homologous enzyme in Drosophila is String), which activates it and allows entry into mitosis. Similar systems have been

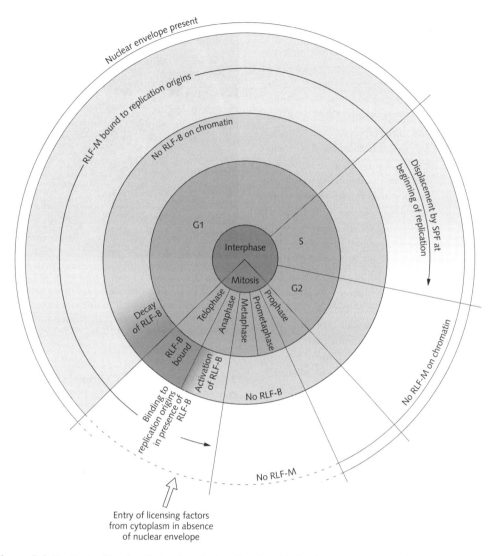

Figure 2.3 Replication licensing during the mitotic cell cycle, with the changes in RLF-B and RLF-M. See text for further explanation.

reported in *S. cerevisiae, Xenopus* and mammals. However, there is evidence for an alternative pathway, involving a Ca^{2+}-calmodulin-dependent-kinase II, in both the fungus *Aspergillus* and in mammalian cells (Roberge, 1992).

2.3 Essentials of mitosis

The essential feature of mitosis is the separation of the two sets of daughter chromosomes,

produced as a result of DNA replication, into two separate and equal groups. This involves several different processes: decatenation (disentanglement) of the newly replicated DNA molecules, and their segregation into sister chromatids; chromosome condensation; attachment of the chromosomes to the mitotic spindle; separation of sister chromatids at the beginning of anaphase, and their segregation into two separate groups; and re-formation of a membrane-bound nucleus at the end of telophase.

2.3.1 Decatenation of DNA

The double-helical nature of the DNA molecule means that newly replicated molecules produced by semi-conservative replication (Section 3.3) are inevitably intertwined, and can only be separated either by untwisting the entire DNA molecule (clearly a physical impossibility, given its great length and the fact that it is complexed with proteins) or by breaking the DNA at intervals and passing one daughter molecule through the resulting gap in the other. This requires the action of an enzyme, topoisomerase II (Topo II; Wang, 1996). The timing of decatenation is not clear. In yeast it occurs mainly immediately after DNA replication (Koshland & Hartwell, 1987), but in mammals it appears to occur much later, mainly in G2 but also during prophase and metaphase (Giménez-Abián et al., 2000). There is a G2 checkpoint to determine whether sister DNA molecules have been sufficiently decatenated by Topo II (Clarke & Giménez-Abián, 2000). Although most textbooks state that early prophase chromosomes are split into two sister chromatids, this is by no means always visible (Flemming, 1965), and modern scanning electron microscopy studies confirm that in at least some mammals early prophase chromosomes are not split (Fig. 2.4) (Sumner, 1991). A high level of Topo II in prophase chromosomes (Sumner, 1996) is consistent with decatenation occurring in early prophase. Evidence from a wide variety of organisms that Topo II is necessary for the separation of daughter chromosomes implies that centromeric DNA remains catenated until the end of metaphase (Section 2.3.3).

Chromosome condensation is necessary for the metaphase and anaphase chromosomes to be of a manageable size to be handled by the cell, without getting entangled with each other, or suffering the risk of breakage through being too long and thin. Nevertheless, in organisms with very small genomes, such as yeasts, the chromosomes are only slightly condensed at mitosis (Ghosh & Paweletz, 1993; Gottschling & Berg, 1998). More details of the condensation process, and its relevance to the structure of chromosomes, are given in Section 6.3.

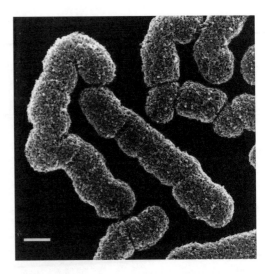

Figure 2.4 Scanning electron micrograph of an early prophase chromosome, showing that it is not yet split into two separate chromatids. Scale bar = 2 μm. Reproduced with permission from Sumner (1991) *Chromosoma* **100**, 410–418, © Springer-Verlag.

2.3.2 Attachment of chromosomes to the spindle, and formation of the metaphase plate

In most eukaryotes, at a phase known as prometaphase, the nuclear envelope breaks down and the spindle is formed. The mechanisms of spindle formation have been reviewed by Andersen (1999). In some organisms, such as yeasts, mitosis occurs without breakdown of the nuclear envelope, and the spindle poles are located in the nuclear envelope (Ghosh & Paweletz, 1993); this is known as closed mitosis.

The spindle consists of microtubules running from one pole of the cell to the other, and from the poles of the cell towards the chromosomes. The minus ends of the microtubules are oriented towards the poles, and the plus ends towards the chromosomes. In somatic cells of animals, the microtubules are assembled round the centrosomes, one at each pole of the cell, in the centre of which lie the centrioles (Marshall & Rosenbaum, 1999). The centrosomes form microtubule-organizing centres, but are not strictly necessary for spindle formation: meiotic cells and plant cells do not have them, and situations have been described in somatic cells of animals in which spindles can form and function without centrosomes (Heald *et al.*, 1996; Waters & Salmon, 1997; Marshall & Rosenbaum, 1999). During mitosis, the protein NuMA (Nuclear protein that associates with the Mitotic Apparatus), which is distributed throughout the nucleus during interphase, becomes attached to the polar regions of the spindle, and this protein, with dynein and dynactin, is needed to stabilize the spindle structure (Compton, 1998).

To function successfully, the spindle microtubules must become attached to the chromosomes, which occurs at a structure called the kinetochore (Chapter 12). Attachment of chromosomes to the microtubules is essentially a random process. Once a chromosome has become attached to a microtubule, it will move to and fro in the cell until it becomes attached to microtubules coming from the opposite pole, whereupon the position of the chromosome becomes stabilized in the middle of the cell, on the metaphase plate.

There is a spindle checkpoint to ensure that the cell cannot proceed to anaphase until all the chromosomes have become properly attached to the microtubules emanating from both poles. This depends on the kinetochores being under tension, which occurs when the chromosome is attached to microtubules from both poles. If the chromosome is only attached to microtubules from one pole or if both kinetochores of the same chromosome are attached to microtubules from the same pole, then those microtubules are not under tension, and the cell cannot proceed to anaphase (Nicklas, 1997). If, however, tension is created artificially by micromanipulation, the cell can progress into anaphase normally. Spindle poisons such as colchicine allow accumulation of mitotic chromosomes at least partly by preventing the development of this tension, so that the chromosomes cannot proceed to anaphase (although they do so eventually in some cells).

Kinetochores contain a protein that is phosphorylated when not under tension, and which becomes dephosphorylated when microtubules are attached and under tension. The kinetochores also contain a kinase and a phosphatase, so that

the whole system for varying the phosphorylation state is present in the kinetochore (Nicklas *et al.*, 1998). Such a system, which has been identified in both mammals and insects, appears to be the basis for the spindle checkpoint.

Several spindle checkpoint proteins have been identified in yeasts, many of which have homologues in vertebrates (Amon, 1999; Gardner & Burke, 2000), and have been localized to unattached kinetochores. These include MAD and BUB proteins, and Cdc20. The spindle checkpoint works by inhibiting the anaphase-promoting complex (APC), or cyclosome, which ubiquitinates various proteins involved in regulating sister-chromatid segregation, thus leading to their destruction (Section 2.3.3). Binding of Cdc20 to the APC is required for its activity, but binding of Mad2 to the Cdc20–APC complex inhibits its activity. In vertebrates, Mad2 is bound to the complex while progression into anaphase is inhibited, but dissociates when all the chromosomes are attached to microtubules. In yeasts, however, Mad1, Mad2 and Mad3 are associated with Cdc20–APC throughout the cell cycle. Probably, therefore, modifications of Mad2, or perhaps additional proteins, are required to inactivate the Cdc20–APC complex (Elledge, 1998; Amon, 1999).

2.3.3   How do the chromosomes separate?

Once all the chromosomes have become attached satisfactorily to the spindle microtubules, the cell is ready to proceed to anaphase, which starts at a fixed time interval after the last chromosome becomes attached (Rieder *et al.*, 1994). The first event in anaphase is the separation of the sister chromatids to form daughter chromosomes. Mechanisms involving severance of both DNA and protein components of the chromosomes have been implicated in this separation, which can occur in the absence of microtubules and is therefore not caused by the pull of the spindle (Ghosh & Paweletz, 1993).

Evidence from a wide variety of species and experimental protocols shows that Topo II is required for chromosome segregation, and therefore that a DNA component of the chromosomes remains to be decatenated (Section 2.3.1) at the start of anaphase. In yeasts, mutants in which Topo II is inactive fail to segregate their chromosomes, but cytokinesis proceeds, producing a *cut* phenotype (e.g. Uemura *et al.*, 1987). In *Drosophila* and mammals, treatment of mitotic cells with a variety of inhibitors of Topo II results in metaphase arrest, delayed passage through mitosis, and induction of polyploid nuclei and endoreduplicated chromosomes, all as expected if anaphase separation of chromosomes is prevented (see references in Sumner, 1998a). Mammalian metaphase chromosomes show a high concentration of Topo II in the centromeric regions, the last parts of the chromosomes to separate, and in the chromosomes of both *Drosophila* (Carmena *et al.*, 1993) and mammals (Bickmore & Oghene, 1996) strands of centromere-specific DNA can be seen still connecting the sister centromeres after the arms have separated, suggesting that decatenation of centromeric DNA sequences by Topo II is necessary to separate chromosomes at anaphase.

Destruction of proteins is also necessary to allow the anaphase separation of chromatids (Nasmyth, 2001; Uhlmann, 2001). For this, the APC (Page & Hieter, 1999) is required. The APC is a complex of nine or more proteins that targets mitotic proteins for destruction by ubiquitinating them. At the metaphase–anaphase transition the APC mediates the destruction of inhibitors of sister-chromatid separation known as securins. Securins are inhibitors of separase, an enzyme that digests cohesin, the protein complex that holds sister chromatids together. Cohesin consists, in the budding yeast *S. cerevisiae*, of the polypeptides Smc1, Smc3, Scc1 (also known as Mcd1 or Rad21) and Scc3 (Nasmyth, 2001; Uhlmann, 2001), and homologous proteins have been found in most organisms that have been studied. The Smc proteins are members of the Structural Maintenance of Chromosomes group of putative ATPases, others of which are involved in the condensin complex responsible for chromosome condensation (Section 6.5). *Xenopus* and humans also have homologues of Smc1, Smc3 and Scc1/Rad21, but have two types of Scc3:

SA1 and SA2. Cohesin binds sister chromatids together immediately after DNA replication, and in budding yeast cohesin remains bound throughout the chromosomes until the metaphase–anaphase transition. In *Xenopus* and humans, however, in which the chromosome arms are more loosely connected than the centromeric regions, cohesin is concentrated at the latter in metaphase (Nasmyth, 2001; Uhlmann, 2001).

At the beginning of anaphase, the Scc1 subunit of cohesin is cleaved by a caspase-like cysteine protease called separase, thus allowing the sister chromatids to separate (Nasmyth, 2001; Uhlmann, 2001). Until securins are destroyed as a result of APC action, they bind to separase and inhibit its action, and by this means the separation process is regulated (Fig. 2.5). However, there appear to be other mechanisms that regulate the cleavage of Scc1 by separase (Nasmyth, 2001). The loss of cohesion between chromosome arms during prophase apparently takes place by a different mechanism. Phosphorylation of cohesin may cause its dissociation from the chromosome arms without any cleavage of Scc1.

Once the sister chromosomes have been separated, the spindle can pull the two groups of daughter chromosomes apart. This occurs in two stages: anaphase A, during which the spindle poles remain the same distance apart, and the two groups of daughter chromosomes move towards their respective poles by shortening of the microtubules at their chromosomal ends; followed by anaphase B, during which further separation of the group of daughter chromosomes is produced by the spindle poles moving further apart.

2.3.4   Telophase – back to the START

Telophase is the final stage of mitosis, when the groups of daughter chromosomes acquire a new nuclear envelope, and the chromosomes decondense. It has not been well studied.

The new nuclear envelope is formed from the membranous vesicles, which are the remnants of the nuclear envelope that disintegrated at prometaphase (Gant & Wilson, 1997). These first become attached to individual telophase chro-

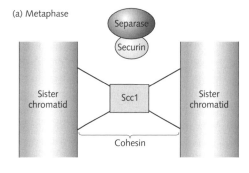

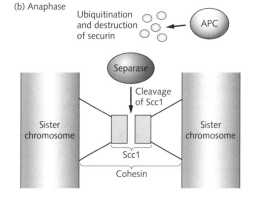

Figure 2.5 The separation of sister chromosomes at anaphase as a result of splitting of cohesins. (a) At metaphase, the sister chromatids are linked together by cohesin, consisting of Smc and Scc subunits. Digestion of cohesin is prevented by binding of securin to separase. (b) At anaphase, the anaphase-promoting complex (APC) targets securin for destruction by ubiquitinating it. As a result, separase can gain access to the Scc1 subunit of cohesin and cleave it, thus separating the sister chromosomes.

mosomes. Lamins, the proteins that line the inner surface of the interphase nuclear envelope, are required for this attachment. The lamin B receptor, which is a component of the chromosome periphery (Section 6.6), promotes targeting of lamins to the chromatin. Subsequently the vesicles fuse, by an unknown mechanism, to form a complete nuclear envelope, and at the same time the chromosomes decondense until they are no longer individually distinguishable. Reassembly of a nucleus involves the protein NuMA, absence of which results in the formation of several micronuclei instead of a single nucleus (Compton & Cleveland, 1994).

2.4   Other cell-cycle events must be co-ordinated with mitosis

Although the main focus of this chapter is the replication and segregation of DNA and chromosomes during the cell cycle, several other processes must be co-ordinated with the chromosomal ones to produce successful cell division. Aspects of spindle formation (Section 2.3.2) and re-formation of the nucleus at telophase (Section 2.3.4) have already been touched upon.

Many of the activities involved in cell division are controlled by the Polo-like kinases (Nigg, 1998), which are of prime importance in co-ordinating the progression of cells through the cell cycle. This group of enzymes takes its name from the Polo kinase of *Drosophila*, although the enzymes are known by different names in different organisms. These enzymes regulate centrosome separation during mitosis, and thus control the formation of a bipolar spindle. They also control the process of cytokinesis – the division of the cytoplasm into two after the nucleus has divided – and septation in yeast. They activate cyclin-dependent kinases, and promote cyclin destruction and exit from mitosis.

The aurora kinases also co-ordinate aspects of chromosome segregation and cytokinesis (Adams *et al.*, 2001). Aurora kinase B forms a complex with the chromosome passenger protein INCENP (Section 6.6). After the metaphase–anaphase transition, this complex recruits ZEN-4 kinase to the midzone, where it bundles microtubules and allows completion of cytokinesis. INCENP probably also targets aurora B kinase to the cell cortex to help form the cleavage furrow.

2.5   Meiosis

Meiosis has been described as two rounds of nuclear division with only one round of DNA synthesis, and in many ways it is regulated in the same ways as mitosis (Sections 2.2 and 2.3). There are, however, some essential differences: the pairing and alignment of homologous chromosomes; the maintenance of cohesion between sister chromatids throughout the first meiotic division; the processes of recombination and crossing-over; the arrest that frequently occurs, often for many years, in female meiosis; and the suppression of DNA synthesis between the first and second meiotic divisions.

The 'classical' view of meiosis was that the cell entered meiotic prophase with homologous chromosomes unpaired (leptotene), that pairing of homologues occurred during zygotene, when the synaptonemal complex (SC) formed, and that recombination occurred within the SC, which was absolutely necessary for this process. Sites of recombination were marked by dense bodies, the recombination nodules. Subsequently, the homologous chromosomes separated somewhat, being held together only at the points where recombination had occurred, the chiasmata, although there was a possibility that the chiasmata might move away from the actual sites of recombination by the process known as chiasma terminalization. Certain features of this scenario have now been questioned, particularly the timing of pairing of homologues, and the role of the SC.

2.5.1 How do homologues get together?

The mechanism by which homologous chromosomes recognize each other at meiosis, come together and form intimate homologous associations (synapsis) is unknown, but it seems to involve several stages, and may differ from one species to another. In some organisms, the distribution of chromosomes in interphase nuclei appears to be essentially random, and there is a real difficulty in understanding how homologues in such nuclei might find each other, especially when there are large numbers of chromosomes. In other species, the interphase nuclei are more organized, and the chromosomes remain in the Rabl configuration (Section 5.2) in which they are aligned with their centromeres towards one pole of the nucleus and their telomeres towards the other. An extreme case is found in organisms such as *Drosophila*, in which the chromosomes are paired in somatic nuclei, and there is therefore no need to bring them together for meiosis.

In many plants, the first stages of association occur during the pre-meiotic interphase, when homologous chromosomes come to lie in close proximity to each other (Sybenga, 1999; Zickler & Kleckner, 1999); the complexities of this process may account for this interphase being so much longer than that of most dividing somatic cells. More precise alignment occurs during leptotene and zygotene, and synapsis occurs during zygotene, when the chromosomes come into intimate association and an SC is formed between them. In many organisms a bouquet is formed by the telomeres clustering together on the nuclear envelope, and this may facilitate the initiation of pairing (Zickler & Kleckner, 1998; Scherthan, 2001). The processes involved in initial pairing of chromosomes are not clear, but might involve chromosomal proteins (Sybenga, 1999) as well as weak DNA–DNA interactions (Stack & Anderson, 2001).

Synapsis often begins at the ends of the chromosomes, and proceeds inwards, but interstitial origins also occur. Synapsis is often not complete: heterochromatic regions of chromosomes often do not synapse, and in some organisms with a restricted distribution of crossing-over the distribution of synapsis and the SC is similarly restricted.

The enzyme Spo11, which also produces the double-strand DNA breaks that lead to meiotic recombination, is also required for synapsis in many organisms, such as the fungus *S. cerevisiae*, the plant *Arabidopsis* and the mouse (Lichten, 2001). This does not seem to be so in the nematode *Caenorhabditis elegans* or in *Drosophila*, however, in which *Spo11* mutants can still undergo synapsis, leading to the suggestion that these species have special 'pairing centres' in their chromosomes (Lichten, 2001; Mitchell, 2001).

2.5.2 The synaptonemal complex – cause or consequence of crossing-over?

Synapsed meiotic prophase chromosomes normally have a synaptonemal complex (SC) between them (Fig. 2.6). Unpaired leptotene chromosomes each contain an axial core; when the homologues synapse, the axial cores become

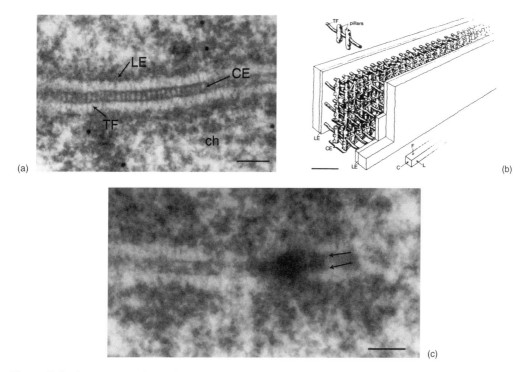

Figure 2.6 The synaptonemal complex (SC) from the beetle *Blaps cribosa*. (a) Frontal section showing the lateral elements (LE), central element (CE) and transverse filaments (TF), surrounded by chromatin (ch). (b) A three-dimensional model of the SC. (c) A recombination nodule in contact with the central element (arrows). Scale bars = 100 nm. (a, b) Reproduced with permission from Schmekel & Daneholt (1995) *Trends in Cell Biology* **5**, 259, © Elsevier Science. (c) Reproduced with permission from Schmekel & Daneholt (1998) *Chromosome Research* **6**, 155–159, © Kluwer Academic Publishers.

the lateral elements of the SC, which are connected by numerous transverse filaments (Zickler & Kleckner, 1999). In the middle, attached to the transverse filaments, is the central element, made up of thickenings known as pillars. There can be three to five layers of filaments and pillars, which are held together by fibrous bridges (Schmekel *et al.*, 1993a, b). This tripartite structure of the SC appears to be universal, although there are detailed structural differences in the central element from one species to another. Numerous proteins have been identified in the SC (Table 2.2), although the functions of many of them are not yet clear. Correlations can be made between the spatial and temporal occurrence of specific proteins, and particular functions that occur at such places and times, but in only a few cases has it been shown that specific proteins are required

for specific processes. Cohesins appear to be major constituents of the axial elements, consistent with the fact that sister chromatids remain intimately linked during the first meiotic prophase (Nasmyth, 2001; cf. Section 2.3.3).

Early studies showed a good correlation between the presence of an SC and the occurrence of crossing-over (Bostock & Sumner, 1978, pp. 319–321; John, 1990, pp. 91–92): if there were no SCs, no crossing-over occurred; and if SCs were confined to specific regions of chromosomes, these were the regions where crossing-over took place. Another feature of SCs that shows a good correlation with crossing-over is the recombination nodules (RNs; Fig. 2.7), which were originally thought to show a correlation in number and position with sites of crossing-over (John, 1990, pp. 166–169). Further study

Table 2.2 Synaptonemal complex proteins.

Location	Name	Species	Comments	Refs
Axial core	SCP2	Rat	?DNA binding	1
	SCP3 (COR1)	Rat	?DNA binding	2
	M_r 52–70K	Lilium		1
	Hop1p	Yeast	Absent in synapsed regions	3
	Red1p	Yeast		3
	Rad51p	Mouse, human	?Early recombination nodules	4, 5
	Rec8	Yeasts, mammals	Cohesin	11
	Meis322	Drosophila	Centromeric cohesin	11
	Atr	Mouse, human	Unsynapsed axes	6
	Atm	Mouse, human	Synapsed axes	6
	Topo II		Late meiotic prophase	9
?	Spo11	S. cerevisiae	Required for DSB ???	7
Central element	SCP1 (SYN1)	Rat	?DNA binding	1, 2
	SC48	Rat		1, 2
	SC65	Rat		11
	Zip1p	Yeast	?Transverse filament component ?DNA binding Required for recombination and interference	1, 2, 8
Ends of paired bivalents	Rap1p	Yeast	Telomeric	9
Initiation sites for synapsis	Zip2	S. cerevisiae	Sites of recombination	10

References: 1, Heyting, 1996; 2, Moens & Spyropoulos, 1995; 3, Smith & Roeder, 1997; 4, Barlow *et al.*, 1997; 5, Moens *et al.*, 1997; 6, Keegan *et al.*, 1996; 7, Keeney *et al.*, 1997; 8, Storlazzi *et al.*, 1996; 9, Klein *et al.*, 1992; 10, Chua & Roeder, 1998; 11, Zickler & Kleckner, 1999.

showed, however, that RNs were more numerous earlier in prophase (Plug *et al.*, 1998; Zickler & Kleckner, 1999) and it was suggested that only those that led to crossing-over would persist to late pachytene. The distribution of these late nodules (LNs) is correlated with sites of crossing-over. The early, more numerous RNs are now often referred to by the less tendentious name of meiotic nodules (MNs), zygotene nodules or early nodules (ENs), and it has been suggested that they might be involved in checking homology before synapsis occurs. A fraction of them appear to transform into late nodules, often with a change in morphology.

Early nodules are commonly spherical or ellipsoidal, and the shape may differ among ENs in the same cell. Late nodules, on the other hand, are usually all of the same shape in any given organism, but may be spherical, ellipsoidal or bar-like. Recombination nodules vary in size

from 30nm to 200nm, and LNs are possibly rather smaller than ENs (Zickler & Kleckner, 1999). Early nodules are found in a variety of positions in relation to SC components (Zickler & Kleckner, 1999). As well as bridging the width of SCs, they can also be associated with unsynapsed axial elements, or can form the only points of contact between otherwise divergent axial elements (Fig. 2.7A). Late nodules, however, are only found in contact with the SC, usually located to one side of it.

Since their discovery, RNs have been assumed to contain enzymes and other proteins required for recombination, and immunocytochemical studies of meiotic chromosomes confirm this. Early nodules have been shown to contain a variety of proteins (Table 2.3), of which two (RAD51 and Dmc1) are homologues of the bacterial RecA protein, which is involved in searching for homology between DNA molecules

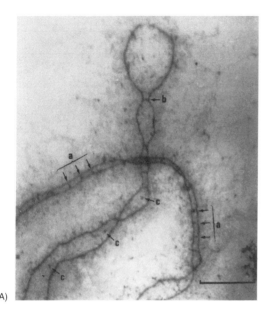
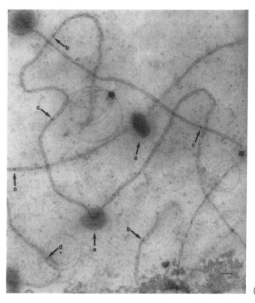

(A)

(B)

Figure 2.7 Recombination nodules from *Allium cepa*. (A) Zygotene nodules: (a) nodules associated with the synaptonemal complexes; (b) nodules at sites of association; (c) nodules midway between axial cores. Reproduced with permission from Albini & Jones (1987) *Chromosoma* **95**, 324–338. © Springer-Verlag. (B) Pachytene nodules: (a) centromeres; (b) distally located late recombination nodules; (c) interstitial late nodules; (d) proximal late nodules. Scale bar = 1 μm. Reproduced with permission from Albini & Jones (1988) *Genome* **30**, 399–410. © National Research Council of Canada.

and in catalysing strand exchange. The protein ATM, also found in early nodules, is involved in detecting DNA breaks and other damage and activating the appropriate cell-cycle checkpoint. One of the proteins found in late nodules, MLH1, is a mismatch repair protein, which is not surprising as recombination is closely related to mismatch repair (Arnheim & Shibata, 1997).

Although the scenario just described fits many known facts, there have long been suggestions that recombination can occur without SCs, or before their formation, and the discovery that certain species can recombine their chromosomes without forming SCs has refocused attention on the function of SCs. The suggestion is that synapsis and SC formation are consequences of recombination, and that the function of SCs is something other than providing a framework for recombination.

Certain organisms, including *Aspergillus nidulans* and the yeast *S. pombe* (Roeder, 1997; Zickler & Kleckner, 1999), have high levels of meiotic recombination but do not form an SC, and an SC is not necessary for recombination in *S. cerevisiae*. *Aspergillus nidulans* and *S. pombe* differ from most organisms in their lack of meiotic interference (Heyting, 1996; Roeder, 1997). Interference appears as a non-random distribution of chiasmata and crossing-over and a restricted number of chiasmata per chromosome. The presence of one crossing-over event interferes with another occurring nearby, so that chiasmata do not form too close to each other. It has therefore been suggested that the function of the SC is to mediate this phenomenon of interference. This is supported by the observation that mutations of *ZIP1*, which encodes a component of the central element of the SC, have only minimal effects on recombination, but abolish interference (Roeder, 1997).

Some caveats must be entered at this stage. First, it is not necessary to assume that the

Table 2.3 Proteins of recombination (meiotic) nodules.

Type	Protein	Function	Refs
Early nodules (meiotic nodules, MNs)	RAD51	On synapsed and unsynapsed chromosomes. RecA homologue. ? Homology searching. Associates with single-stranded DNA at sites of DSBs	1, 2, 3, 4
	RPA*	Single-stranded DNA-binding protein. Only on synapsed chromosomes	4
	ATR*	Presynaptic MNs	4
	ATM*	Postsynaptic chromosomes. Detection of DNA damage and cell-cycle control	4
	Dmc1	RecA homologue. Recognition of homologous DNA and catalysis of strand exchange	5
Late nodules (recombination nodules, RNs)	MLH1	Sites of crossing-over. Mismatch repair protein	4, 6
	RPA	See above	4

*Inferred to be in ENs by fluorescence microscopy studies, but not yet confirmed by electron microscopy.
References: 1, Anderson *et al.*, 1997; 2, Barlow *et al.*, 1997; 3, Moens *et al.*, 1997; 4, Plug *et al.*, 1998; 5, Pittman *et al.*, 1998; 6, Barlow & Hultén, 1998.

sequence of meiotic events is identical in all organisms. Second, simple organisms with small genomes, such as *Aspergillus* and *S. pombe,* may be able to operate with a simplified meiotic system that would not work efficiently in more complex organisms with larger genomes. Third, the existence of organisms such as the female silkworm (*Bombyx*), which have no crossing-over but form good SCs, indicates that SC formation is not necessarily a consequence of recombination. The *Spo11* mutants of *Drosophila* and *Caenorhabditis* that lack recombination also form normal SCs (Lichten, 2001).

As in mitosis, there are checkpoints to control progression through meiosis. The pachytene checkpoint is specific to meiosis, and ensures that meiosis does not proceed if synapsis and recombination are incomplete (Roeder, 1997).

2.5.3   Recombination, crossing-over and chiasmata

We have already discussed the timing of recombination in relation to other events in meiosis, and the role or roles of the SC and recombination nodules in the process, and it is clear that there is still much to be learnt at the chromosomal level. On the other hand, a good deal is known about the process of homologous recombination at the molecular level, and although much of this knowledge has been derived from studies of bacteria, particularly *Escherichia coli*, there is good reason to suppose that the recombination process in eukaryotes is generally similar, and uses enzymes homologous to those found in bacteria (Shinagawa & Iwasaki, 1996).

There are essentially three stages in recombination at the molecular level: formation of double-strand breaks (DSBs) in the DNA, formation of Holliday junctions and resolution of Holliday junctions to produce a crossing-over (Figs 2.8 & 2.9). Numerous proteins are involved in each of these processes, and defects in any one of them may disrupt recombination and the progress of the cell through meiosis. In the yeast *S. cerevisiae,* DSB formation is catalysed by the protein Spo11, and exonuclease action digests the resulting 5′ termini to yield single-stranded tails about 600 nucleotides long (Keeney *et al.*, 1997). Double-strand break formation is not random, but occurs preferentially in hot-spots, which in *S. cerevisiae* are located preferentially in promoters of genes but are not confined to specific sequences (Smith & Nicolas, 1998). These regions consist of nuclease-sensitive chromatin (Haber, 1997), which are regions of high acces-

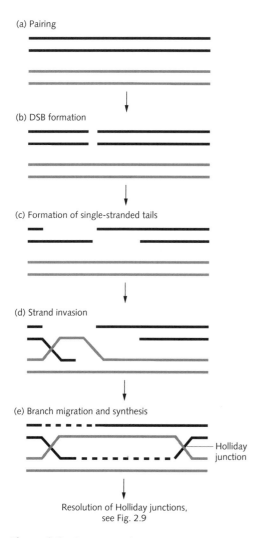

(a) Pairing

(b) DSB formation

(c) Formation of single-stranded tails

(d) Strand invasion

(e) Branch migration and synthesis

Holliday
junction

Resolution of Holliday junctions,
see Fig. 2.9

Figure 2.8 The process of crossing-over. See text for further explanation.

sibility. In mammals also, recombination occurs preferentially in G+C-rich, gene-rich regions (Zickler & Kleckner, 1999), which are also nuclease-sensitive. Sites of crossing-over and chiasma formation are not randomly distributed throughout the chromosomes (John, 1990, pp. 47–65).

Following the formation of DSBs and single-stranded tails, the latter are postulated to invade a homologous double-stranded DNA molecule (Fig. 2.8). The single-stranded segment pairs with its complementary strand from the double-stranded DNA, forming a Holliday junction. The

site of the junction can move along the paired molecules (branch migration), and gaps in the first DNA molecule (the one in which the DSBs were formed) can be filled using the other DNA molecule as a template. The whole process is similar to the repair of DSBs (induced by radiation, etc.) in non-meiotic cells (Section 3.6.5), and in fact meiotic recombination may well be derived from the repair process. However, whereas repair in non-meiotic cells involves recombination between sister DNA molecules, meiotic recombination occurs mainly between non-sister homologues, which are not identical because they will carry different alleles for many genes. Just as there is a DNA damage checkpoint in G2 of the mitotic cycle, meiotic prophase cells have a recombination checkpoint that depends on detection of DSBs (Arnheim & Shibata, 1997; Page & Orr-Weaver, 1997). The final stage in the recombination process is the resolution of the Holliday junctions, which must be cut either by a resolvase or a topoisomerase (Smith & Nicolas, 1998). Depending on how the junctions are cut, the recombinant molecules will either exchange flanking markers (crossing-over) or not exchange them (Fig. 2.9). In either case, gene conversion can occur. The same enzymatic pathways are involved in each case.

Once recombination has been completed, the cell can continue its passage through meiosis. As it passes into diplotene and diakinesis, the homologous chromosomes in each bivalent separate, except at the chiasmata, although sister chromatids remain closely apposed, so that the appearance is quite unlike anything seen during mitosis (Fig. 2.10). Chromatids can, in suitable preparations, be seen crossing over from one chromosome to the other (Fig. 2.11), and they are clearly a manifestation, at the chromosomal level, of the recombination that has occurred at the DNA level. Nevertheless, it was believed for many years that the chiasmata might not coincide exactly with sites of crossing-over, but were subject to a process of terminalization. However, experiments using bromodeoxyuridine to label sister chromatids differentially (Section 3.4) show that sites of exchange coincide exactly, at the chromosomal level, with the position of chias-

(a) Resolution without exchange of flanking markers

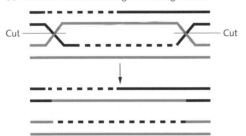

(b) Crossing-over with exchange of flanking markers

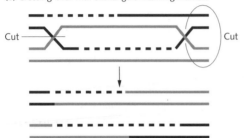

Figure 2.9 Resolution of Holliday junctions at the end of crossing-over. (a) Resolution without exchange of flanking markers. (b) Crossing-over, with exchange of flanking markers. See text for further explanation.

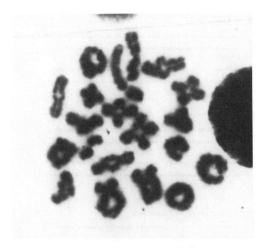

Figure 2.10 Human chromosomes at diakinesis. Micrograph kindly provided by R.M. Speed.

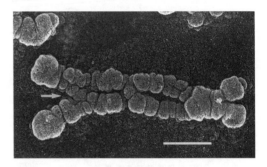

Figure 2.11 Scanning electron micrograph of a bivalent with three chiasmata from a spermatocyte of the locust *Schistocerca gregaria*. Crossing-over can clearly be seen at one chiasma (arrowed). Scale bar = 5 μm. Reproduced with permission from Wolf *et al.* (1994) *Journal of Submicroscopic Cytology and Pathology*, **26**, 79–89.

mata (John, 1990, pp. 73–77), so that direct evidence for terminalization is still lacking.

Although chiasmata originate as a consequence of crossing-over, they also have an essential function in ensuring proper chromosome segregation at the first meiotic metaphase. As the only regions that hold homologous chromosomes together in late meiotic prophase, they not only prevent premature disjunction, but also help to ensure that the kinetochores of the homologous chromosomes that comprise a bivalent are oriented towards opposite poles of the cell. The structure of the kinetochores at the first meiotic metaphase is still not clear: it has been suggested that they remain undivided at this stage, to ensure that there is no problem of sister chromatids becoming attached to microtubules emanating from opposite poles, although there does not seem to be any good evidence for this (John, 1990, pp. 40–42). As in mitosis, there is a spindle checkpoint to ensure that all the chromosomes are correctly attached to the spindle before anaphase can occur (Page & Orr-Weaver, 1997)

(Section 2.3.2). The meiotic cohesin system is similar to that in mitotic cells (Section 2.3.3), but there are specific meiotic cohesin subunits. Polypeptide Scc1 is replaced by Rec8, and the somatic variants of Scc3 (SA1 and SA2) are replaced by a different form of Scc3 known as STAG3 (Nasmyth, 2001). In yeast, the onset of meiotic anaphase is associated with destruction of securin (Nasmyth, 2001; cf. Section 2.3.3).

2.5.4 Meiotic arrest

In the mitotic cell cycle, the main variable in the length of the cell cycle is the length of G1, and once the cell is committed to divide, it proceeds through S, G2 and mitosis without delay. Similarly, male meiosis is generally a continuous process, designed to produce vast quantities of spermatocytes with the minimum delay. Female germ cells, however, can arrest in meiosis, often for years, at a variety of different stages, according to the species (John, 1990, pp. 105–109). In many invertebrates this arrest is at metaphase I, whereas in most vertebrates it is at metaphase II. The stimulus for completion of meiosis is usually fertilization.

Meiotic arrest also occurs commonly at diplotene. In mammals, female meiosis starts in the embryo, and proceeds as far as diplotene, when the chromosomes become diffuse and the cells are referred to as being in the dictyate stage. This arrest is under hormonal control, and the oocyte recommences growth and passage through meiosis in response to luteinizing hormone in adult life. Thus in large mammals with a long period of immaturity, the oocytes may be arrested in the dictyate stage for many years. It has been suggested that prolonged meiotic arrest in human oocytes could be a cause of aneuploidy (Section 17.2). The oocytes of amphibia and many other organisms also spend a prolonged period in diplotene, but it is misleading to describe this as meiotic arrest, because this is the stage when lampbrush chromosomes (Chapter 14) are formed and undergo intense RNA synthesis.

2.5.5 Meiosis – the final stages

The first meiotic division (meiosis I) differs from mitotic division in that whole chromosomes segregate to opposite poles of the cell, and the division is said to be reductional, because the number of chromosomes in each daughter cell is reduced. Mitosis and meiosis II, on the other hand, are equational divisions, because the number of chromosomes in each daughter cell is the same as in the parental cell (although each consists of only one chromatid instead of two). These differences are, however, simply consequences of the way the chromosomes and their kinetochores are arranged prior to division. (There are, however,

organisms with holocentric chromosomes [Section 12.5] that have an inverted meiosis in which the first division is equational and the second is reductional; John, 1990, pp. 93–96.)

The most interesting feature of the later stages of meiosis is, however, the absence of DNA synthesis between the first and second chromosomal divisions. This is obviously crucial to one of the main purposes of meiosis, because DNA replication would restore the DNA content of the cell to the diploid level, with an exponentially increasing level of polyploidy in subsequent generations after fertilization. Although many details of the regulation of this process remain unknown, a cyclin–Cdk system is used to suppress DNA replication between meiosis I and II (Picard *et al.*, 1996). Unfertilized starfish eggs are unable to inactivate MAP kinase, and this prevents them from proceeding to embryogenesis until the egg is fertilized.

2.6 Accuracy is ensured in cell division

Growth and maintenance in eukaryotes are almost invariably dependent on mitosis and cell division (but see Chapter 15, Polytene chromosomes), and in general multicellular organisms must use meiosis in the course of reproduction. Absolute accuracy is essential for a successful outcome in these processes, and there is abundant evidence that the consequences of failure are disastrous (Chapter 17). Systems of checkpoints have therefore evolved to ensure that cells cannot proceed to the next stage of chromosomal division until all the preparations have been completed satisfactorily. Chromosomes are scanned, by mechanisms that are still far from being fully understood, for the presence of such features as DNA damage, incomplete replication or non-attachment to the spindle. The efficacy of such systems is all the more remarkable when we consider that they can apparently detect one break in 6×10^9 bases of DNA (in humans and other mammals), or one unattached kinetochore among a hundred. As a result, the failure rate in chromosomal division is generally extremely low.

DNA, the genetic code

3

3.1 Stability and variability of DNA

Deoxyribonucleic acid is the basis of all eukaryotic genetic systems, and as such it needs to be stable in quantity and sequence. In practice there is extensive variation in both the amount of DNA and its sequence. This variation is due largely to sequences that have little or nothing to do with genes (Section 3.3). Nor is DNA a particularly stable substance. It is constantly being damaged, and therefore needs to be repaired constantly. The DNA repair process is not always perfect and can sometimes lead to mutations (Section 3.6), and DNA is also subject to epigenetic modification in the form of methylation (Section 3.5). Methylation is involved in the control of gene expression (see also Chapter 9), but it can also be a cause of mutations. In this chapter, then, we shall consider the paradox that DNA needs to be stable to perform its genetic functions properly, but that in practice it is variable and mutable.

3.2 The amount of DNA in nuclei, and the C-value paradox

Humans, and mammals generally, have about 6–7 pg or approximately 2 m of DNA in every diploid nucleus. Some organisms, particularly simple ones, have very much less DNA; others, such as some plants, and lower vertebrates such as lungfish, newts and salamanders, have many times as much (Table 3.1). This immense variation in the amount of DNA per nucleus, which is clearly not related to the complexity of the organism, is the C-value paradox. No-one would argue that a lily is 250 times more complicated than *Drosophila*, or that a newt is over tenfold more complicated than a mammal. Great variation in genome size can hide significant similarities: a high proportion of syntenic genes in the pufferfish *Fugu* are also syntenic in humans, in spite of the great difference in the sizes of their genomes (McLysaght *et al.*, 2000).

Part of the answer to the C-value paradox is that only a fraction of the genome consists of genes and their associated DNA sequences such as introns, promoters and enhancers. Numbers of genes are much less variable than the total amount of DNA, and vary from about 5800 to 39 000 in genomes sequenced so far (Table 3.2). In humans, the fraction of DNA that is transcribed into RNA – the genes and their introns – is at most about 10% of the total, and of this RNA only a fraction is translated into protein. The rest of the genome consists of a large variety of sequences, many of them repetitive (Section 3.3), that in general have no known function and are often regarded as junk. Large genomes contain much more repetitive DNA than small ones (Bennett, 1995; Hancock, 1996; Bennetzen *et al.*, 1998), and the genes are much closer together in small genomes (Elgar, 1996; Bennetzen *et al.*, 1998; Table 3.2).

Table 3.1 Amounts of DNA in diploid nuclei of selected animals and plants.

Species	Picograms	Megabase pairs*	Length†	Ref.
Fungi				
Saccharomyces cerevisiae	0.05	45	15 mm	1
Plants				
Arabidopsis thaliana	0.2	183	62 mm	2
Oryza sativa (rice)	0.93	898	30.5 cm	2
Glycine max (soybean)	2.37–2.48	2160–2260	73.4–76.8 cm	3
Zea mays (maize)	5	4565	1.55 m	6
Allium cepa	40	36 520	12.42 m	6
Lilium 'Enchantment'	88	80 345	27.32 m	6
Fritillaria davisii	225	205 425	69.84 m	2
Protozoa				
Trypanosoma brucei	0.097	89	30.2 mm	8
Animals				
Insects				
Bombyx mori (silkworm)	1.04	950	32.3 cm	6
Locusta migratoria	12.7	11 595	3.94 m	6
Drosophila melanogaster	0.36	330	11.2 cm	6
Apis mellifica	0.34	310	10.5 cm	6
Fish				
Torpedo ocellata	15.0	13 695	4.66 m	4
Petromyzon marinus (lamprey)	4.2	3835	1.30 m	5
Fugu rubripes (puffer fish)	0.88‡	800	27.2 cm	7
Salmo trutta (brown trout)	5.9	5387	1.83 m	5
Amphibia				
Xenopus laevis	6.3	5752	1.96 m	6
Triturus cristatus	70	63 910	21.73 m	1
Trachemys scripta (turtle)	5.3	4840	1.65 m	5
Bird				
Gallus domesticus	2.5	2283	77.6 cm	5
Mammals				
Mus musculus	6.0	5478	1.86 m	1
Homo sapiens	7.0	6391	2.17 m	5

*Calculated as 913×10^6 base pairs per picogram of DNA (see Ref. 6).
†Calculated as 0.34 nm per base pair.
‡Calculated from number of base pairs.
References: 1, Adams *et al.* (1992); 2, Uozu *et al.* (1997); 3, Chung *et al.* (1998); 4, Stingo *et al.* (1989); 5, Tiersch *et al.* (1989); 6, Rasch (1985); 7, Elgar (1996); 8, Borst *et al.* (1982).

3.3 Repetitive DNA – sequences with a function, or just junk?

Repetitive DNA is classified into tandem repeats and interspersed repeats. Tandem repeats consist of the same sequence repeated thousands or millions of times in tandem, thus forming discrete blocks of DNA. Interspersed repeats, on the other hand, consist of specific sequences dispersed around the genome, but not forming tandemly repeated blocks.

Many types of repetitive DNA have been

Table 3.2 Total numbers of genes estimated from whole genome sequencing of various species.

Species	No. of genes	Genes/million bases
Yeast *Saccharomyces cerevisiae*	5800	483
Arabidopsis thaliana	25498	221
Nematode *Caenorhabditis elegans*	19099	197
Drosophila melanogaster	13601	117
Homo sapiens	32000–39000	12–15

Data from Bork & Copley (2001).

regarded as 'junk' or 'selfish' DNA, which are not essential to the functioning of the organism (see also Section 7.3.1), and the properties of many repetitive sequences may lead to their accumulation in the genome, independently of any selective advantage or disadvantage (Charlesworth *et al.*, 1994). Exceptions to the rule that repetitive DNA seems to be without function are found at the telomeres (Section 13.2) and certain sequences in *Drosophila* heterochromatin (Section 7.4.2).

3.3.1   Tandem (satellite) repeats

Tandem repeats are often referred to as satellites, because the first ones discovered had a different DNA base composition and density from 'main-band' DNA, and therefore formed a 'satellite' band during density gradient centrifugation. Tandem repeats are now all commonly referred to as satellites, regardless of whether they have a distinctive base composition. Three classes of satellites are recognized: 'classical' satellites minisatellites and microsatellites.

3.3.1.1   'Classical' satellites

Classical satellites often have a repeat unit consisting of hundreds or even thousands of nucleotides, although the repeat is only 7 bp in some *Drosophila* satellites, 6 bp in some mammals and as short as 2 bp in a crab (Beridze, 1986). These very short repeat units are as short as or shorter than those of mini- and microsatellites, but classical satellites differ from mini- and microsatellites in their quantity (they sometimes make up more than half of the genome), and they form cytologically visible blocks of heterochromatin on

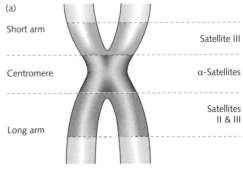

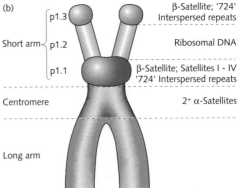

Figure 3.1 Arrangement of satellite DNAs on chromosomes: (a) human chromosome 10; (b) a generalized human acrocentric chromosome.

chromosomes (Section 7.3.1). Several different types of satellite DNA can occur together in the same block of heterochromatin (Choo, 1990; Eichler, 1999), often with interspersed (Section 3.3.2) and unique sequences (Fig. 3.1). Classical satellites often have a distinctive base composition: not only may they be richer in A+T or G+C than main-band DNA but, particularly when the repeat unit is very short, there can be a considerable difference in mean base composition between the two DNA strands.

Certain satellites, such as human alpha-satellite and mouse minor satellite, may have an important role in centromere organization (Section 12.2.3). Otherwise there is no evidence that classical satellite DNAs have any function, and the amount of satellite can vary between and within individuals of a species without phenotypic effect (Section 7.4), although such variations may affect features such as growth rate and size in plants (e.g. Rayburn et al., 1985; Section 16.5).

3.3.1.2 Minisatellites and microsatellites

In addition to the 'classical' satellites, there are also minisatellites or VNTR ('variable number tandem repeats'), which are tandem repetitions of short sequences of between 10 and 100 base pairs forming arrays of about 0.5–30 kb, and microsatellites, which are repeats of very short sequences, no more than six bases long and often as short as two or three bases. They have been found in all organisms studied and are distributed throughout the chromosomes, although microsatellites may be less common in coding regions and at telomeres (Hancock, 1999). The commonest dinucleotide repeat in humans and *Drosophila* is $(CA)_n$, but in plants $(GA)_n$ and $(AT)_n$ are commoner.

Different individuals in a species have different numbers of repeats in a particular mini- or microsatellite, and as a result heterozygosity for these sequences is high. Estimated mutation rates for mammals are in the region of 10^{-3} per locus per generation, but seem to be only 6×10^{-6} in *Drosophila* (Hancock, 1999). This great quantity of variation, combined with the relative stability of such sequences, makes it possible to use minisatellites for 'DNA fingerprinting', which can be used for the analysis of linkage, in population genetics, for proving paternity and for forensic analysis. Microsatellites are valuable genetic markers, and have been essential tools in constructing genetic maps in humans and other organisms.

Apart from these practical applications, minisatellites are of great interest as they appear to be involved in gene conversion and meiotic recombination (Jeffreys et al., 1998). There is evidence that both micro- and minisatellite sequences can bind chromosomal proteins, and may form gene regulatory elements and parts of coding sequences. Variation in repeat number can have significant phenotypic effects (Kashi & Soller, 1999).

A particular type of microsatellite is the repetition of trinucleotides that is found in association with certain genetic diseases, such as Fragile X syndrome and Huntington's disease. In normal individuals, the genes involved have a low number of copies of the trinucleotide, whereas in patients with the disease the number of copies is higher. The trinucleotide sequences may be within the coding sequence of the gene, so that the extra trinucleotides cause the gene to produce a toxic product, or external to the coding sequence, resulting in loss of function. These sequences and the diseases they cause are discussed in more detail in Section 17.5.

3.3.1.3 Mechanisms of variation of satellite DNA arrays

It is characteristic of arrays of satellite DNAs, whether they be 'classical', mini- or microsatellites, that the length of the arrays can vary, often dramatically. There are three mechanisms by which this can happen, which depend on the repetitive nature of the sequences. In unequal crossing-over (Fig. 3.2a), two arrays can pair out of register and then exchange, so that one becomes longer and the other shorter (Kurnit, 1979; Hancock, 1999). However, unequal crossing-over occurs most readily with long tandemly repeated sequences, and it is more likely that mini- and microsatellites vary as a result of gene conversion, in which there is a unidirectional transfer of sequences (Fig. 3.2b), or by replication slippage. In replication slippage (Fig. 3.2c), either the template or the newly synthesized strand of DNA can become detached from its partner, fold up and reattach out-of-register, so that the newly synthesized strand is different in length from that of the template (Hancock, 1999). Instability of microsatellites is found in certain types of cancer (Atkin, 2001; Section 17.9.1).

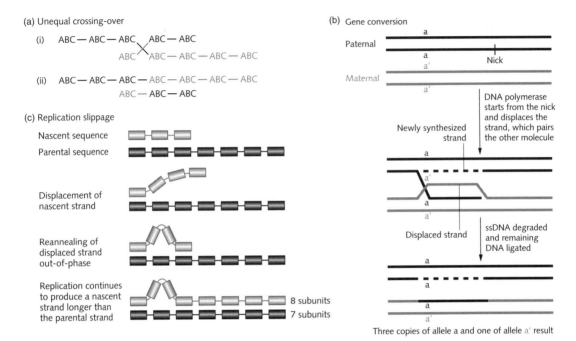

Figure 3.2 (a) Unequal crossing-over: as a result of repeated sequences (represented by ABC) pairing out of phase, crossing-over produces a longer ABC sequence and a shorter one. (b) Gene conversion. (c) Replication slippage: as a result of the nascent DNA strand becoming displaced and then annealing out-of-phase, the nascent strand becomes longer than the template strand. After Hancock (1996) with permission.

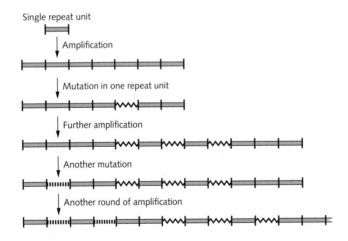

Figure 3.3 Evolution of satellite DNAs by repeated mutation and amplification. Mutations can include base changes, deletions, etc. The size of the unit of amplification can be a single repeat or multiple repeat units.

The satellite repeat units in a species are not entirely homogeneous, but consist of a number of 'families'. A change in one or more bases can occur in a single repeat unit, which may then be amplified to form a subunit that comprises a significant fraction of the total satellite. In several species (e.g. rat, mouse; Beridze, 1986) the individual satellite repeat unit consists of a number of subrepeats that have diverged from a single sequence. Thus satellites appear to be produced by repeated mutation and amplification (Fig. 3.3). Many different alpha-satellites have evolved on

Table 3.3 Classification and properties of interspersed DNA repeats.

Class*	Size	Number in human genome (estimated)	Proportion of human genome	Examples		
				Humans	*Drosophila*	Plants
DNA transposons	2–3 kb (autonomous) 80–3000 bp (non-autonomous)	404 900	2.7%	MER1-Charlie MER2-Tigger Zaphod Tc2 *Mariner*	*P*-elements *hobo*	Ac, Ds, Mu
LTR-containing	6–11 kb	697 300	7.9%	ERV Class I ERV-K & -L Etns; IAP; MaLR	TY1-*copia* *gypsy*	*Ta1*
LINEs	6–8 kb	1 371 100	18.9%	L1–3 CR1_Hs HAL1 BovB or BDDF (bovine)	*I; R2*	
SINEs	100–300 bp	1 841 000	12.5%	*Alu*; MIR; MIR3 B1 (mouse)		
Others		9600	0.1%	Pseudogenes		

*LINE, long interspersed nuclear element; LTR, long terminal repeat; SINE, short interspersed nuclear element.
Data from: Charlesworth *et al.* (1994); Kazazian (1998); Smit (1999); Li *et al.* (2001); IHGSC (2001).

different chromosomes in humans and chimpanzees by unequal crossing-over and concerted evolution (Willard, 1998). Nevertheless, there must also be mechanisms for maintaining homogeneity of satellite sequences, even between different individuals and chromosomes in the same species. For example, all the mouse chromosomes except the Y contain similar major satellite sequences, and many other examples could be cited where the same satellite occurs on different chromosomes in a species (Miklos & Gill, 1982; Beridze, 1986). How this homogeneity is maintained is not clear, although gene conversion (Fig. 3.2b) seems most likely. There is evidence for the association of regions of different human chromosomes that contain the same satellite DNAs (Schmid *et al.*, 1983), indicating that exchange between satellites on different chromosomes might be possible. Such exchange could be involved in a homogenization mechanism.

3.3.2 Interspersed repetitive sequences

Interspersed repeats are so called because they do not form tandemly repeated blocks, but are intermingled with other sequences around the genome. They fall into several classes, and together they can form a substantial part of the genome – about 45% or more in humans (IHGSC, 2001; Li *et al.*, 2001) and 50% in maize (Bennetzen *et al.*, 1998) – and the different amounts in different species explain, to a considerable extent, the C-value paradox. Interspersed repeat elements of all types are widespread in protozoa (protista), fungi, plants and animals, both invertebrates and vertebrates (Bennetzen *et al.*, 1998; Plasterk *et al.*, 1999; Arkhipova & Meselson, 2000). They are either actual mobile elements (transposons or retrotransposons) or sequences derived from mobile elements.

A classification of interspersed repeats is given in Table 3.3. Many DNA transposons (Fig. 3.4)

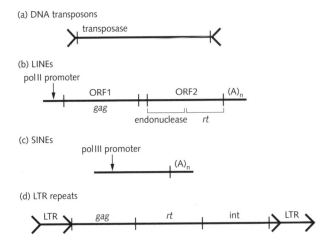

(a) DNA transposons

(b) LINEs

(c) SINEs

(d) LTR repeats

Figure 3.4 Structure of interspersed repeat elements. (a) DNA transposases, which consist of a transposase gene flanked by inverted repeats (arrowheads). (b) LINEs (long interspersed nuclear elements), which have a pol II promoter, two open reading frames (ORFs) of which ORF2 contains an endonuclease and a reverse transcriptase (rt), and a poly(A) tail. (c) SINEs (short interspersed nuclear elements), which have a pol III promoter and a poly(A) tail, but no ORFs. (d) LTR (long terminal repeat) repeat elements, which have long terminal repeats at each end and contain the *gag* gene plus genes for reverse transcriptase (rt) and integrase (int).

encode a transposase that enables the sequence to move by a 'cut-and-paste' mechanism (Plasterk et al., 1999), although other, non-autonomous transposons have lost the transposase sequence. Horizontal transmission of DNA transposons between species has occurred repeatedly between species that are not closely related (IHGSC, 2001), for example between insects and humans (Hartl, 1996).

The remaining types of interspersed elements are all retrotransposons that replicate through RNA intermediates. Long interspersed nuclear elements (LINEs; Fig. 3.4b) contain two open reading frames, coding an endonuclease and a reverse transcriptase; there is also a polymerase II promoter. In the cytoplasm, LINE RNA binds its own proteins and then migrates to the nucleus, where the endonuclease makes a single-stranded nick in the DNA, from which the reverse transcriptase primes transcription from the 3′ end of the LINE RNA. Reverse transcription is often incomplete, and only proceeds for about 1 kb in humans (IHGSC, 2001), and in fact only a small proportion of human LINEs are completely unmutated and active (Kazazian, 1998).

Short interspersed nuclear elements (SINEs; Fig. 3.4c) are mainly derived from transfer RNA (tRNA) in humans, but the important Alu sequences are derived from 7SL signal recognition particle RNA (IHGSC, 2001). Mice have SINEs derived from both tRNA and from 7SL. The

SINEs have a polymerase III promoter but lack any genes, and so rely on LINEs for transposition. Processed pseudogenes similarly lack the machinery for retrotransposition, being complementary DNA copies of fully processed messenger RNAs that therefore lack promoters (Weiner, 2000).

Long terminal repeat (LTR) retrotransposons (Fig. 3.4d) contain *gag* and *pol* genes, which encode protease, reverse transcriptase, RNase H and integrase, and retrotranspose by a retroviral mechanism, with reverse transcription occurring in the cytoplasm in a virus-like particle.

In humans, neither DNA transposons nor LTR retrotransposons are now significantly active in transposition, while the activity of L1 and *Alu*, the only active LINEs and SINEs, is declining (Smit, 1999; IHGSC, 2001). In contrast, transposon activity is much higher in the mouse genome (Kazazian, 1998; Smit, 1999; IHGSC, 2001). Organisms such as *Drosophila*, *Caenorhabditis* and *Arabidopsis* have much smaller proportions of interspersed repeats than mammals (they also have much smaller genomes; Table 3.1), and in the latter two species DNA transposons are the dominant class (IHGSC, 2001). In *Drosophila*, and also in maize, transposon activity is much greater again than in mouse (Smit, 1999), and it is estimated that the maize genome has doubled in size over the last 3 million years simply by retrotransposition (Smit, 1999).

Interspersed repeats do not only add bulk to the genome. Recombination between retrotrans-

Table 3.4 Some properties of repeated genes.

Gene*	Species	No. of copies
Histones	*Drosophila*	~100
	Xenopus laevis	20–50
	Homo sapiens	10–20
rRNA	*Drosophila*	120–240
	Triturus (Amphibia)	3900–5460
	Mammals	100–300
5S RNA	*Drosophila*	~160
	Xenopus laevis	>9000
	Homo sapiens	2000
tRNA	*Drosophila*	600–900
	Xenopus	~7000
	Homo sapiens	1310
Globins	*Homo sapiens*	8 (plus pseudogenes)

*rRNA, ribosomal RNA; tRNA, transfer RNA.

posons can produce disease-causing deletions, and insertions of retrotransposons have been estimated to cause about 1 in 600 disease-causing mutations in humans and as many as 1 in 10 in mice (Kazazian, 1998). On the other hand, transposons have been adopted by their host genomes to produce new genes, and it is estimated that nearly 50 human genes may have been derived in this way (Smit, 1999; IHGSC, 2001); these include telomerase (Section 13.3), the centromeric protein CENP-B (Section 12.4.1) and the Rag1 and Rag2 recombinases involved in VDJ recombination. In *Drosophila melanogaster* transposable elements are used to form the telomeres (Section 13.2).

3.3.3 Repeated genes

Some genes, especially those for ribosomal RNA, 5S ribosomal RNA, transfer RNAs and histones, are repeated (Table 3.4). These are often genes whose products are required in large quantities in cells, and the multiple copies of these genes are kept identical by an unknown mechanism, possibly gene conversion. Repeated genes are often, but not always, concentrated in specific regions of chromosomes, and in the case of 18S and 28S ribosomal genes tend to produce specific chromosomal structures, the

secondary constrictions (Section 11.1). Homeobox genes form a number of related families, members of which are arranged in a specific sequence on chromosomes (Finnerty & Martindale, 1998; Ferrier & Holland, 2001). Other families of genes with multiple copies, such as those for globins, do not form clusters on the chromosomes, and each gene has a distinct function.

3.4 DNA replication

The first stage of DNA replication is the separation of the two DNA strands ('unwinding') so that each is available as a template for synthesis; this is effected by a DNA helicase. The helicase activity is thought to be in the Mcm4, Mcm6 and Mcm7 subunits of the Mcm complex that is associated with the initiation of replication (Section 2.2.2.1; Labib & Diffley, 2001; Takisawa *et al.*, 2001). Because DNA synthesis can only proceed in the $5' \rightarrow 3'$ direction, only one strand of DNA (the 'leading' strand) can be synthesized continuously, whereas the other (the 'lagging' strand) is synthesized in short segments, the Okazaki fragments, back towards the replication origin (Fig. 3.5). Many aspects of the synthesis of the leading and lagging strands are similar, but

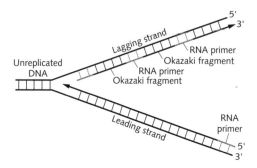

Figure 3.5 Diagram of DNA replication, showing RNA primers and Okazaki fragments.

the lagging strand requires extra processing to join up the Okazaki fragments.

All DNA synthesis starts from an RNA primer of 8–9 nucleotides, which is synthesized by the primase activity of DNA polymerase alpha. After the RNA primer has been synthesized, DNA polymerase alpha adds a short length of initiator DNA (iDNA) to the primer. Replication factor C (RF-C) attaches itself to the iDNA, and PCNA (proliferating cell nuclear antigen) binds to the RF-C, forming a 'sliding clamp' round the duplex DNA. The PCNA binds DNA polymerase delta or epsilon, displacing DNA polymerase alpha from the DNA template, and these DNA polymerases complete the DNA synthesis (Waga & Stillman, 1998).

To complete DNA synthesis, the RNA primer for the leading strand and the primers for each Okazaki fragment on the lagging strand have to be removed, and the gap filled by DNA. RNase H is a class of enzymes that degrades RNA strands hybridized to DNA. The RNA is digested endonucleolytically to produce oligoribonucleotides that dissociate easily from the DNA. A single ribonucleotide remains attached to the first deoxyribonucleotide of the iDNA, and this is removed by a 5' → 3' nuclease known as FEN1/RTH1. The resulting gap is filled by DNA synthesis using DNA polymerase delta (or epsilon), and this DNA is joined to the adjacent Okazaki fragment by DNA ligase.

A single chromosome is far too big to be replicated as a single unit, and in fact replication takes place from numerous origins simultaneously. There is nevertheless a consistent temporal sequence in DNA replication, with some segments consistently being replicated early, and others later, in the S phase (Sections 2.2.2.1, 7.1 and 10.2.2). In the yeast *Saccharomyces cerevisiae*, replication is initiated at specific sequences – the autonomously replicating sequences (ARS) (Gilbert, 2001; Méchali, 2001) – but not all ARS are necessarily used. However, in higher eukaryotes no specific initiation sequences have been identified (Gilbert, 2001; Méchali, 2001); instead, initiation appears to occur anywhere within quite large stretches of DNA (Section 2.2.2.1).

The replication of chromosomal DNA is semi-conservative, that is, each daughter molecule of double-stranded DNA consists of an 'old' strand and a newly synthesized strand. This can be demonstrated at the chromosomal level by the following experiment. Cells are grown in the presence of bromodeoxyuridine (BrdU), an analogue of thymidine that can be incorporated into DNA instead of thymidine and can be detected either by its modification of the staining properties of the chromosome or immunocytochemically (Box 3.1). After two cycles of replication, one sister chromatid has incorporated more BrdU than the other, so they can be distinguished (Fig. 3.6).

3.5 5-Methylcytosine – epigenetic modification of DNA

5-Methylcytosine is important because it is a post-synthetic modification of DNA that is nevertheless heritable. It is therefore an epigenetic phenomenon, that is, a stable change in the course of development. It is a potent source of mutations, its distribution in different genomes is highly variable and in some cases it is associated with repression of gene activity. In animals, cytosine is normally methylated only when it occurs in the dinucleotide CpG, although in plants trinucleotides having the sequence CNG (where N is any base) can also be methylated. The DNA methyltransferase Dnmt1 (maintenance methylase) has a very strong preference for

Box 3.1 Immunocytochemistry

Immunocytochemistry is the use of the specific interaction between antigens and antibodies to localize substances at the cellular level. Firstly, an antibody is raised against the antigen of interest. This primary antibody is raised in a species different from that of the cells being examined (e.g. rabbit antibody for human antigens). When the antigen is incubated with the cell preparation under appropriate conditions, the antigen binds specifically to the antigen *in situ* (Fig. 1). To identify the location of the antigen in the cell, the antigen has to be labelled by a very sensitive method: fluorescence, a histochemical enzyme reaction or (for electron microscopy) colloidal gold. The primary antibody can be labelled directly, but it is more sensitive and flexible to use a species-specific labelled secondary antibody (e.g. anti-rabbit antibody in the example above). A species-specific labelled antibody can be used to label a wide variety of primary antibodies raised in the same species.

Antibodies are now available commercially against a vast range of protein epitopes (parts of molecules that are recognized by antibodies). For labelling nucleic acids, especially DNA, antibodies are used against various modified nucleotides, which can be incorporated into the DNA molecule at S phase. Such nucleotides include bromodeoxyuridine – a thymidine ana-

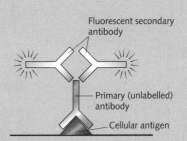

Figure 1 The principle of immunocytochemistry. A primary unlabelled antibody is allowed to bind to the antigen in the cell. The site of binding of the primary antibody is recognized by binding a labelled species-specific secondary antibody to the primary antibody.

logue that is widely used for studies of DNA replication – and digoxygenin-labelled nucleotides – which are widely used for *in situ* hybridization (Box 5.1). Nucleotides (and antibodies) can also be labelled with biotin, a vitamin that is very strongly bound by the egg-white protein avidin or the similar bacterial protein streptavidin. The biotin–avidin or streptavidin reaction can be used for highly specific labelling in the same way as an immunocytochemical reaction.

For further reading and practical information, see Polak & van Noorden (1997).

---

methylating cytosines in CpG dinucleotides where the complementary CpG in the other strand of the DNA is already methylated, so that the pattern of methylation is inherited from one cell generation to the next. Lack of Dnmt1 results in embryonic lethality (Hendrich, 2000). The Dnmt1 is targeted to replication foci through binding to PCNA (Verreault, 2000; Section 4.2.2). However, demethylase (Wolffe *et al.*, 1999) and DNA methyltransferases Dnmt3a and Dnmt3b (Reik *et al.*, 1999) are responsible for the demethylation and *de novo* methylation that occur in gametogenesis and early development (Sections 8.4.3 and 9.3), and absence of

Dnmt3 activity is lethal (Okano *et al.*, 1999). Both Dnmt3a and Dnmt3b have different specificities: Dnmt3b methylates centromeric satellite DNA, and its deficiency leads to ICF syndrome (Section 17.7).

There is a strict temporal sequence of demethylation and methylation during development in mammals (Yoder *et al.*, 1997). During preimplantation development there is a decline in methylation, followed by an increase starting at implantation, with the adult level reached during gastrulation. Different classes of DNA show different patterns of methylation, however: CpG islands remain unmethylated, except in the

(a)

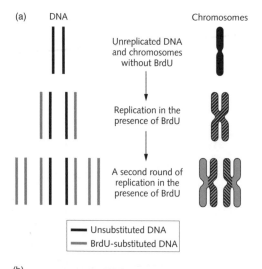

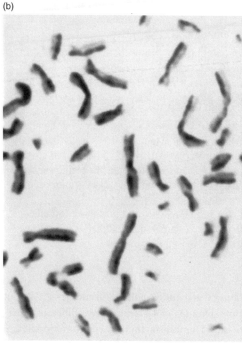

Figure 3.6 Detection of semi-conservative replication of chromosomes. After one cycle of DNA replication, BrdU is incorporated into one of the two strands of DNA in both chromatids. After two cycles of replication, one daughter molecule has both strands that contain BrdU, and therefore the chromatid that contains this molecule stains weakly, while the other daughter molecule contains one molecule substituted with BrdU and the other without BrdU, so that it is coloured more strongly. (a) Diagram. (b) Micrograph.

inactive X chromosome (Section 8.4.3), and imprinted genes have their own distinctive patterns of demethylation and methylation (Section 9.3). The L1 and IAP retrotransposons (Section 3.3.2) are never demethylated in males, but are inherited from the mother in an unmethylated state and are methylated at gastrulation. The *Alu* sequences, on the other hand, are inherited fully methylated from the mother, but unmethylated from the father, the latter being methylated again during gastrulation.

The amount of methylation of cytosines can range from undetectable as in *Caenorhabditis*, extremely low as in *Drosophila* (Gowher et al., 2000) and other arthropods, 2–7% in mammals (Colot & Rossignol, 1999) and up to 33% in some plants (Adams et al., 1992, p. 121). There are also great differences in 5-methylcytosine distribution between invertebrates and vertebrates. In the former, methylation is confined to a minor fraction of the genome (up to 30%, but often much less); many genes are found in the unmethylated fraction, but other genes in invertebrates are methylated (Tweedie et al., 1997). On the other hand, methylation occurs throughout the vertebrate genome, the exceptions being the CpG islands associated with the promoters of many genes (Jones & Takai, 2001); CpG islands are also found in plants (Gardiner-Garden & Frommer, 1992; Gardiner-Garden et al., 1992) and fungi. The CpG islands are regions that contain a higher concentration of the dinucleotide CpG than the rest of the genome. In fact, the CpG concentration is not unusually high in the islands, but is what would be expected from the DNA base composition in those regions. There is good evidence for a deficiency of CpG in the remainder of the genome in mammals, and a corresponding excess of the dinucleotides TpG and its complementary sequence CpA as a result of deamination of the 5-methylcytosine (Bird, 1987). 5-Methylcytosine is a rather unstable base, and is quite easily deaminated to produce thymine (Fig. 3.7), which, because it is a base normally present in DNA, is not recognized as a mutation. Cytosine can also become deaminated, but because the product of deamination is uracil, which does not normally occur in DNA, it is recognized and repaired

(Section 3.6.2). The CpG islands tend to become methylated in certain situations such as in X-chromosome inactivation (Section 8.4.3) and in imprinting (Section 9.3), although methylation is not a general mechanism of transcriptional control. Changes in methylation of genes are also associated with cancer (Section 17.9.3). Repeated DNA sequences, including multicopy transgenes and satellite DNAs in heterochromatin (Colot & Rossignol, 1999), are often heavily methylated,

and demethylation of heterochromatic DNA sequences results in decondensation of the heterochromatin (Sections 7.3.1 and 17.7). The general significance of methylation has not yet been established, but it has been proposed that inactivation of transposons (interspersed repetitive sequences) is an important function (Yoder et al., 1997; Vaucheret & Fagard, 2001). Failure of methylation in a hybrid marsupial leads to widespread activation of retroviral elements and chromosome remodelling (O'Neill et al., 1998).

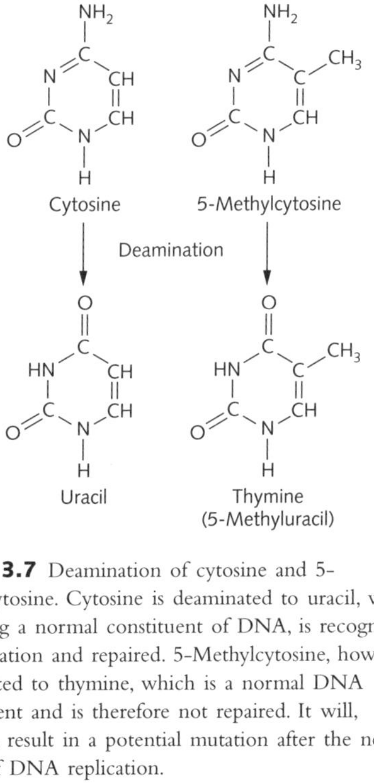

3.6 DNA damage and repair

Deoxyribonucleic acid is not an especially stable molecule, and can be damaged by a variety of agents. These include ionizing (e.g. X-rays, gamma-rays) and non-ionizing (UV) radiation and a wide variety of chemicals, not only pollutants but also therapeutic drugs and many natural products. Even the normal cellular environment, which contains free oxygen radicals, can cause DNA damage. The types of damage produced include breaks in the DNA strands and various types of damage to the bases (Table 3.5). Unless these types of damage are repaired, they can lead to mutations, which in turn may lead to cell death or cancer, or to visible chromosome damage. There are several types of DNA repair: 'mismatch' repair, excision repair, photorepair, double-strand break repair and translesion DNA synthesis. Deficiencies in DNA repair result in several human diseases characterized by the

Figure 3.7 Deamination of cytosine and 5-methylcytosine. Cytosine is deaminated to uracil, which, not being a normal constituent of DNA, is recognized as a mutation and repaired. 5-Methylcytosine, however, is deaminated to thymine, which is a normal DNA component and is therefore not repaired. It will, however, result in a potential mutation after the next round of DNA replication.

Table 3.5 Types of damage to DNA caused by different agents.

Type of damage	Agents causing damage	Repaired by
Pyrimidine dimers	UV light	Photorepair
Alkylation of bases	Chemicals	Excision repair
Base damage	UV light	Excision repair
Base damage	Reactive oxygen	Excision repair
Base damage	Spontaneous hydrolysis	Excision repair
Depurination	Chemicals	Excision repair
Single-strand breaks	X-rays, gamma-rays	Excision repair
Double-strand breaks	X-rays, gamma-rays	Double-strand break repair
Mispaired bases	Replication errors	Mismatch repair

Table 3.6 DNA glycosylases and their specificities in base excision repair.

DNA glycosylase	Substrate	Species
Uracil	Uracil	*Saccharomyces cerevisiae; Homo sapiens*
3-Methyladenine	3-Methyladenine 7-Methylguanine Hypoxanthine 7,8-Dihydro-8-oxoguanine	*Saccharomyces cerevisiae; Homo sapiens;* *Bos taurus*
8-Oxoguanine (fapy)	2,5-Amino-5-formamido-pyrimidine 7,8-Dihydro-8-oxoguanine	*Saccharomyces cerevisiae; Homo sapiens*
Thymine glycol (Endonuclease III)	5-Hydroxycytosine Urea Thymine glycol	*Saccharomyces cerevisiae; Homo sapiens;* *Mus domesticus; Bos taurus*
A-G mismatch	Adenine 5-Hydroxycytosine	*Homo sapiens*
G-T mismatch	T-G; U-G	*Homo sapiens*
Hydroxymethyluracil	Thymine glycol	*Homo sapiens*
Formyluracil	5-Formyluracil	*Homo sapiens*

Data from: Demple & Harrison (1994); Seeberg *et al.* (1995).

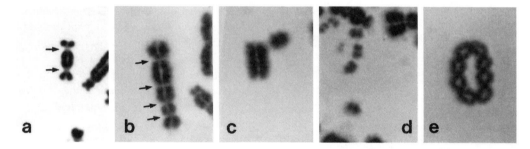

Figure 3.8 Different types of chromosome damage: (a) a dicentric chromosome – the two centromeres are arrowed; (b) a multicentric chromosome – the centromeres are arrowed; (c) an acentric fragment – the chromatids lie parallel throughout their length – there is no centromeric constriction; (d) minute chromosomes; (e) ring chromosome.

development of cancers or chromosomes breakage (Section 17.4).

The types of chromosomal damage that result from damage to the DNA include chromosome and chromatid breaks, dicentric and multicentric chromosomes, acentric fragments, minute chromosomes, pericentric and paracentric inversions, isochromosomes and ring chromosomes (Box 3.2; Fig. 3.8). If the chromosomal damage is caused during the G1 phase of the cell cycle, the damage may be replicated with the rest of the DNA during S phase, so that it appears in both chromatids at the following metaphase: this is 'chromosome-type damage'. If the damage occurs during G2, then only one chromatid is affected, resulting in 'chromatid-type damage'. There is good evidence that a single break in a DNA molecule is sufficient to cause a chromatid or chromosome break. Other kinds of chromosome damage are the result of one broken chromosome end fusing with another broken chromosome end. Unlike the normal chromosome ends, the telomeres (Section 13.4) – freshly broken chromosome ends – are 'sticky' and may fuse with any other sticky ends that they encounter, either from different chromosomes to

Box 3.2 Types of chromosome damage and their formation

Different types of chromosome damage, and the mechanisms by which they are produced, are illustrated in Fig. 1.

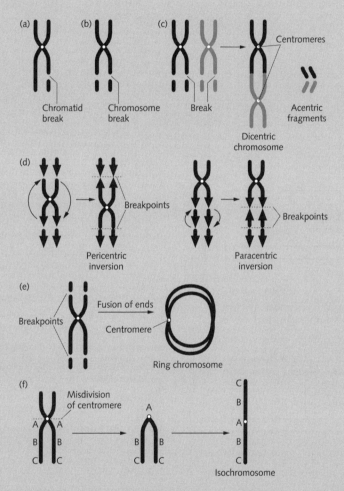

Figure 1 Formation of different types of chromosome damage.
(a) A *chromatid break* is the result of breaking the DNA in one chromatid, normally in the G2 phase after DNA replication.
(b) A *chromosome break* is the result of breaking the DNA in G1, before it has been replicated. The break is therefore 'replicated' at S phase, and so appears at identical sites in the sister chromatids.
(c) *Dicentric chromosomes* and *acentric fragments* can be formed when two chromosomes are broken, and subsequently fuse with each other. If both components of the fused chromosome contain a centromere, a dicentric chromosome is formed. At anaphase there is a possibility that the two centromeres on the same chromatid will be pulled in opposite directions. The chromatid will then form a bridge between the daughter cells and will break again between the centromeres. If the broken daughter chromosomes then fuse to form more dicentric chromosomes, this breakage–fusion–bridge cycle can be repeated indefinitely.
Formation of a dicentric chromosome also results in the production of acentric fragments, which need not fuse

Continued on p. 38

Box 3.2   *Cont.*

with each other. Acentrics can also be produced when the end of a chromosome is broken off, independently of the formation of a dicentric. Because acentrics have no centromeres, they cannot attach to the spindle and so are usually lost at cell division.

(d)   *Inversions* are formed when two breaks occur in the same chromosome, and the chromosomal segment in the middle rotates before it fuses back into the chromosome. If the middle segment contains the centromere, the inversion is *pericentric*; if it does not contain a centromere it is *paracentric*.

(e)   *Ring chromosomes* are formed when both ends of the chromosome are broken off, and the newly exposed 'sticky' ends fuse together.

(f)   *Isochromosomes* are chromosomes with homologous arms, that is, arms that are structurally and genetically identical and are mirror images of each other (i.e. their sequences run in opposite directions from the centromere). They can arise in various ways. Two identical acrocentric chromosomes may fuse at their centromeric ends, or a translocation may occur that has the same effect. Another mechanism is centromere misdivision, in which transverse breakage of the centromere occurs.

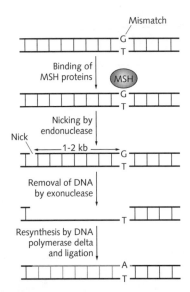

Figure 3.9 Mismatch repair (proof-reading of newly synthesized DNA).

form di- and multicentrics (Fig. 3.8a,b), or from the same chromosome to form ring chromosomes (Fig. 3.8e) and isochromosomes.

3.6.1   Mismatch repair

Mismatch repair (Fig. 3.9) is a method for correcting the sequence of newly replicated DNA (Buermeyer *et al.*, 1999; Kolodner & Marsischky, 1999). The fidelity of DNA replication is good,

but not perfect, and it is therefore necessary to have a method of 'proof-reading'. The repair system recognizes normal but mismatched bases, and small loops produced by the insertion of too many or too few bases during replication (IDLs, insertion/deletion loops). It is not known how such mismatches are recognized, or how the faulty strand rather than the correct one is selected for repair. In eukaryotes, three MSH proteins have been identified that form different heterodimers: MSH2-MSH6 recognizes base mismatches and 1 bp IDLs, while MSH2-MSH3 recognizes 2–4 bp IDLs. Other proteins, known as Mlh1 (Msh1 in *Saccharomyces cerevisiae*) and PMS1 and PMS2, form a complex with the MSH proteins to form a higher order complex that is directed to the replication fork by PCNA. The binding to PCNA may allow discrimination of the newly synthesized DNA strands. When the complex is assembled, repair is initiated by nicking the DNA at a distance of up to 1–2 kb from the mismatched site, followed by degradation of the DNA by exonuclease I until the mismatch is reached and removed. A new DNA strand is then synthesized by DNA polymerase delta (Sancar, 1999).

As well as repairing mismatches that occur during replication, the mismatch repair system also corrects mismatches that result from recombination. Mutations in the mismatch repair system occur commonly in hereditary non-polyposis colon cancer (HNPCC; Section 17.9.1).

3.6.2 Base excision repair

The other types of repair deal with the damage produced after the DNA has been synthesized. In excision repair, the damage (generally referred to as a lesion) is removed from the DNA, and a new DNA strand is synthesized to fill the gap, using the other strand of the DNA molecule as a template (Thoma, 1999).

In base excision repair (McCullough *et al.*, 1999), a single damaged base is recognized, removed and replaced. The lesion in the DNA molecule is recognized by a specific DNA glycosylase, of which several have been recognized (Table 3.6). The main endogenous lesions corrected by base excision repair are AP (apurinic/apyrimidinic) sites and uracil (produced by deamination of cytosine). Other lesions excised are those produced by oxygen and by alkylation. Certain base mismatches are also corrected. Removal of the damaged base leaves an AP site, from which the deoxyribose phosphate is removed by AP-endonuclease, which cleaves 5′ to AP sites, and AP lyase, which cleaves 3′ to AP sites, although the latter does not seem to be essential (Seeberg *et al.*, 1995). Finally, the gap is filled by DNA polymerase beta (in mammalian cells) or DNA polymerase delta (in *S. cerevisiae*), and the new nucleotide is ligated to the rest of the DNA (Fig. 3.10).

3.6.3 Nucleotide excision repair

Nucleotide excision repair (NER) is less specific than base excision repair, as it does not require a specific enzyme for each type of lesion, but uses the same mechanism to remove a wide variety of lesions. It occurs throughout eukaryotes, but is not essential for viability, and the occurrence of certain human diseases deficient in NER (xeroderma pigmentosum, Cockayne's syndrome and trichothiodystrophy; Section 17.4) has been valuable in elucidating the mechanisms of NER.

In NER, a stretch of nucleotides around the lesion is removed and then resynthesized. The

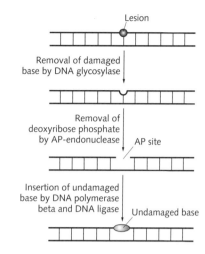

Figure 3.10 Base excision repair.

first stage is recognition of the lesion (Fig. 3.11; de Laat *et al.*, 1999; Thoma, 1999). Unlike base excision repair, this is a non-specific process that depends on detection of a distortion in the DNA, which is recognized by the XPC-hHR23B complex (for a listing of corresponding yeast and rodent NER proteins, see Table 3.7). These recruit the transcription factor TFIIH, followed by XPA and RPA, which remodel the chromatin to produce a more open conformation, while DNA helicases XPB and XPD separate the DNA strands. The DNA is cut 3′ and 5′ to the lesion, by XPG and XPF-ERCC1, respectively, and the damaged strand of DNA, about 27–29 bp long (Sullivan, 1995), is removed. The gap is then filled by new synthesis catalysed by DNA polymerase delta and epsilon, and the new DNA joined to the old by DNA ligase. All these processes take place in chromatin, of course, rather than on naked DNA, and remodelling of nucleosomes is needed before the lesion can be recognized and corrected (Smerdon & Conconi, 1999; Thoma, 1999).

The sequence described in the previous paragraph refers to what has become known as global genome NER (GG-NER), but actively transcribing DNA is repaired by a slightly different mechanism, transcription-coupled NER (TC-NER). Transcribed genes are repaired more quickly than non-transcribed DNA, and lesions

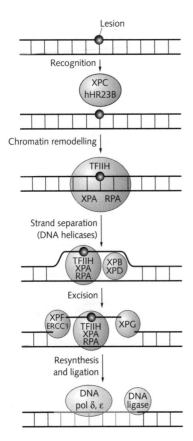

Figure 3.11 Nucleotide excision repair.

are removed from the transcribed strand more quickly than from the non-transcribed strand (Thoma, 1999). The repair deficiency in Cockayne's syndrome (Section 17.4) involves TC-NER. When RNA polymerase II stalls at a lesion, the proteins CSA and CSB bind, and the polymerase backs away from the lesion but does not release the RNA. Both XPA and TFIIH are then recruited to the site, the repair complex is assembled and, after the repair has been completed, transcription resumes.

3.6.4 Photorepair

When DNA is irradiated with UV light, various types of damage are produced, of which the principal ones are cyclobutane pyrimidine dimers (CPDs) and pyrimidine (6-4) pyrimidone photoproducts (6-4PPs) (Fig. 3.12). The CPDs are formed between any two adjacent pyrimidines in the same strand of DNA: TpT, TpC, CpT or CpC. The 6-4PPs are usually formed between TpC and CpC. Although such lesions can be removed by nucleotide excision repair (Section 3.6.3), many organisms have a special photorepair system for removing ultraviolet damage (Thoma, 1999). Specific photolyases bind

Table 3.7 Nucleotide excision repair (NER) proteins in mammals and yeasts.

| | NER protein | | |
Process	Human	Rodent	Yeast
Recognition of DNA lesion	XPC		Rad4
	hHR23B		Rad23
Chromatin remodelling	TFIIH		TFIIH
	XPA		Rad14
	RPA		Rfa
DNA unwinding (DNA helicases)	XPB	ERCC3	Rad25
	XPD	ERCC2	Rad3
Incision (3′)	XPG	ERCC5	Rad2
Incision (5′)	XPF	ERCC4	Rad1
	ERCC1	ERCC1	Rad10

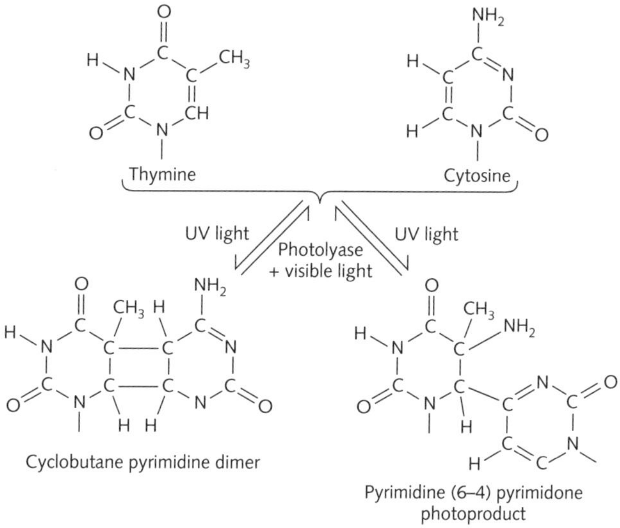

Thymine

Cytosine

UV light

Photolyase + visible light

UV light

Figure 3.12 Photorepair: formation and repair of cyclobutane pyrimidine dimers and pyrimidine (6-4) pyrimidone photoproducts. The pyrimidines involved (cytosine and thymine) are usually on the same strand of DNA.

Cyclobutane pyrimidine dimer

Pyrimidine (6–4) pyrimidone photoproduct

to CPDs or 6-4PPs and, in the presence of visible light as an energy source, restore the original dinucleotides. The CPD photolyases are found in fungi, plants, invertebrates and many vertebrates (but not in humans), and 6-4PP photolyases have been reported in *Drosophila*, silkworms, *Xenopus laevis* and rattlesnakes.

3.6.5 Double-strand break repair

Double-strand breaks (DSBs) in DNA present particular problems for cells. On the one hand, if unrepaired they can lead to chromosome breaks, which may be lethal; on the other hand, they may be more difficult to repair than the lesions previously considered because there is no intact complementary strand of DNA that can be used as a template. Double-strand breaks are caused by ionizing radiation (X-rays and gamma-rays) and by certain chemicals; they are repaired by at least three different mechanisms, and deficiencies in DSB repair can lead to cancer or to ataxia telangiectasia and Nijmegen breakage syndrome (Section 17.4).

The simplest way to repair DSBs is by non-homologous end joining (NHEJ) (Kanaar *et al.*, 1998; Karran, 2000). No homology between the adjacent ends is required, and the repair is not error-free. The first stage of NHEJ is the binding

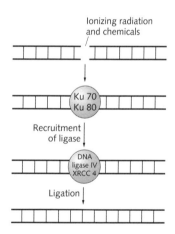

Ionizing radiation and chemicals

Ku 70 Ku 80

Recruitment of ligase

DNA ligase IV XRCC 4

Ligation

Figure 3.13 Double-strand break repair by non-homologous end joining.

of a heterodimer of the proteins Ku70 and Ku80 to the broken ends, which not only holds the ends together but prevents their nucleolytic degradation. The Ku heterodimer activates DNA-dependent protein kinase, which facilitates end joining by a heterodimer of DNA ligase IV and XRCC4 (Fig. 3.13).

Homologous recombination (Fig. 3.14) repairs DSBs by a recombinational mechanism (Kanaar *et al.*, 1998; Karran, 2000). A nuclease digests the DNA on either side of the break to produce

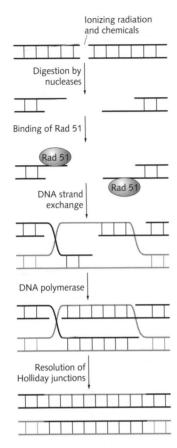

Figure 3.14 Double-strand break repair by homologous recombination.

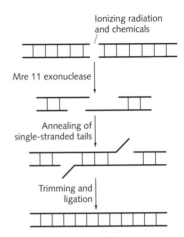

Figure 3.15 Double-strand break repair by single-strand annealing. This mechanism results in a variable loss of DNA on either side of the break.

potentially capable of leading to chromosome rearrangements, this is rarely seen.

Homologous recombination can be responsible for the repair of about 50% of DSBs in mammalian cells (Richardson *et al.*, 1998), and the proportion of DSBs repaired by NHEJ is also believed to be substantial. There is, however, a third mechanism of DSB repair: single-strand annealing (Karran, 2000). Like homologous recombination, single-strand annealing relies upon homology, but it does not involve recombination. The proteins Rad50 and Mre11 form a complex that binds to the DSB (Fig. 3.15). Protein Mre11 has $3' \rightarrow 5'$ exonuclease activity that digests away short lengths of one strand on each side of the break, and these single-stranded segments then start to search for homology between themselves. Once annealing of the single-stranded tails has occurred, they are trimmed to size and ligated.

3.6.6   Translesion DNA synthesis

Normally DNA synthesis is blocked when the polymerase reaches a lesion, and cannot continue until the lesion has been repaired. There are, however, polymerases that can continue past lesions, and although strictly this process is not repair, it does provide a means of replicating

3′ overhangs onto which Rad51 protein polymerizes. The resulting nucleoprotein filaments then search for homologous double-stranded DNA. The DNA strand exchange produces a joint molecule from the intact and damaged DNA molecules. The two strands of the intact molecule are used as a template to synthesize the missing segments of the damaged DNA, and finally the Holliday junctions are resolved to produce two separate DNA molecules again.

Because of the large quantities of interspersed repetitive elements in mammalian genomes, recombination between non-homologous chromosomes is possible, and in fact occurs, during the repair of DSBs (Richardson *et al.*, 1998). Although homologous recombination is faulty DNA that has hitherto escaped repair systems and has evaded the DNA repair checkpoint (Section 2.2.3). Two different DNA polymerases can perform translesion synthesis (Bridges, 1999): DNA polymerase zeta, encoded by *REV3* in *Saccharomyces cerevisiae* and humans, carries out error-prone synthesis and actually introduces mutations into the DNA; however, DNA polymerase eta, encoded by *RAD30* in *S. cerevisiae*, can insert the correct bases opposite a cyclobutane thymine dimer, and a similar polymerase may be deficient in the human disease xeroderma pigmentosum variant (XPV).

3.7 DNA is dynamic

Deoxyribonucleic acid is clearly not stable to the point of being inert. Over the course of evolution it has had to change, although a comprehensive set of repair mechanisms has limited the changes, but not eliminated them. Indeed, mutational changes are the basis of evolution. A changed DNA sequence may sometimes lead to a favourable change in the organism's phenotype that gives it a selective advantage. This point has been understood for many years, since the nature of gene mutation became known. It is only more recently that the epigenetic modification of DNA by methylation of cytosine has been understood as a means of modulating gene activity in certain special cases, although its general function has still not been elucidated.

Finally, it now appears that a large proportion of an organism's genome is made up of a variety of repetitive sequences that show a good deal of autonomy in their behaviour. Such sequences can multiply and transfer themselves from one genome to another. They are largely responsible for the C-value paradox. Even here, though, the organism probably exerts more control than might appear at first sight: different organisms maintain small or large genomes apparently by controlling the quantity of their repetitive sequences. The incorporation of the DNA into chromatin, as described in the next chapter, is essential both for maintenance of DNA and for the control of its functions.

Websites

DNA C-values
DOGS (Database of Genome Sizes):
http://www.cbs.dtu.dk/database/DOGS/
Angiosperm DNA C-values Database:
http://www.rbgkew.org.uk/cval/database1.html

DNA structure
Nucleic Acid Database (NDB):
ndbserver.rutgers.edu/NDB/

Interspersed repetitive elements
Repbase:
www.girinst.org/Repbase_Update.html

DNA methylation
DNA Methylation Society:
dnamethsoc.server101.com/

DNA repair
www.nih.gov/sigs/dna-rep/whatis.html
mcbio.med.buffalo.edu/RPN530/DNA_Repair.html

Assembly of chromatin

4.1 Introduction

We have seen in Chapter 3 that the length of DNA in the nucleus of each cell is enormous – approximately 2 m in every mammalian nucleus, for example (see Table 3.1) – so it must obviously be packed in some manageable form. This packing is accomplished by combining the DNA with a variety of proteins, which also enable the DNA to carry out its functions.

The packing of DNA into chromatin and chromosomes is achieved in a number of distinct stages, each with its own characteristic packing ratio (the ratio of the length of the DNA to that of the structure into which it is compacted) (Table 4.1). The higher levels of packing differ somewhat between interphase chromatin and metaphase chromosomes, the latter being more highly compacted (Sections 6.3 and 6.4). However, the lower levels of packing, whether in interphase or metaphase, are, with few exceptions, the same and form the subject of this chapter.

4.2 The nucleosome fibre

The first stage of compaction of the DNA fibre is produced by winding it round a body consisting of eight histone molecules, two each of H2A, H2B, H3 and H4 (Kornberg & Lorch, 1999; www.average.org/~pruss/nucleosome.html). The histones are small basic proteins that consist of a central, highly structured region, the histone-fold domain, and C- and N-terminal tails that lack secondary structure (Fig. 4.1). Specific amino acid residues in the N-terminal tails can be modified by acetylation (Section 4.2.4), phosphorylation, methylation, ADP-ribosylation or ubiquitination (Section 4.2.6). These modifications have important and specific effects on the properties of the histones and of the chromatin in which they are incorporated, and form a 'histone code' that regulates the behaviour of DNA in chromatin (Jenuwein & Allis, 2001).

Each histone octamer has approximately 1.7 turns of DNA wrapped round it, and each nucleosome is connected to the next by a stretch of linker DNA, the length of which varies from one species or cell type to another (Table 4.2). Digestion of chromatin with nucleases produces different sizes of particles, depending on the conditions and length of digestion. Initially, fragments containing 200 bp of DNA are produced, and then successively smaller fragments as digestion continues. Different names have been applied to different types of nucleosomal structures with different lengths of DNA attached (Fig. 4.2). The histone octamer with 146 bp of DNA wrapped round it is the *core particle*. The same structure with a molecule of histone H1 apparently holding together the DNA where it enters and leaves the core particle, and which contains 168 bp of DNA, is known as the *chromatosome*. The complete structure, including on average 35 bp of linker DNA, making a total of about 200 bp of DNA, is the *nucleosome sensu stricto*. Formation of the nucleosomal fibre com-

Table 4.1 Packing ratios of DNA in chromatin and chromosomes.

Structure	Length per cell*	Breadth	Packing ratio
DNA molecule	2 m ($2 \times 10^6 \mu$m)	2 nm	1
Nucleosome fibre	0.28 m ($2.8 \times 10^5 \mu$m)	10 nm	7
Solenoids	0.04 m ($4 \times 10^4 \mu$m)	30 nm	50
Loops	1 mm ($10^3 \mu$m)	0.26 μm (260 nm)	2000
Chromosomes	200 μm	2 μm (2000 nm)	10 000

*In humans and most other mammals.

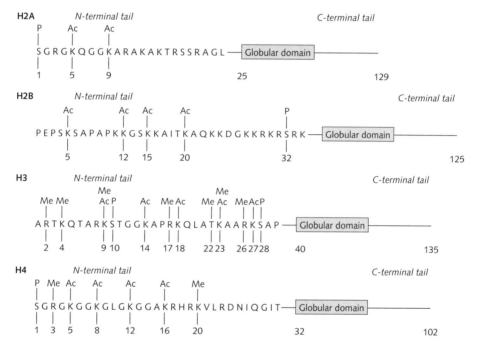

Figure 4.1 The structure of core histone molecules (H2A, H2B, H3 and H4). These consist of an apolar, globular, histone-fold domain, an unstructured highly basic N-terminal tail and a short basic C-terminal tail. The N-terminal tails contain: lysine (K) residues that can be acetylated (Ac); arginine (R), lysine (K) and threonine (T) residues that can become methylated (Me); and serine (S) residues that can become phosphorylated (P). Certain lysine residues in H3 can become either acetylated or methylated, but not both simultaneously. The numbers below each drawing are the amino-acid numbers. Data from Cheung *et al.* (2000), Strahl & Allis (2000) and Zhang & Reinberg (2001).

pacts the DNA sevenfold, and produces a chromatin fibre approximately 10 nm in diameter, corresponding to that seen by electron microscopists when specimens are prepared under low ionic strength conditions. Further dispersion of the chromatin leads to the 'beads-on-a-string' structure; this, however, is probably an artefact that rarely, if ever, occurs in nature.

The structure of the nucleosome has been determined at high resolution (Luger *et al.*, 1997;

Rhodes, 1997) (Fig. 4.3). The central parts of the histone molecules form the core of the nucleosome, while the amino-terminal tails extend outwards and are involved in interactions between adjacent nucleosomes. The DNA follows a bent path round the histone core, and is held strongly by electrostatic bonds between the basic amino acids of the histones and the phosphates of the DNA. Although the length of DNA wrapped round the histone octamer is usually regarded as

Table 4.2 Lengths of DNA per nucleosome.

Species	Cell type	DNA/nucleosome (bp)
Yeast		163,165
Aspergillus		154
Neurospora		170
Physarum		171,190
Sea urchin	Gastrula	218
	Sperm	241
Chick	Oviduct	196
	Erythrocyte	207,212
Rabbit	Cortical neuron	162
	Cortical glia	197
Rat	Bone marrow	192
	Kidney	196
Human	HeLa	183,188

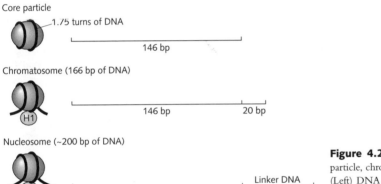

Figure 4.2 The structure of the core particle, chromatosome and nucleosome. (Left) DNA wrapped round a histone octamer. (Right) Linear DNA from the different structures.

constant, in fact it varies in length between about 100 and 170 bp. The nucleosome is therefore a much more variable and dynamic structure than suggested by the 'standard' 146 bp model (van Holde & Zlatanova, 1999). The nucleosomal histones are generally stable, but a fraction of H2B appears to be continually exchanged (Kimura & Cook, 2001).

4.2.1   Linker histones

Linker histones (histone H1 and its variants, and some other related proteins — Table 4.3, and see below) bind to the DNA that links adjacent nucleosome core particles, and also to the core particles themselves, thus stabilizing the nucleosomal fibre and playing a part in the further condensation of the 10 nm fibre to form the 30 nm fibre (Section 4.3). For discussion of the possible modes of binding of linker histones, see van Holde & Zlatanova (1996), Travers (1999) and Wolffe & Hayes (1999). The linker histones are much more variable in structure than the core histones, and some are radically different from the 'standard' type found in higher eukaryotes. Histone H1 typically consists of a central globular domain, with C- and N-terminal tails that are rich in basic amino acids and can bind strongly

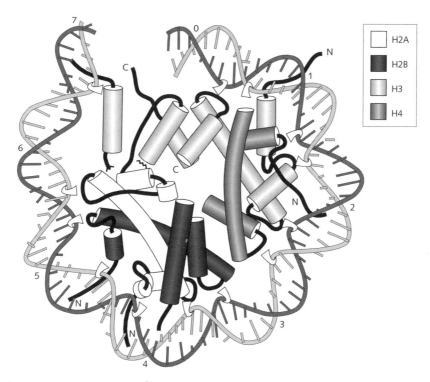

Figure 4.3 Nucleosome structure at 2.8 Å resolution; half the nucleosome is shown, with one turn of DNA round it. Points of contact between DNA and histones are indicated by hooks. Reprinted with permission from Rhodes (1997) *Nature* **389,** 231–232. © Macmillan Magazines Limited.

to DNA. However, the linker histone of *Tetrahymena* has no globular domain, being similar to the C-terminal domain of a typical H1 histone, while in the yeast *Saccharomyces cerevisiae* there is an H1-like protein that has two globular domains (Wolffe *et al.*, 1997); it is not clear if the latter acts as a linker histone. Even among vertebrates, there may be several different types of H1 in the same species, which may be present simultaneously, or at different developmental stages, or in particular cell types (Table 4.3). For example, histone H5 in birds is a linker histone specific to the inactive chromatin of the erythrocyte nuclei.

Linker histones are not essential for assembly of chromatin, and cells lacking them are viable (Wolffe *et al.*, 1997); instead, they may have a role in gene regulation (Zlatanova & van Holde, 1998; Wolffe & Hayes, 1999). Like core histones, histone H1 is in dynamic equilibrium, and part

of it is constantly being exchanged between different sites on chromatin (Misteli *et al.*, 2000).

4.2.2 Assembly of nucleosomes

Assembly of nucleosomes on DNA occurs during S phase, immediately after DNA synthesis (Verreault, 2000; Mello & Almouzni, 2001), and requires two different processes. Existing nucleosomes are assembled on one or other of the daughter strands of DNA, but of course this provides only half as many nucleosomes per unit length of DNA as there were before replication. New nucleosomes are therefore also assembled, using histones that have been synthesized in the cytoplasm during the same S phase. First a tetramer of H3 and H4 is deposited, and then two dimers consisting of H2A and H2B are added. Acetylation of the N-terminal tails of H3 and H4 is necessary for their deposition; H4 must

Table 4.3 Linker histones.

Name	Species	Comments
H1a	Mammals	In dividing cells
H1b	Mammals	In dividing cells
H1c	Mammals	
H1d	Mammals	
H1e	Mammals	In non-dividing cells
H1oo	Mammals	Oocyte-specific (Tanaka et al., 2001)
H1^0	Mammals	In non-dividing cells
H1t	Mammals	Testis-specific (Lin et al., 2000)
MDBP-2-H1	Chicken	Methylated DNA binding protein-2-H1: truncated H1 subtypes (Schwarz et al., 1997)
H5	Fish, amphibia, birds	In nucleated erythrocytes
B4 (H1M)	Xenopus	Oocytes and embryonic cells
Cs-H1	Sea urchin	Oocyte-specific cleavage stage histone H1 (Tanaka et al., 2001)
H1	Sea urchin	Six different forms, expressed at different stages of development
H1-1	Euplotes crassus	Macronuclear (Ray et al., 1999)
H1-2	Euplotes crassus	Macronuclear (Ray et al., 1999)
H1-1	Arabidopsis	Ascenzi & Gantt (1999)
H1-2	Arabidopsis	Ascenzi & Gantt (1999)
H1-3	Arabidopsis	Induced by drought stress (Ascenzi & Gantt, 1999)

For more detailed information on histones, see: http://genome.nhgri.nih.gov/histones/

be acetylated specifically on lysine residues 5 and 12. Deposition of H3 and H4 on the DNA is mediated by the protein CAF1 (chromatin assembly factor 1), which is localized at sites of replication in S-phase nuclei by binding to proliferating cell nuclear antigen (PCNA). However, it does not appear to be an essential protein, as mutants of CAF-1 show relatively minor defects. One alternative pathway may use the chaperone Asf-1 (Mello & Almouzni, 2001). Once H3 and H4 have been assembled on the DNA, H2A and H2B are added in a process mediated by Nap1 (nucleosome assembly protein 1). Finally, a maturation process occurs, in which the nucleosomes become regularly spaced on the DNA, H3 and H4 are deacetylated and histone H1 is added.

High levels of DNA methylation are correlated with low levels of histone acetylation, at least in mammalian chromosomes, and particularly in constitutive heterochromatin (see Chapter 7). The maintenance methylase Dnmt1 is targeted to replication foci through binding to PCNA (Fig. 4.4) (Verreault, 2000). The Dnmt1 in turn binds the deacetylase HDAC1, which is therefore in the correct position to deacetylate the newly deposited acetylated histones H3 and H4.

4.2.3 How can replication and transcription take place on the nucleosomal fibre?

Nucleosomes can present an obstacle to the free passage of protein complexes required for replication and transcription, and to the separation of the DNA strands that is required for these processes. How then can the processes of replication (Section 3.4), transcription and DNA repair (Section 3.6) occur, which require the access of bulky protein complexes to the DNA and the separation of the DNA strands? Clearly, the interactions between the DNA and the nucleosomal histones need to be loosened. In some cases, the binding of transcription factors alone is sufficient to displace the DNA, at least partly, from the histone octamer (Felsenfeld,

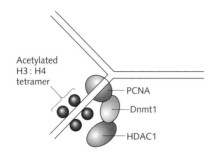

Acetylated
H3 : H4
tetramer

PCNA

Dnmt1

HDAC1

Figure 4.4 Deacetylation of histones deposited at replication forks containing methylated CpG dinucleotides. The maintenance methyltransferase Dnmt1 associates with PCNA at the replication fork, and in turn binds the histone deacetylase HDAC1 to facilitate deacetylation at sites containing methylated DNA. The positions of the different molecules are arbitrary and do not necessarily reflect their three-dimensional arrangement *in vivo*.

1996; Felsenfeld *et al.*, 1996), but in other cases more complex mechanisms are involved. These involve histone modifications and chromatin remodelling complexes (Kornberg & Lorch, 1999).

4.2.4 Histone acetylation

Histone acetylation – more specifically, acetylation of lysines 9 and 14 on the N-terminal tail of histone H3, and of lysines 8 and 16 on the N-terminal tail of H4 – is associated with activation of genes (Davie, 1998; Kuo & Allis, 1998; Luger & Richmond, 1998; Struhl, 1998). Acetylation reduces interactions between the histone and the DNA, leading to a much looser structure. At the chromosomal level, regions rich in genes are hyperacetylated (Section 10.2.2), whereas inactive regions such as heterochromatin (Section 7.3.2) and the inactive X chromosome in female mammals (Section 8.4.3) are hypoacetylated. Acetylation is reversible, carried out by enzymes known as histone acetyltransferases (HATs; Marmorstein & Roth, 2001) and can be removed by histone deacetylases (HDACs; Khochbin *et al.*, 2001). Histone acetyltransferase activity is associated with various transcriptional co-activators (Boyes *et al.*, 1998; Davie, 1998;

Kornberg & Lorch, 1999; Jenuwein & Allis, 2001), which associate with sequence-specific transcription factors and thus provide specificity for the process. Acetyltransferases can also acetylate transcription factors (Boyes *et al.*, 1998), so that acetylation is required at several levels to promote transcription.

4.2.5 Chromatin remodelling

Acetylation of histones is not always sufficient to make the DNA available for transcription. Several chromatin remodelling complexes have been identified (Table 4.4), each of which is specific for particular promoters and controls the transcription of only a small fraction of the genome (e.g. SWI/SNF affects less than 6% of the genes in *S. cerevisiae*; Sudarsanam & Winston, 2000). The remodelling complexes contain between 2 and 15 subunits, one of which is ATPase that, in the presence of ATP, can disrupt nucleosome structure (Gregory & Hörz, 1998; Lemon & Freedman, 1999; Kornberg & Lorch, 1999; Tyler & Kadonaga, 1999; Peterson & Workman, 2000). Some facilitate binding of transcription factors to the chromatin, while others promote transcription. Some are also involved in chromatin assembly. The ISWI remodelling complexes, such as CHRAC and ACF, induce regular spacing of nucleosomes on the DNA (Flaus & Owen-Hughes, 2001). Another protein, FACT (facilitates chromatin transcription), acts after initiation of transcription to allow transcription to continue past nucleosomes (Orphanides *et al.*, 1998). Thus although remodelling of nucleosomes at promoter regions of genes allows initiation of transcription, a second mechanism is required to allow it to continue.

4.2.6 Other modifications of histones

Histones can be modified in other ways, with various effects (Table 4.5) (Wolffe & Hayes, 1999; Strahl & Allis, 2000; Jenuwein & Allis, 2001), although in general these modifications have not been studied as intensively as acetylation. These modifications, like acetylation, generally modify the structure of the chromatin, with conse-

Table 4.4 Chromatin remodelling complexes.

Complex	Species	No. of subunits	Function
SWI/SNF family			Disrupts nucleosomal structure
SWI/SNF	S. cerevisiae	11	
RSC	S. cerevisiae	15	
Brahma	D. melanogaster	~7	
hSWI/SNF	Humans	~10	
ISWI family			Proper spacing of nucleosomes
ACF	Drosophila	2	Facilitates binding of activators to chromatin
CHRAC	Drosophila	5	Increases accessibility of chromatin
NURF	Drosophila	4	Mediates binding of transcription factors
ISW1	S. cerevisiae	4	Disrupts nucleosomes
ISW2	S. cerevisiae	2	Nucleosome spacing
RSF	Humans	2	Facilitates transcription
Mi-2/CHD family			
Mi-2	Xenopus	6	
NuRD	Humans	~7	

Data from Armstrong & Emerson (1998) and Tyler & Kadonaga (1999).

Table 4.5 Modifications of histones.

Modification	Histone(s) modified	Effects/function
Acetylation	H3, H4	Activation of genes
ADP-ribosylation	Core histones	Local disruption of chromatin structure to facilitate DNA repair
Methylation	H3, H4	Repression of transcription
Phosphorylation	H1	?Chromatin condensation
	H3	Gene activation; chromatin condensation
	H2A, H4	?Nucleosome assembly
Ubiquitination	H2A (and H2B)	?Disruption of chromatin structure to facilitate transcription

For more detailed information on modifications of histones, see Sullivan *et al.* (2000) and http://genome.nhgri.nih.gov/histones/

quences for transcription, DNA repair or chromatin condensation.

The function of histone methylation has long been a mystery, but it is now known that methylation of lysine 9 of histone H3 represses transcription (Rea *et al.*, 2000). The acetylated lysine 9 is first deacetylated; a histone H3-specific methyltransferase, encoded by SUV39H1, then methylates this lysine, which then recruits chromatin protein HP1 (Section 7.3.2) to repress transcription. This can lead to heterochromatin

formation (Zhang & Rheinberg, 2001). On the other hand, methylation of lysine 4 of H3, either alone or in combination with other sites, appears to lead to transcriptional activation (Zhang & Rheinberg, 2001). In fission yeast, heterochromatin contains H3 methylated at lysine 9, while in euchromatin H3 is methylated at lysine 4 (Noma *et al.*, 2001).

The ADP-ribosylation of core histones occurs when DNA repair (Section 3.6) is activated. The introduction of the acidic residues presumably neutralizes the basic groups of the histones and destabilizes the nucleosome structure to allow access of DNA repair enzymes.

Phosphorylation is a major modification of histones, occurring mainly on histones H1 and H3. Phosphorylation of H1, which occurs mainly on specific amino acid sequences in the histone tails, tends to neutralize the positive charge and disrupt the chromatin structure, which allows activation of transcription (Wolffe & Hayes, 1999). Other situations have been described in which the relationship between phosphorylation and condensation is rather different. In *Tetrahymena*, H1 becomes dephosphorylated in transcriptionally inactive condensed chromatin, and the same is true of histone H5 (an H1 variant) in bird erythrocytes. Newly synthesized H5 is phosphorylated, but becomes dephosphorylated as the nuclei condense (Wolffe, 1998).

Phosphorylation of H3 at serines 10 and 28 has also been associated with chromosome condensation (Cheung *et al.*, 2000). This phosphorylation starts at the pericentric heterochromatin, spreads throughout the chromosomes in G2 and is lost in anaphase (Hendzel *et al.*, 1997). Phosphorylation on H3 is also associated with activation of gene expression (Cheung *et al.*, 2000). Paradoxically, serine 10 is again involved. Phosphorylated H3 may be preferred to unphosphorylated H3 by several transcription-associated histone acetyltransferases.

Ubiquitin is a polypeptide, 76 amino acids long, that becomes attached to many proteins to target them for destruction. However, there is no evidence that this is the reason for its attachment to certain histones, but rather that it is another modification associated with transcription, and is therefore assumed to disrupt chromatin structure. Histone H2A is the principal histone that is ubiquitinated, to form a protein known as A24, and whereas only about 4% of H2A molecules are ubiquitinated in inactive chromatin, about 50% are modified in transcriptionally active genes (Wolffe & Hayes, 1999). Histone H2B is also ubiquitinated, but at a much lower level.

Unmodified histones then are the building blocks, along with DNA, of the basic 10 nm nucleosomal fibre, but they are incompatible with the many activities that chromatin has to take part in. Thus mechanisms have evolved to loosen the nucleosomal structure so that the DNA is more accessible, or to condense it to prevent transcription. Modifications to the histone tails form a 'histone code' that is recognized by different proteins involved in remodelling chromatin (Fig. 4.5), and different histone-modifying enzymes may act in concert with remodelling complexes (Jenuwein & Allis, 2001).

4.2.7 Deviant histones and deviant nucleosomes?

Most chromatin is composed of nucleosomal fibres based on the four standard core histones and a linker histone, usually H1, all of which can be modified to allow transcription and other activities. Linker histones are much more variable than the core histones (Section 4.2.1), but there are situations in which more radically different histones and histone-like molecules can be found (Table 4.6) (Wolffe, 1995; Wolffe & Pruss, 1996). It is perhaps not appropriate to describe as histones those transcription regulatory factors that use the histone-fold domain, although it is no doubt functionally significant that they make use of similar structures, which are probably highly relevant to their binding to DNA. However, it does seem plausible that proteins such as H2A.Z, CENP-A at mammalian centromeres and MacroH2A on inactive X chromosomes are variants designed to induce a specific type of chromatin structure in specific chromosomal regions or at particular phases of development.

4.2.8 Other chromatin proteins – the HMG proteins

So far we have discussed the histones and the way they are complexed with DNA to form nucleosomes. The nucleosomal structure implies that there are proteins that can destabilize this structure to permit replication (Section 3.4), transcription (Section 4.2.3) and DNA repair (Section 3.6). But there are other proteins that are important components of chromatin, many of which probably remain to be discovered.

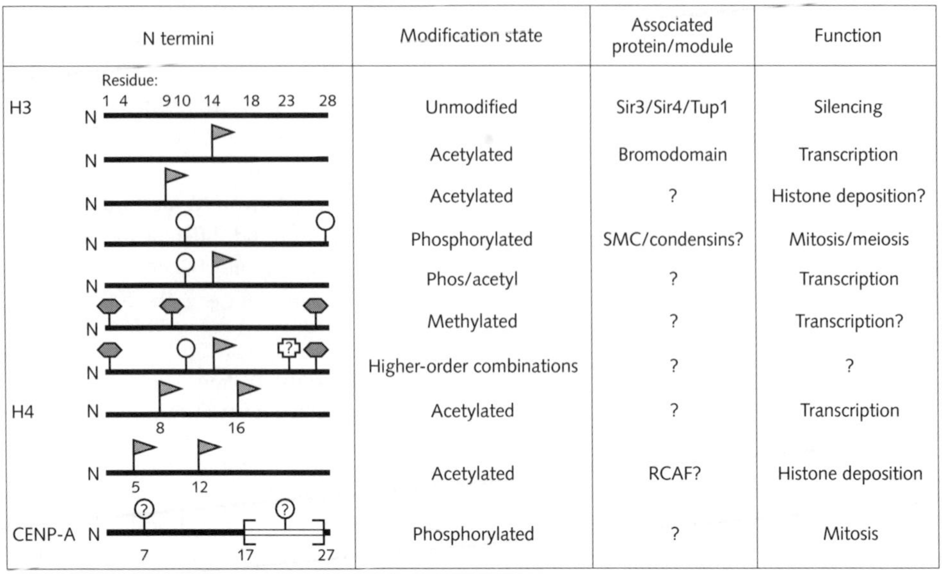

Figure 4.5 The histone code. Patterns of histone modifications at specific residues are associated with specific functions, often in combination with other proteins. Reprinted with permission from Strahl & Allis (2000) *Nature* **403**, 41–45. © Macmillan Magazines Limited.

Table 4.6 Histone deviants.

Name	Variant of	Comments
HNF3 (hepatocyte nuclear factor 3)	H1	Sequence-specific transcriptional regulation
H2A.Z	H2A	Essential for early development in *Drosophila*
Macro H2A (mH2A)	H2A	Concentrated on mammalian inactive X chromosome (Costanzi & Pehrson, 1998)
Macro H2A-Bbd	H2A	Excluded from mammalian inactive X chromosome (Chadwick & Willard, 2001a)
Gamma-H2AX	H2A	Associated with double-strand DNA breaks and meiotic crossing-over (Hunter *et al.*, 2001)
HMf		In Archaebacteria. Histone-fold domain without tails; can wrap DNA round itself
CENP-A	H3	Centromeric protein (Section 12.4.1)
HCP-3	H3	*Caenorhabditis* kinetochore protein (Buchwitz *et al.*, 1999)
CSE-4		*S. cerevisiae* homologue of CENP-A
Transcriptional regulatory proteins	H2A, H2B, H3, H4	Use histone-fold domain to bind to DNA and induce protein–protein interactions
gH2A	H2A	
gH2B	H2B	Male gametic cells of *Lilium longiflorum* (Ueda *et al.*, 2000)
gH3	H3	

For more detailed information on histones, see Sullivan *et al.* (2000) and http://genome.nhgri.nih.gov/histones/

Table 4.7 HMG proteins.

Class	Molecular weight	Name	Comments
HMGB (HMG1/2; HMG box proteins)	~29 kDa	HMGB1	Abundant structural protein (10^6 molecules/mammalian nucleus)
		HMGB2	Structural
		HMGB3	Expressed in mouse embryo, not in adult (Vaccari et al., 1998)
		HMG-D	Drosophila homologue of HMG1
HMGN (HMG14/17; nucleosomal binding proteins, NBD)	10–12 kDa	HMGN1	= HMG14
		HMGN2	= HMG17
HMGA (HMGI(Y); AT-hook proteins, ATH)		HMGA1a	Structural; binds to alpha-satellite DNA
		HMGA1b	Structural
		HMGA2	Structural

For more information on the classification of HMG proteins, and correlations with the older nomenclature, see Bustin (2001) and www.informatics.jax.org/mgihome/nomen/genefamilies/hmgfamily.shtml

Chief among these other proteins are the HMG (high mobility group) proteins (Bustin & Reeves, 1996), which have low molecular weights and therefore high mobility on electrophoretic gels. Unlike the histones, the HMG proteins contain significant numbers of acidic amino acids, as well as basic amino acids. They are grouped into three classes (Table 4.7). As well as the HMG proteins, there are several proteins that have structural similarities to the HMGB class, in particular an amino acid sequence known as the HMG domain, which when bound to DNA molecules bends them significantly (Grosschedl et al., 1994). However, whereas the HMGB proteins have multiple HMG domains and low DNA sequence specificity, those proteins with only a single HMG domain recognize specific DNA sequences.

The main HMG proteins are abundant nuclear proteins: for example, a typical mammalian nucleus contains 10^6 molecules of HMGB1, about one-tenth of the quantity of core histones (which are extremely abundant proteins). Both HMGB1 and HMGB2 can bind to the linker DNA in chromatin and, unlike histone H1, generally appear to stimulate rather than inhibit transcription by promoting a looser chromatin structure (Zlatanova & van Holde, 1998). In fact, HMGB1 appears to be a transcriptional regulator (Wolffe, 1999). However, in both Drosophila and Xenopus, early embryos have chromatin enriched in HMG proteins, lack histone H1 and are transcriptionally inactive; transcription only begins when the amount of HMG proteins in the chromatin is reduced, and the H1 histones increased (Spada et al., 1998; Zlatanova & van Holde, 1998). In mouse embryos HMGB1 and histone H1 increase together during embryogenesis, and there is no indication that one substitutes for the other. Protein HMGB3, on the other hand, is highly expressed in mouse embryos, but is virtually absent from adult chromatin (Vaccari et al., 1998). At present, the significance of these changes in the concentrations of HMG proteins during development is not clear. Unlike H1, HMGB1 does not remain attached to metaphase chromosomes, and indeed most of it is perhaps not bound to chromatin even in interphase (Falciola et al., 1997).

Both HMGN1 and HMGN2 are abundant proteins that appear to bind directly to nucleo-somal cores, as a dimer of either HMGN1 or of HMGN2 (Bustin & Reeves, 1996). These two proteins can facilitate loosening of the nucleo-somal structure to allow transcription (Wolffe, 1998), and could be a cause of the nuclease hypersensitivity associated with regions that contain actively transcribed genes. However, because HMGN1 and HMGN2 are released from chromosomes during mitosis, while mitotic chromosomes retain nuclease hypersensitivity, they evidently cannot be the only factor respon-sible for the maintenance of a more open state of chromatin (Hock et al., 1998).

Proteins of the HMGA group also interact with nucleosomal DNA to promote conforma-tional changes that facilitate transcription, which have been studied particularly for the human interferon-β promoter (Grosschedl et al., 1994; Wolffe, 1998). These proteins contain the AT-hook motif (Pro-Arg-Gly-Arg-Pro), which binds to sites in the minor groove of DNA consisting of runs of A+T bases and induces structural alter-ations in the DNA. The HMGA proteins are components of a protein complex associated with DNA enhancers called the enhanceosome (Bianchi & Beltrame, 2000).

4.3 Packing nucleosomes into solenoids

The 10 nm nucleosomal fibre has been described in detail, partly because a lot is known about this level of packing of DNA, and partly because variations and alterations in structure at this level are vital to the various processes in which DNA and chromatin are involved. However, this is only the lowest level of packing, producing a packing ratio of only seven compared with the ratio of 10 000 found in a condensed chromosome, and 10 nm fibres are only rarely seen in preparations of chromatin for electron microscopy that are designed to minimize artefacts.

In fact, the most commonly seen chromatin fibre, both in interphase chromatin and in mitotic chromosomes, has a diameter of approx-imately 30 nm. The most widely accepted model of the packing of the 10 nm fibre into a 30 nm fibre is the solenoid, proposed by Finch & Klug (1976), in which the nucleosomal fibre is wound in a regular helix, with about six nucleosomes per turn and a hole down the middle (Fig. 4.6). The linker histones probably face inwards, towards the central hole, and help to stabilize the solenoid (Widom, 1989; Bartolmé et al., 1994). Coiling of the nucleosomal fibre into a solenoid would result in a total packing ratio in the region of 40–50. Nevertheless, other models for com-paction of the nucleosomal fibre into a higher order structure exist. More irregular aggregations

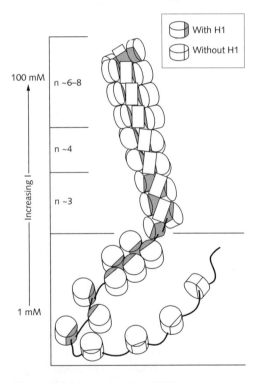

Figure 4.6 The organization of DNA and nucleosomes into a solenoid. At low ionic strength and in the absence of histone H1 no organized structure is formed, but addition of H1 at low ionic strength produces a zigzag structure. At higher ionic strength this reorganizes itself into the solenoid. Reproduced from Thoma et al. (1979) *Journal of Cell Biology* **83**, 402–427. © Rockefeller University Press.

of nucleosomes have been proposed, which have been named 'superbeads' (Zentgraf & Franke, 1984). Another proposed structure is a 30 nm fibre based on a zigzag nucleosomal fibre (Woodcock & Dimitrov, 2001). Scanning force microscopy reveals irregular structures rather than a uniform 30 nm fibre (van Holde & Zlatanova, 1996). Although it is possible to attribute some of these variations to differences in preparation and methods of analysis, it would not be surprising if there were significant variation in the higher order packing of the nucleosomal fibre. It has already been emphasized that the nucleosomal fibre itself must undergo changes to a looser structure, more accessible to the protein complexes needed for replication, transcription and DNA repair, and the same must undoubtedly be true of the 30 nm fibre. Although a regular solenoid may indeed be the predominant structure in inactive regions of chromatin, it would hardly be surprising to find less regular structures in transcriptionally active regions that must unfold to allow access to the DNA.

4.4   Yet more packing

As mentioned in Section 4.3, the packing ratio produced by formation of a 30 nm chromatin fibre is in the region of 40–50, which is still at least 200-fold less than that found in a fully condensed metaphase chromosome. Although it was once postulated that the complete condensation of metaphase chromosomes might be achieved simply by further levels of coiling into thicker and thicker chromatin fibres, it is now generally held that the next level of packing above the 30 nm fibre involves an arrangement of loops that radiate out from a 'core' or 'scaffold'. This will be described in Section 6.3. At least in metaphase chromosomes, there appears to be yet another level of packing beyond this. The 30 nm fibre also appears to be arranged in loops in interphase chromatin, but it is not clear at present whether the mitotic scaffold exists in a recognizable form in the interphase nucleus, and the interphase nuclear matrix (Section 5.4) may well be an unrelated structure.

4.5   Other ways to pack DNA

The organization of DNA into nucleosomal fibres, followed by higher levels of packing as just described, is almost universal among eukaryotes and is therefore evidently an efficient way of packing DNA that is compatible with the diverse functions that DNA participates in. There are, nevertheless, two situations, one involving a particular developmental stage and the other a specific group of organisms in which DNA is not wrapped round nucleosomes. These are, respectively, the spermatozoa of many organisms and a group of Protista known as dinoflagellates.

4.5.1   Spermatozoa

It is characteristic of the spermatozoa of many organisms that their chromatin is highly compacted, is transcriptionally inactive, usually has a very distinctive shape and in many cases their DNA and chromosomes are very highly ordered (e.g. Watson et al., 1996 and references therein). Electron micrographs of, for example, mammalian sperm heads show that the nucleus is filled with dense, structureless chromatin (Fig. 4.7), quite unlike that normally seen in somatic interphase nuclei or even in condensed metaphase chromosomes. In many organisms, the normal somatic histones are replaced by special sperm histones, but others, including many ver-

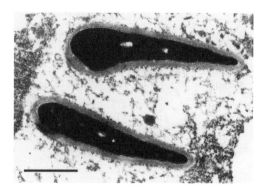

Figure 4.7 Transmission electron micrograph of a section through two human sperm heads from the testis. Note the dense chromatin in which no structure is visible, except for small vacuoles. Scale bar = 1 μm.

tebrates, replace histones completely with very small, very basic proteins known as protamines (Wolffe, 1998). Fish protamines consist of about 30 amino acids, of which roughly two-thirds are arginine. Protamines of eutherian mammals differ from those of most other vertebrates in additionally containing high concentrations of cysteine, which form disulphide cross-links and stabilize the structure of the sperm head even more (Balhorn, 1982). Protamines do not form nucleosomal structures – indeed, it has been calculated that there would not be sufficient space in the sperm head for them to do so – but instead stabilize a hexagonal array of DNA molecules, occupying the spaces between adjacent molecules, neutralizing the charges on adjacent DNA molecules and cross-linking them together (Raukas & Mikelsaar, 1999). The DNA–protamine complexes are a special adaptation to packing sperm DNA securely in a minimal volume in a situation in which the DNA is totally inert.

4.5.2 Dinoflagellates

The dinoflagellates are a group of Protista with a nuclear and chromosomal organization quite unlike that found in any other eukaryotes. The chromosomes do not appear to condense or disperse during the different stages of the cell cycle, and in electron micrographs have a curious, feathery appearance (Fig. 4.8). The DNA of dinoflagellates is not compacted into nucleosomes, but forms a complex with basic proteins that are quite different from histones (Vernet *et al.*, 1990). An interesting feature of dinoflagellate chromosomes is that they contain substantial quantities of transition metals (Mn, Fe, Ni, Cu and Zn) (Kearns & Sigee, 1980), which may well be essential constituents of these curiously constructed chromosomes. Otherwise, little is known of the chromatin of these intriguing little organisms.

4.6 Summary

In this chapter the complexing of DNA with specific proteins to compact it into a manageable form has been described: first the formation of

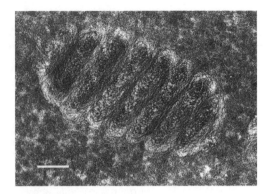

Figure 4.8 A longitudinal section of a late G1 chromosome from the dinoflagellate *Crypthecodinium cohnii*. Scale bar = 0.2 μm. Reprinted with permission from Bhaud *et al.* (2000) *Journal of Cell Science* **113**, 1231–1239, © The Company of Biologists.

the 10 nm chromatin fibre by forming a complex with histones, and then the coiling or folding of this into a 30 nm fibre, reducing the length of the DNA by some 40–50-fold. These are therefore only the basic stages of chromatin compaction. This chromatin structure has to be modified to allow replication, transcription and DNA repair to take place, and this is achieved not only by a variety of modifications of the histones, but also by the binding of a large variety of chromosomal proteins, which open up the chromatin structure and allow the DNA to function. Different types of histones, histone-related proteins or non-histones are found in particular chromosomal regions or at different developmental stages. We must not think of a single, static way of packing DNA into chromatin, but of a dynamic chromatin system that modifies itself according to the current needs of the cell.

Websites

Histones
http://genome.nhgri.nih.gov/histones/

HMG protein nomenclature
www.informatics.jax.org/mgihome/nomen/genefamilies/hmgfamily.shtml

Nucleosomes
www.average.org/~pruss/nucleosome.html

The chromosomes in interphase

5.1 Interphase nuclei: sites of chromosome activity

The interphase nucleus is where chromosomes spend most of their time and carry out most of their functions, especially transcription and, in growing and dividing cells, DNA replication. An understanding of the organization of chromosomes in interphase is therefore of prime importance; however, individual chromosomes are not (with a few exceptions, such as polytene chromosomes, Chapter 15, and the inactive X chromosome in female mammals, Section 8.4.3) visible at this stage, and study of the interphase nucleus by traditional methods gives few clues to its organization. An electron micrograph of a typical nucleus (Fig. 5.1) shows four main features: a double membrane (the nuclear envelope) segregating the nuclear contents from the cytoplasm; the nucleolus (described in detail in Chapter 11); and regions of densely and weakly stained chromatin. The denser chromatin is often concentrated against the nuclear envelope. The more and less electron dense materials are often equated with heterochromatin and euchromatin, respectively, but this is misleading: the densely stained chromatin can certainly comprise a greater proportion of the nuclear content (sometimes all of it, as in nucleated erythrocytes) than heterochromatin, as defined by more traditional criteria (see Chapter 7), and it seems better to refer to the different regions of the interphase nucleus as condensed and dispersed chromatin. What are certainly not visible are discrete chromosomes. Nevertheless, application of appropriate methods, especially fluorescence *in situ* hybridization (FISH; Box 5.1) and immunocytochemistry, together with sophisticated microscopic and image processing methods that allow reconstruction of three-dimensional images and spatial correlation of signals representing different nuclear components, has resulted in an enormous increase in our understanding of the organization of interphase nuclei.

In this chapter we shall consider the spatial arrangement of chromosomes within the nucleus, how they are made up from 30 nm fibres (as described in Section 4.3), the question of the nuclear matrix and the sites of replication and transcription within the nucleus. Various nuclear organelles such as the coiled bodies will be described here, but not the most prominent of such organelles, the nucleolus, which has a chapter to itself (Chapter 11). The nuclear envelope and its pores will only be considered insofar as they are relevant to the other topics of this chapter; briefly, the nuclear envelope consists of a double membrane, on the inside of which is a network of lamins and other proteins (Wilson, 2000). The envelope is interrupted at intervals by the nuclear pores – structures of great complexity through which molecules and small particles can pass in and out of the nucleus (Stoffler *et al.*, 1999; Talcott & Moore, 1999; Allen *et al.*, 2000; Ryan & Wente, 2000). Special machinery interacts with import and export signals to transport proteins and RNAs through the nuclear pores

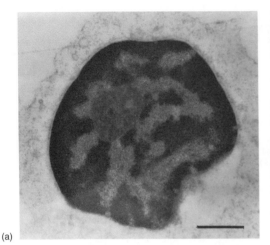
(a)

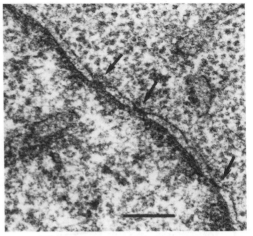
(b)

Figure 5.1 (a) Electron micrograph of a nucleus from a mammalian cell, showing condensed and dispersed chromatin and the nucleolus. Scale bar = 1 μm. (b) Electron micrograph of a portion of a nucleus, showing the double nuclear envelope and nuclear pores (arrows). Scale bar = 0.2 μm.

(Görlich, 1998; Nakielny & Dreyfuss, 1999; Michael, 2000).

5.2 How are the chromosomes arranged in the nucleus?

Examination of nuclei by electron microscopy at sufficiently high power shows only that they are full of chromatin fibres, which are more densely packed in some places than in others, but show no discontinuities such as would be expected if the chromosomes were clearly separated. Nevertheless, chromosome painting (FISH using probes specific for a particular chromosome; Box 5.2) and related methods show that interphase chromosomes remain as discrete objects, generally larger and less regularly shaped than metaphase chromosomes and with rather less sharp outlines (Fig. 5.2a). Frequently, the two homologues are on opposite sides of the nucleus. The space occupied by an individual interphase chromosome is generally referred to as the chromosome territory (Cremer & Cremer, 2001). Within a particular cell type, chromosomal position in the nucleus is more or less fixed, and the chromo-

somes only make small movements (Zink & Cremer, 1998). Larger scale movements are associated with changes in the functional state of the cells, however.

Evidence for a higher level of organization of the chromosomes in nuclei is also available, and indicates that this differs greatly between species. Certain chromosomes, because of their function (or lack of it), lie in specific regions. Thus the inactive X chromosome in female mammals forms a compact structure – the Barr body – against the nuclear envelope (Fig. 5.2b). Chromosomes that bear nucleolus organizer regions (NORs; Chapter 11) are inevitably associated with the nucleolus if the NORs are active; in a species such as humans, with five pairs of chromosomes bearing NORs (not all of which are usually active, however), these chromosomes are therefore closely associated with each other. Certain human chromosomes seem to be more closely associated with each other in quiescent (non-cycling) nuclei than would be expected by chance (Nagele et al., 1999), although the significance of such associations is far from clear. In human cells, it has been reported that the most gene-rich chromosomes are nearer the centre of

Box 5.1   Fluorescence *in situ* hybridization (FISH)

Fluorescence *in situ* hybridization (FISH) is the labelling of specific DNA sequences *in situ* (in chromosomes or nuclei) with a fluorescently labelled complementary nucleic acid, so that the location of these sequences can be seen under a microscope. It can be used for labelling whole genomes (GISH, Box 5.3), repeated sequences or single genes, for chromosome painting (Box 5.2) or for comparative genome hybridization (CGH, Box 17.1).

The first stage in FISH is preparation of the probe. This can be produced by PCR (polymerase chain reaction) or as a sequence cloned in a cosmid, BAC (bacterial artificial chromosome) or YAC (yeast artificial chromosome). The probe is labelled either directly with a fluorochrome or with a hapten such as biotin or digoxygenin. The hapten can be detected after hybridization using an antibody labelled with a fluorochrome. This procedure is more sensitive and versatile than direct labelling. The labelled nucleotides can be incorporated directly during PCR or by nick translation.

The FISH process is carried out by denaturing the DNA in a chromosome preparation, incubating it with the probe and, if a biotin- or digoxygenin-labelled probe has been used, incubating with fluorescently labelled avidin or anti-digoxygenin, respectively (Fig. 1). Full protocols for FISH are found in standard laboratory manuals (e.g. Craig, 1999; Schwarzacher & Heslop-Harrison, 2000; Saunders & Jones, 2001).

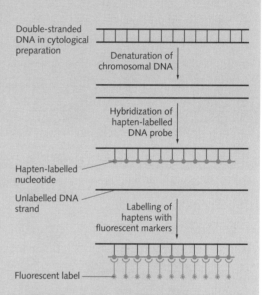

Figure 1 Diagrammatic representation of the main stages in fluorescence *in situ* hybridization (FISH).

Box 5.2   Chromosome painting

Chromosome paints are FISH (Box 5.1) probes that are designed to label only a single chromosome pair in a genome. They are used for identifying that chromosome in mitosis, meiosis or in interphase (Fig. 5.2a), in hybrid cells, for evolutionary studies (Section 16.3.1; Fig. 16.1) and for studying translocations involving the labelled chromosome (Ried *et al.*, 1998; Bridger & Lichter, 1999).

Chromosome paints are produced from flow-sorted or microdissected chromosomes, or from hybrid cells that contain only a single chromosome from the species of interest. Using microdissection, paints can be produced for a specific part of a chromosome, such as the short or the long arm. Human chromosome paints are now available commercially from several manufacturers. Chromosome paints have also been produced for a wide variety of mammals.

To avoid cross-hybridization by DNA from other chromosomes, hybridization is carried out in the presence of an excess of $C_{o}t1$ DNA, that is, the DNA that anneals most rapidly and therefore contains the repetitive sequences that are most likely to be common to different chromosomes. For this reason, blocks of heterochromatin, which contain highly repetitive DNA (Section 7.3.1), are usually not labelled by chromosome paints.

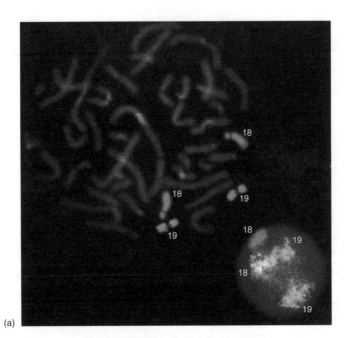

(a)

Figure 5.2 Micrographs showing the distribution of various chromosomal components in the nucleus. (a) Chromosome painting of interphase and metaphase chromosomes, showing the distinct territories occupied by chromosomes 18 and 19 in the nucleus. Micrograph kindly supplied by Wendy Bickmore. (b) Barr body (arrowed) lying against nuclear envelope in a human female buccal epithelial cell. (c) The Rabl configuration in an interphase nucleus from a wheat root tip: c, centromeres; t, telomeres. Reproduced with permission from Dong and Jiang (1998) *Chromosome Research* **6**, 551–558, © Kluwer Academic Publishers. (d) Scattered centromeres and telomeres in an interphase nucleus from a rice root tip: c, centromeres; t, telomeres. Reproduced with permission from Dong and Jiang (1998) *Chromosome Research* **6**, 551–558, © Kluwer Academic Publishers.

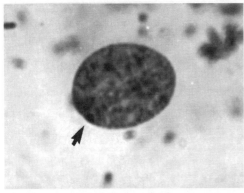

(b)

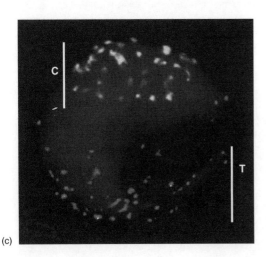

(c)

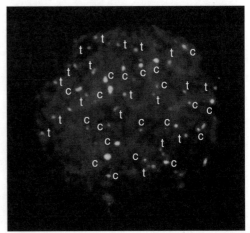

(d)

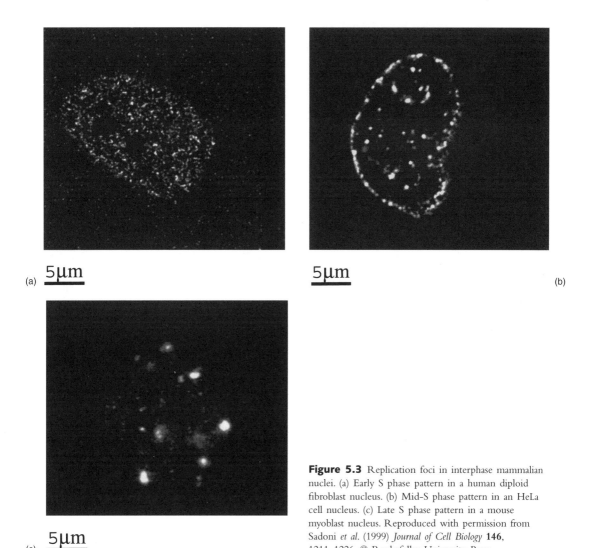

(a) 5μm

5μm   (b)

(c) 5μm

Figure 5.3 Replication foci in interphase mammalian nuclei. (a) Early S phase pattern in a human diploid fibroblast nucleus. (b) Mid-S phase pattern in an HeLa cell nucleus. (c) Late S phase pattern in a mouse myoblast nucleus. Reproduced with permission from Sadoni *et al.* (1999) *Journal of Cell Biology* **146**, 1211–1226. © Rockefeller University Press.

the nucleus and the gene-poor chromosomes tend to be closer to the nuclear periphery (Boyle *et al.*, 2001). Similarly, in chicken cells the gene-rich microchromosomes are centrally located whereas the gene-poor macrochromosomes are concentrated towards the nuclear periphery (Habermann *et al.*, 2001). On the other hand, it has been reported that larger chromosomes tend to be closer to the nuclear periphery, and smaller ones nearer the middle (Heslop-Harrison *et al.*, 1993), or vice versa (Cremer *et al.*, 2001). It seems certain that a variety of factors, including

cell type and transcriptional activity, are important in determining chromosomal distribution in nuclei. In some cases, transcriptionally inactive genes may localize to centromeric heterochromatin, but become repositioned when activated (Brown *et al.*, 1999).

In *Drosophila*, specific sites on chromosomes are found in specific parts of the nucleus, varying by only 0.5 μm in average position from cell to cell; certain chromosomal regions are associated consistently with the nuclear envelope (Marshall *et al.*, 1997).

Box 5.3   Genomic *in situ* hybridization (GISH)

Genomic *in situ* hybridization (GISH) is a variant of FISH in which a whole genome is labelled to distinguish it from the other parental genome in a hybrid. It is particularly useful for studies of plants, many of which are hybrids (Bennett, 1995).

It is necessary to use a labelled probe for only one parental genome. To prevent non-specific labelling, unlabelled DNA from the other parental species is added to the hybridization mixture at a much higher concentration than the labelled probe. One parental genome might be labelled yellow by fluorescein isothiocyanate (FITC) using the labelled probe, while the other parental genome might be counterstained orange-red by a DNA-specific fluorochrome such as propidium.

Larger scale arrangements of chromosomes in nuclei have also been found. In hybrid plants, the chromosome sets from each parental species can be labelled distinctively using GISH (genomic *in situ* hybridization, Box 5.3), and it has been found that they remain separate throughout the cell cycle for many years (Leitch *et al.*, 1991; Heslop-Harrison *et al.*, 1993; Bennett, 1995). In scale insects, the paternal genome becomes heterochromatinized and is eliminated in males (Nur, 1990), and this happens as a single mass of chromosomes, rather than individual chromosomes scattered among non-heterochromatinized ones. In this case, one of the parental genomes may be imprinted (Chapter 9) and thus marked for elimination.

Perhaps the best known and highest degree of organization of chromosomal distribution in interphase nuclei is found in those organisms that show the Rabl configuration of chromosomes. At anaphase, the chromosomes inevitably move towards the poles of the cells with their centromeres leading and their telomeres trailing. In 1885, Rabl found a similar arrangement of chromosomes in prophase nuclei, and deduced that it must have been maintained throughout interphase. The ability to label centromeric and telomeric regions of chromosomes distinctively, particularly using FISH, has amply confirmed that the Rabl configuration occurs in the cells of many plants and animals (e.g. Fussell, 1975; Marshall *et al.*, 1997; Abranches *et al.*, 1998; Dong & Jiang, 1998) (Fig 5.2c), but by no means all.

Mammalian somatic cells, for example, do not in general show a Rabl configuration. As noted in Section 2.5.1, the Rabl configuration might be expected to facilitate pairing of homologues at meiosis, yet it is far from being universal: in many organisms, centromeres and telomeres are scattered throughout the nucleus and are not necessarily associated with the nuclear envelope (Fig. 5.2d) (Dong & Jiang, 1998). Evidently, chromosomes or groups of chromosomes may have defined locations in the nuclei of particular organisms, but there is no universal arrangement, and the functional advantages of any particular arrangement may be difficult to discern.

5.3   Where do replication and transcription take place?

As we shall see later (Sections 7.1, 8.4.3 and 10.2.2), different chromosomal regions replicate their DNA at different times during the S phase, and genes tend to be concentrated in particular regions of chromosomes. It would therefore be reasonable to expect that sites of replication and transcription in interphase nuclei would not be uniformly distributed, and that the distribution of replication sites would change as the nuclei moved through S phase. It certainly turns out that sites of replication and transcription are non-uniformly distributed in interphase nuclei, although perhaps not in such a simple way as might be expected by extrapolation from the structure of mitotic chromosomes.

Sites of replication and of transcription can be identified by labelling newly synthesized DNA or RNA, respectively, with an appropriately labelled (usually fluorescent) precursor. In general, such sites appear as a large number of discrete foci throughout the nuclei, but closer examination shows that they are generally found at the surface of chromosome territories (Kurz et al., 1996; Verschure et al., 1999; Cremer & Cremer, 2001), although this may not invariably hold true (Abranches et al., 1998). However, it has been argued that the finding of transcription sites throughout the chromosome territories could be an artefact. Contrary to expectation, the volumes of the chromosome territories of the active and inactive X chromosomes in female mammals are similar; the difference lies in the much more convoluted surface of the active X (Eils et al., 1996). Thus transcription sites that appear to be within the body of a chromosome territory might in fact lie in an invagination of its surface. The view that the surfaces of the chromosome territories and the spaces between them form the site for transcription, messenger RNA (mRNA) processing and transport is known as the interchromosome domain model (Cremer & Cremer, 2001).

In mouse fibroblasts there are about 10 000 replication sites per nucleus, and each must contain, on average, six replicons, that is, independent replication units of DNA (Ma et al., 1998; Zink et al., 1998). The pattern of active replication sites changes as the cell progresses through S phase: early, middle and late replication patterns can be recognized, just as would be expected from the study of metaphase chromosomes (Section 10.2.2) (Kill et al., 1991; O'Keefe et al., 1992; Berezney & Wei, 1998). In immortalized or transformed cell lines, the earliest nuclear replication patterns consist of hundreds of small foci throughout the interior of the nucleus (Fig. 5.3a). Next, labelling appears round the nucleoli and the periphery of the nucleus, until these are the main sites by mid-S phase (Fig. 5.3b). In late S, replication is restricted to a

few large foci in the interior of the nucleus as well as at the periphery (Fig. 5.3c) (Sadoni et al., 1999). However, the pattern of early replication may be changed by the process of transformation, and in primary fibroblasts DNA synthesis starts at a small number of sites surrounding the nucleus (Kennedy et al., 2000). Patterns of DNA replication in the nucleus therefore appear not to be fixed, but can be regulated. Both the timing and positioning of early and late replicating domains are established early in G1 (Dimitrova & Gilbert, 1999).

The DNA polymerases are immobilized in 'replication factories', which not only contain newly replicated DNA, but also the enzymes and other proteins required for DNA replication (Cook, 1999; Leonhardt et al., 2000). The replication factories appear to be attached to the nuclear matrix (Section 5.4), and can be recognized as dense bodies in electron micrographs. In each early S phase replication factory in human cells there are probably about 40 active replication forks.

Sites of transcription also appear as foci throughout the nucleus, each of which is responsible for the transcription of several genes (Jackson et al., 1993), but the sites of transcription are distinct from the sites of replication (Wei et al., 1998). Sites of transcription mediated by RNA polymerase I are necessarily in the nucleolus (Chapter 11), because this enzyme transcribes ribosomal DNA. Enzyme RNA polymerase II is responsible for transcription of most genes, while RNA polymerase III transcribes 5S RNA and tRNA; sites of activity of both of these are found throughout the nucleus (Fig. 5.4). The enzymology of RNA polymerase action, and the proteins involved, has recently been reviewed by Paule and White (2000). In HeLa cells, there are in the region of 8000 sites using pol II, each containing between 8 and 15 transcription units (Cook, 1999), and about 2000 sites containing pol III, each with about five molecules of this enzyme (Pombo et al., 1999). Like replication factories, transcription factories are probably attached to the nuclear matrix. Splicing of precursor messenger RNA (pre-

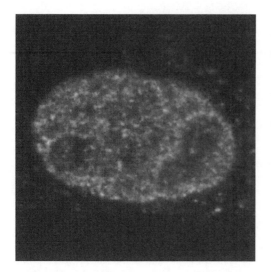

Figure 5.4 Transcription sites in an interphase HeLa nucleus. Reproduced with permission from Pombo *et al.* (1999) *EMBO Journal* **18**, 2241–2253. © European Molecular Biology Organization.

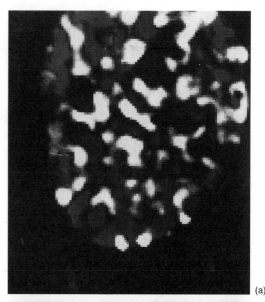

(a)

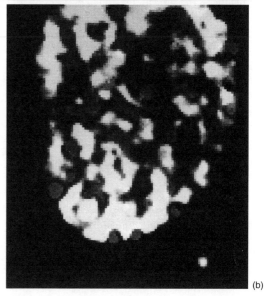

(b)

Figure 5.5 Transport of mRNA in interchromosome domains of a rat myoblast nucleus: (a) chromatin; (b) newly synthesized mRNA. There is no overlap between the chromatin and the mRNA, indicating that the mRNA is transported in the interchromosome domain. Reproduced with permission from Politz *et al.* (1999) *Current Biology* **9**, 285–291. © Elsevier Science.

mRNA) appears to occur mainly at the time of transcription. Sites of transcription are often quite closely associated with discrete foci – the splicing factor compartments (SFCs) – which contain about 150 proteins, including splicing factors (Misteli, 2000). Morphologically the SFCs correspond to interchromatin granules and perichromatin fibrils (Section 5.5.1).

Products of DNA replication obviously become parts of chromosomes, and remain in the nucleus. Transcription products, on the other hand, must be exported to the cytoplasm. This movement of gene products is believed to occur through narrow spaces, the interchromosome domains, which connect with the pores in the nuclear envelope (Fig. 5.5). Messenger RNA can apparently diffuse freely through the interchromatin space to reach the cytoplasm (Daneholt, 1999; Politz *et al.*, 1999).

5.4 The nuclear matrix

As mentioned in the previous section, foci of replication and transcription remain in interphase nuclei after extraction of most of the nuclear content, implying that they are firmly attached to some nuclear structure. This structure is referred to as the nuclear matrix. It has been, and still is, a controversial structure ever since its existence was announced (Hancock, 2000; Pederson,

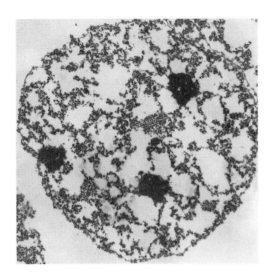

Figure 5.6 Electron micrograph of a nuclear matrix from rat liver, prepared by extraction. Reproduced with permission from Berezney *et al.* (1995) *International Review of Cytology* **162A**, 1–65. © Academic Press.

2000). Early studies using different methods of extraction produced a variety of images in electron micrographs of nuclei, from almost empty nuclei to nuclei containing extensive networks of material. Typically, such a nucleus consists of an envelope, the remains of the nucleolus (the 'nucleolar matrix') and fibrous and granular material that occupies the areas of less dense chromatin (Fig. 5.6). Another point of uncertainty is the relationship between the interphase nuclear matrix and the mitotic chromosome scaffold or core (Section 6.3). The mitotic scaffold is a proteinaceous structure that runs along the centre of the chromatids, and from which loops of 30 nm chromatin fibres radiate. The nuclear matrix, on the other hand, seems to be associated with the surfaces of the chromosome territories. Nevertheless, there are some important similarities between the interphase matrix and the mitotic scaffold. Topoisomerase II is an important protein of both structures (Nelson *et al.*, 1986), and the DNA sequences that attach to the scaffold (scaffold attachment regions, SARs) seem to be the same as those that bind to the interphase matrix (matrix attachment regions, MARs) (Nelson *et al.*, 1986; de Belle *et al.*,

1998). Yet there is a paradox here: if replication and transcription factories are components of the interphase matrix, and function by passing molecules of DNA through themselves, one should not expect to find any specific DNA sequence associated with the matrix.

A radial loop model of the interphase chromosome, in which loops of chromatin fibres would loop out from, and return to, the interphase matrix in a similar way to that found in mitotic chromosomes (Section 6.3), is only one of the proposals for interphase chromatin structure that have been made (Zink *et al.*, 1998; Belmont *et al.*, 1999). Another model is a giant-loop, random-walk structure, with loops of several megabase pairs in size (Yokota *et al.*, 1995). It has also been proposed that interphase chromosomes form structures with successive levels of coiling or folding that produce progressively thicker fibres. There is insufficient evidence to choose between the different models at present, but it would seem most economical to use a similar structure during interphase to the well established loops-and-scaffold structure found at mitosis.

The observed structure of the matrix depends on the type of cell from which it has been obtained, and on the method used to prepare it (Lewis *et al.*, 1984; Verheijen *et al.*, 1988). As a result, a matrix may be completely absent, or may consist of a dense network of fibres throughout the nucleus. Adult chicken erythrocyte nuclei show no matrix when extracted with procedures that reveal a matrix in other types of nuclei (Verheijen *et al.*, 1988), but when they are artificially reactivated they enlarge, the chromatin decondenses, proteins are taken up from the cytoplasm, RNA synthesis begins and a matrix develops. It seems therefore that the nuclear matrix is a dynamic structure, probably absent from transcriptionally inactive nuclei and likely to vary in structure according to the metabolic state of the cell. It would not, therefore, be possible to give a description of the matrix that would be applicable to all types of cell.

If the matrix is the site of processes that take place at the surface of the chromosome territories, such as DNA replication and RNA transcription, then the matrix would form a network between the chromosome territories, and indeed certain images of matrices are consistent with this idea (Fig. 5.6). The same region is, however, occupied by the interchromatin domains (Cremer & Cremer, 2001), which form channels along which the products of RNA synthesis are transported to the cytoplasm and therefore cannot be completely occupied by a relatively solid structure such as a proteinaceous matrix. Razin and Gromova (1995) have proposed a solution to this apparent paradox, in which the matrix forms channels leading to the nuclear pores (Fig. 5.7). Loops of chromatin could be attached to the matrix, the DNA transcribed into RNA and packaged as ribonucleoprotein (RNP) and the RNP could then diffuse along the channels and out of the nucleus.

5.5   Other nuclear structures

As well as the major components of the nucleus – nuclear envelope, chromatin, nuclear matrix and nucleolus – there are various structures that form a less prominent part of the nucleus, but which are nevertheless essential for its functioning. These include perichromatin fibrils and interchromatin granules (Section 5.5.1) and a

number of different structures that are collectively known as nuclear bodies but otherwise do not have a lot in common.

5.5.1   Perichromatin fibrils and interchromatin granules

Perichromatin fibrils simply represent the newly synthesized RNA at the surface of the chromatin (Fakan, 1994). Interchromatin granules (Fig. 5.8) form irregular clusters of granules – the interchromatin granule clusters (IGCs), each of which is about 20–25 nm in diameter – that consist of ribonucleoprotein (Thiry, 1995; Misteli & Spector, 1998; Sleeman & Lamond, 1999). By light microscopy it is not possible to distinguish the perichromatin fibrils and the IGCs, and together they are known as splicing factor compartments (SFCs; Misteli, 2000). The SFCs contain a wide variety of proteins (Mintz *et al.*, 1999), which include pre-mRNA splicing factors and ribosomal proteins. At least some of these proteins are in dynamic equilibrium, and move in and out of the SFCs rapidly (Phair & Misteli, 2000). Pre-mRNA splicing generally occurs at the time of transcription, and both processes take place at the periphery of the SFCs (Misteli, 2000) but not in their interior. Newly synthesized RNA only appears in the interchromatin gran-

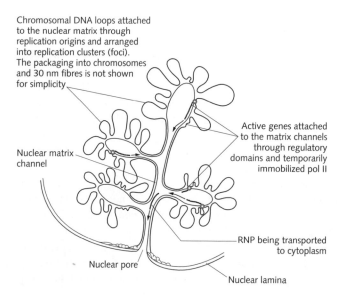

Chromosomal DNA loops attached to the nuclear matrix through replication origins and arranged into replication clusters (foci). The packaging into chromosomes and 30 nm fibres is not shown for simplicity

Nuclear matrix channel

Active genes attached to the matrix channels through regulatory domains and temporarily immobilized pol II

RNP being transported to cytoplasm

Nuclear pore

Nuclear lamina

Figure 5.7 A model of the nuclear matrix as a system of channels. Redrawn by permission of Wiley-Liss, Inc., a subsidiary of John Wiley and Sons, Inc., from Razin & Gromova (1995) *Bioessays* **17**, 443–450. © John Wiley.

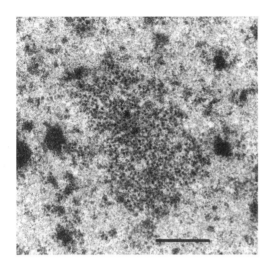

Figure 5.8 Interchromatin granules. Scale bar = 0.5 µm. Reproduced with permission from Thiry (1995) *Histology & Histopathology* **10**, 1035–1045. © Jiménez Godoy.

ules after several hours, and they may in fact contain stable mRNAs that are not exported to the cytoplasm. The splicing proteins are probably only stored in the interchromatin granules, but are not active there.

5.5.2 Coiled bodies (Cajal bodies)

Scattered throughout the nuclei are a number of foci of certain nuclear proteins, recognizable by immunofluorescence, including proteins involved in transcription and in processing of RNA (Matera, 1999; Gall, 2000). Most such nuclear bodies are, as yet, poorly characterized; the best known of such bodies are the coiled bodies and the promyelocytic leukaemia (PML) bodies, which are described below in more detail.

The coiled bodies, also known as Cajal bodies (Matera, 1999; Gall, 2000), have a very respectable history, having been described first by Ramon y Cajal in 1903 as accessory bodies. Electron microscopists described a body about 0.5 µm in diameter (Fig. 5.9) that appeared to consist of randomly coiled threads and was therefore named the coiled body. This was subsequently shown to be the same as Ramon y Cajal's accessory body. A nucleus contains one or

a few coiled bodies. Apart from their structure, the distinguishing feature of coiled bodies is that they contain a protein of about 80 kDa known as coilin (Bellini, 2000), although this protein is not restricted to coiled bodies and is possibly not an essential component of them (Gall, 2000). They also contain the three eukaryotic RNA polymerases and factors required for transcription and processing of different types of RNA (Gall, 2000, 2001). Coiled bodies contain several small nuclear RNPs (snRNPs) but they appear to be mature, and therefore the coiled bodies do not seem to be the sites for processing snRNPs. Possibly coiled bodies are organelles essential for the sorting and directing not only of snRNPs but also of snoRNPs (small nucleolar RNPs) (Matera, 1999).

Coiled bodies associate with specific gene loci (Matera, 1999; Gall, 2000); interestingly, the spheres of amphibian lampbrush chromosomes (Section 14.2), which are analogous to coiled bodies, have long been known to associate with specific sites, and are indeed well-known markers for mapping lampbrush chromosomes. In HeLa cells, 30–40% of coiled bodies, as well as the spheres of amphibian lampbrush chromosomes, contain U7 snRNP, which catalyses the removal of the 3′ end of histone pre-mRNAs before the mature histone mRNA is exported to the cytoplasm (Gall, 2000). Coiled bodies are also associated with genes for U1, U2, U3 and other snRNAs. It has been proposed that coiled bodies are the sites where the transcription machinery is assembled. The different RNA polymerases would form complexes with their transcription and processing factors in the coiled bodies, and the complexes would then be transported to the sites of transcription (Gall, 2000).

5.5.3 PML bodies

Promyelocytic leukaemia bodies (also known as ND10 or PODs) are structurally distinct from coiled bodies (Matera, 1999; Maul *et al.*, 2000). They consist of an outer layer containing the PML protein and an inner core that lacks this protein. The whole structure is about 0.5 µm in diameter. Little is known about the function of

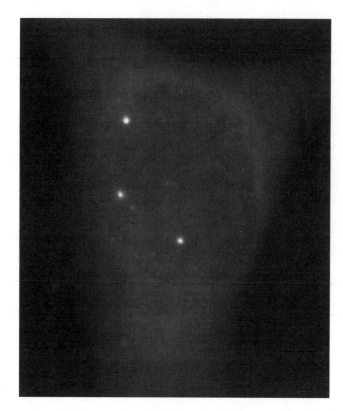

Figure 5.9 Coiled bodies in a mammalian cell nucleus. Reproduced with permission from Matera (1999) *Trends in Cell Biology* **9**, 302–309. © Elsevier Science.

PML bodies, although it has been suggested that they may be involved in cell-cycle regulation and apoptosis. The main interest in PML bodies is that they are modified in disease. Specifically, they are disrupted in patients with the 15;17 translocation, which leads to acute promyelocytic leukaemia (Section 17.9.1). This translocation results in fusion of the PML protein with the retinoic acid receptor α. The fusion protein not only fails to localize in the PML bodies, but also prevents wild-type PML from localizing in them.

Acute promyelocytic leukaemia is one of many situations where a chromosomal change produces a change in a gene that has clear consequences at the cellular and organismal level. Although many chromosomal changes are likely to be lethal, we shall encounter many more situations in which a pathological state can be attributed directly to an alteration visible at the chromosomal level.

5.6 Interphase nuclei are highly organized and dynamic

Although traditionally interphase nuclei have been regarded as rather amorphous, it has now become clear that they are highly organized and contain many different compartments. Apart from the condensed and dispersed chromatin and the nucleoli, whose existence has been known for a very long time, special compartments have been recognized in which processes such as DNA replication, RNA transcription and RNA processing take place. These include perichromatin fibrils, interchromatin granules, SFCs, coiled or Cajal bodies and PML bodies. We now have a good idea of the functions of these compartments, although there can be no doubt that there is still much to be learnt. Owing to methods such as FISH, which allow the identification of individual chromosomes and parts of chromosomes in interphase nuclei, we can now locate the positions of chromosomes in inter-phase nuclei, and study their relationships with each other. It turns out that chromosomal posi-tion is significantly non-random, and is related to transcriptional activity and possibly other factors. An understanding of the interphase nucleus is now emerging: it consists of numerous compart-ments, each with their own functions and each showing highly dynamic behaviour in response to both normal and abnormal physiological states of the cell.

Website

A website that gives additional information on the organization of the nucleus, particularly the three-dimensional distribution of chromosomes in the nucleus, and interactions between chromatin and the nuclear envelope, is:

Sedat lab: util.ucsf.edu/sedat/sedat.html

Structure of mitotic and meiotic chromosomes

6.1 Chromosomes of dividing and interphase cells compared

We have seen in the previous chapter (Chapter 5) that in interphase the chromosomes are made up of 30 nm chromatin fibres (that is, the fibres are not coiled or folded in any way to make a thicker fibre). One model for the higher order organization of chromatin fibres in interphase nuclei is that they form loops attached to a scaffold or skeleton, the nuclear matrix. Similarly, in both mitotic and meiotic chromosomes there is a proteinaceous structure to which loops of 30 nm chromatin fibres are attached. Nevertheless, there are substantial differences in the organization of chromosomes in interphase, in mitosis and in meiosis. The most obvious difference is perhaps that the chromosome structure is more compact and the shape better-defined in dividing cells than in interphase chromosomes. In addition, there are many differences between the matrix of interphase chromosomes, the scaffold of mitotic chromosomes and the synaptonemal complex (SC) of meiotic cells. Although it might seem logical that these structures would be homologous (and indeed they do have certain components in common), both their structure and composition are clearly related to the functions that they perform. It is still far from clear whether a single structure is adapted to perform the different skeletal functions of the nuclear matrix (Section 5.4), the mitotic chromosome scaffold and the meiotic SC (Section 2.5.2), or

whether each structure is formed anew at the appropriate stage of the cell cycle. Similarly, there appear to be differences in the way in which the chromatin fibres are attached to the skeletal elements of the chromosomes at different stages of the cell cycle. In this chapter, we shall first consider the arrangement of the chromatin loops and the chromosome scaffold in mitotic chromosomes – the situation about which most is known – and then go on to describe what is known about the situation in meiotic chromosomes.

There is a steady process of condensation throughout prophase, both in mitosis and meiosis, to produce the fully condensed metaphase chromosomes that are the main subject of study by cytogeneticists. This process of condensation is also the subject of this chapter. It must be remembered that the packing ratio of a fully contracted mammalian mitotic metaphase chromosome is in the region of 10 000 (Table 4.1), and the organization of chromosomes into loops attached to a scaffold may be insufficient to achieve this. An extra level of condensation would therefore be required, but details of this remain controversial (Section 6.4). In yeasts, however, the degree of compaction of mitotic chromosomes is at least several-fold less than in mammals (Yanagida, 1990), suggesting that this final level of condensation might not be required. Nevertheless, mechanisms of chromosome condensation in yeasts and vertebrates have many features in common (Section 6.4).

6.2 Making a mitotic chromosome

Certain features of mitotic chromosomes now seem so obvious that it is worth taking a little space to consider why they are not axiomatic, and what the evidence is for these features. As well as the 'loops-and-scaffold' structure, to be described in detail below (Section 6.3), it is accepted that there is a single DNA molecule in each chromosome (or in each chromatid after the DNA has been replicated at S phase), and that there is a fixed order of genes and other structural features on the chromosome. Early electron micrographs of whole mounts or sections of metaphase chromosomes showed an apparently disorganized tangle of chromatin fibres, with little indication of any particular organized structure (Fig. 6.1). Indeed, DuPraw (1970) proposed a 'folded-fibre' model of the chromosome, in which the chromatin fibres were folded in different ways throughout the body of the chromosome in different physiological conditions, implying a good deal of randomness in the structure of chromosomes. However, a variety of evidence shows that chromosomes are reproducibly and systematically organized.

6.2.1 One chromosome, one DNA molecule

Mitotic chromosomes, whether at metaphase or in the less-contracted prophase state, are very much thicker than a DNA molecule or even a 30 nm chromatin fibre. Because, as already remarked (above and Fig. 6.1), chromatin fibres appeared to run in all directions in chromosomes, there was no evidence from direct observations of chromosomes to show whether they were composed of many DNA molecules (polynemy) or of only a single DNA molecule (uninemy). In fact, there is now abundant evidence, mostly obtained by methods other than direct observation, that unreplicated mitotic or meiotic chromosomes are unineme (Sumner, 1998b; Zimm, 1999). Molecules of DNA replicate semi-conservatively, and so do chromosomes (Section 3.4, Fig. 3.6), and the most economical explanation of this is a single DNA molecule per

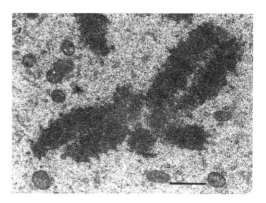

Figure 6.1 An electron micrograph of a cross-section through metaphase chromosomes from a Chinese hamster ovary cell. No substructure is visible except for a fine granulation, representing cross-sections through chromatin fibres. Scale bar = 1 μm.

unreplicated chromosome. It is possible to estimate the sizes of very large molecules such as DNA by procedures such as ultracentrifugation and viscoelastic retardation. It is difficult to prepare intact DNA molecules from chromosomes of eukaryotes, because the DNA is so easily broken, but with care it can be done for organisms with small chromosomes. Measurements of such molecules are consistent with one molecule per chromosome.

Irradiation of chromosomes with X-rays can cause chromosome breakage. The energy required to break DNA is well known, and it has been shown that the energy of X-rays required to produce a single chromosome break is the same as that required to break just one DNA molecule.

Lampbrush chromosomes (Chapter 14) are meiotic rather than mitotic, but these too provide evidence for the uninemy of chromosomes. The first piece of evidence was that the kinetics of the digestion of the axial fibre of lampbrush chromosomes by DNase were consistent with a single DNA molecule per chromatid, and this was reinforced when it became possible to make accurate measurements of the width of the axial fibre.

Although it is now accepted that mitotic and meiotic chromosomes are unineme, there are nevertheless certain chromosomes, the polytene chromosomes (Chapter 15), that are polyneme.

Such chromosomes, of course, do not divide. When one contemplates the problems that might be caused by trying to ensure an equal division of chromosomes consisting of thousands of parallel DNA molecules, it becomes clear that uninemy is the most efficient solution to the problem of chromosome segregation.

6.2.2   Chromosomes have a fixed linear order

The DNA of a chromosome does not merely consist of a single molecule per chromatid or per unreplicated chromosome, but it is also arranged in a pattern that is essentially fixed. The single DNA molecule does not follow a random course throughout the body of the chromosome (as postulated by DuPraw's 'folded-fibre' model, Section 6.2). Even before it became accepted that DNA was the genetic material, it was clear that chromosomal structures, especially centromeres but also secondary constrictions, normally occurred at the same sites on any particular chromosome, although such constancy of structure would not necessarily have been attributed to DNA. Stronger evidence for a regular pattern of DNA organization came from the study of chromosome banding patterns (Section 10.2), which in many cases reflect DNA base composition. Now that numerous genes and other DNA sequences have been localized on chromosomes by fluorescence *in situ* hybridization (FISH), it is clear that specific DNA sequences occupy specific sites on chromosomes, and that the order of genes on chromosomes corresponds to that in the DNA molecule. Much of the remainder of this chapter is concerned with describing a structure that not only compacts the DNA and chromatin fibres substantially, but also anchors the fibres in such a way that a linear order of DNA sequences is maintained in the chromosomes as well as on the DNA molecule.

6.3   Loops and scaffolds

A tangle of chromatin fibres would hardly be expected to maintain a strict linear order, and by the early 1970s a number of proposals had been made that chromosomes contained a linear core that could maintain the shape of the chromosome and keep the DNA in a fixed order (for references, see Sumner, 1998b; Stack & Anderson, 2001). However, it was not until after 1977, when Laemmli and his colleagues published their first studies on dehistonized chromosomes (Adolph *et al.*, 1977; Paulson & Laemmli, 1977), that it became generally accepted that chromosomes consisted of a core structure, usually called the scaffold, from which loops of chromatin radiated (Fig. 6.2). Initially there were many doubts about the reality of the chromosome scaffold, for a variety of reasons. Firstly, such a structure could not normally be detected in intact chromosomes. This is no longer a valid criticism, as the scaffold can be stained with silver (Fig. 6.3a), or immunolabelled for topoisomerase II (one of its main constituents, see below) (Fig. 6.3b). Secondly, the images of the scaffold were highly variable, and differed from one preparation to another; a loose network of fibres was most commonly seen (Fig. 6.2), rather than a discrete structure running along the centre of each chromatid, as might have been expected. The precise structure of the chromosome scaffold remains a matter of controversy, and will be discussed further below. Thirdly, it was argued that the 'scaffolds' seen in histone-depleted chromosomes were merely

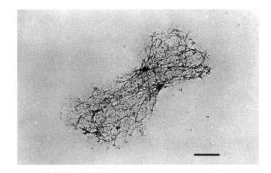

Figure 6.2 A dehistonized chromosome, showing loops, scaffold and two concentrations of material in the centromeric region that are believed to represent the kinetochores. Scale bar = 2 μm. Reproduced with permission from Hadlaczky *et al.* (1981) *Chromosoma* **81**, 537–555. © Springer-Verlag.

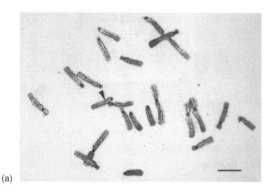

(a)

(b)

Figure 6.3 Chromosome scaffolds. (a) Scaffolds of the plant *Lilium longiflorum* stained with silver. Arrowhead indicates a region where the scaffold appears to be double. Scale bar = 10 μm. Reproduced with permission from Stack (1991) *Genome* **34**, 900–908. © National Research Council of Canada. (b) Scaffolds of Chinese hamster ovary chromosomes immunolabelled with topoisomerase II: (left) ethidium fluorescence of DNA, showing the whole chromosomes; (right) topoisomerase II immunofluorescence restricted to the centre line of each chromatid. Reproduced with permission from Sumner (1998) *Advances in Genome Biology* **5A**, 211–262. © JAI Press.

non-specific aggregations of proteins that occurred during the process of preparation. This has largely been refuted by the finding that scaffolds contain specific proteins (see below).

In spite of these uncertainties, it is now possible to give a reasonable, though still incomplete account of chromosome structure at this level. (See Stack & Anderson, 2001, for more detailed discussion of chromosome structure.) Loops of chromatin radiate out from a central structure running along the length of the chromosome. Average properties of the loops have been summarized by Pienta & Coffey (1984), and are listed

in Table 6.1. A number of points are immediately apparent. Firstly, there is great variability: some of this is likely to be due to uncertainties in measurement, but it could also indicate that not all the loops are of the same size. Secondly, the number of loops is of the same order as the number of genes (between about 32 000 and 39 000 genes per human cell; Bork & Copley, 2001), suggesting a possible correlation between loops and genes. Thirdly, the average length of a loop of chromatin is in the region of 0.5 μm; because it is a loop, it would only project 0.25 μm from the central core, but because loops project on both sides of the core, this would produce a structure about 0.5 μm in diameter, which is rather less than the diameter of a fully condensed chromosome (for which Pienta & Coffey quote a diameter of 0.85 μm). No allowance has been made in this calculation for the thickness of the core or scaffold. As the scaffold appears to be a relatively diffuse structure, it probably occupies little space and would add little to the diameter of the chromatid. It is possible to make an estimate of the packing ratio that would be produced by the formation of loops. Assuming that the space occupied at the core is the same as the thickness of the chromatin loops (30 nm or 0.03 μm), the packing ratio for a single loop would be 0.52 μm ÷ 0.03 μm, which equals 17.333. Estimates for the number of loops radiating from a single point on the core are in the region of 17 (Pienta & Coffey, 1984), giving a total packing ratio of about 295. Together with the packing ratio of about 40 of DNA in 30 nm fibres, this gives a total of about 12 000, which is similar to the estimated figure of about 10 000 for a condensed chromosome (Table 4.1). Given certain assumptions, therefore, the loops-and-scaffold model would be adequate to produce the final level of condensation required to form a condensed metaphase chromosome. We shall, however, reconsider this point further on (Section 6.4).

It should be emphasized that a model in which the loops and scaffold form a structure 0.5 μm or more in diameter is not universally accepted, and indeed is not supported by any direct observations. Several authors have

Table 6.1 Properties of average loops in chromosomes (ranges in parentheses).

Base pairs/loop	63 000 (30 000–100 000)
Length of DNA/loop	21.4 μm (10–34 μm)
Length of chromatin loop	0.52 μm (0.25–0.83 μm)
Number of loops/chromosome set (human)	95 000 (60 000–200 000)

After Pienta & Coffey (1984).

described structures with a diameter of 0.2–0.25 μm, which must condense further to form the fully condensed metaphase chromosome (El-Alfy & Leblond, 1988; Hao *et al.*, 1990; Manuelidis & Chen, 1990; Rattner, 1992). How this further level of condensation might be accomplished is considered later (Section 6.4).

The model just presented for the loops is an average one, and does not address the questions of whether loops are all the same size, whether they are attached at fixed points to the scaffold or can move and change their length and whether they have any kind of regular structure. It should be remembered that, in interphase, chromatin fibres are apparently drawn through replication and transcription 'factories', and are not static (Section 5.3). In fact, there is good evidence that chromatin loops in metaphase chromosomes are not all the same size, and that specific types of DNA sequences can form loops of specific sizes. Thus the genes for 18S + 28S rRNA (ribosomal genes) have notably small loops, each of which consists of a single copy of the repeated sequence (Keppel, 1986; Marilley & Gassend-Bonnet, 1989; Bickmore & Oghene, 1996). On the other hand, there appear to be no differences in the average loop size in euchromatic and heterochromatic segments of mammalian chromosomes (Bickmore & Oghene, 1996). Average sizes of loops appear to differ by only small amounts between gene-rich and gene-poor regions of chromosomes (Craig *et al.*, 1997), although such observations could nevertheless conceal large size differences between specific loops.

6.3.1 Scaffold attachment regions (SARs)

Specific DNA sequences, known as scaffold attachment regions (SARs), bind the loops to the metaphase chromosome scaffold. The SARs are DNA sequences that remain attached to scaffolds after exhaustive nuclease digestion, and are A+T-rich (Gasser *et al.*, 1989). They contain 70–75% of A+T base pairs but, except that they often (but not always) contain a topoisomerase II (Topo II) cleavage sequence, there appear to be no highly conserved SAR sequences. Nevertheless, SARs from one species (e.g. *Drosophila*) will bind to scaffolds from very distantly related species (e.g. mammals and yeast) (Amati & Gasser, 1988; Mirkovitch *et al.*, 1988). The Topo II cleavage site appears to be highly significant because Topo II is a major component of the scaffold (see below). It should be noted that A+T-richness alone is not sufficient for interaction with the scaffold.

The SARs are several hundred base pairs long – *Saccharomyces cerevisiae* SARs vary from about 300 to 1500 bp, and are spaced <3 kb to 140 kb apart (Gasser *et al.*, 1989) – providing further evidence that the loops may vary greatly in size, although it is not known if every SAR is necessarily attached to the scaffold. Loop size may determine directly certain features of chromosome structure. For example, the loops that contain ribosomal genes are particularly small (Keppel, 1986; Marilley & Gassend-Bonnet, 1989; Bickmore & Oghene, 1996), and the chromosomal regions that contain these genes are narrower, forming a secondary constriction. Similarly, the centromeric constriction could be a consequence of the tendency of certain centromeric DNA sequences (alphoid DNA in humans) to associate with the scaffolds rather than form loops (Bickmore & Oghene, 1996). The constriction formed by a segment of yeast DNA inserted into mouse chromosomes has also been interpreted as a consequence of shorter loops in yeast DNA than in the host DNA (McManus *et al.*, 1994). However, other explanations are possible, based on differential chromosome condensation (Sections 11.2 and 12.2.1).

The SARs do not occur in coding sequences, and in fact often appear to flank genes and form the boundaries of nuclease-sensitive regions associated with active genes. They have also been associated with origins of replication, and sizes of replicons are in fact similar to the sizes of chromatin loops (Buongiorno-Nardelli *et al.*, 1982). Loops delineated by SARs might therefore be functional units of chromosome organization. However, it must be recognized that there is sometimes a good deal of confusion in the literature between the metaphase chromatin loops, which are apparently relatively static, and the interphase loops, which, as we have seen (Section 5.3), are wound in and out of replication and transcription factories that form part of the interphase nuclear matrix.

6.3.2 Structure and composition of the scaffold

The nature of the scaffold itself is probably the most uncertain feature of the loops-and-scaffold model of chromosome structure. It can appear as a continuous straight line along the middle of each chromatid, as when stained with silver (Fig. 6.3a) or immunolabelled to show Topo II (Fig. 6.3b). However, the Topo II labelling is not always continuous, but may instead form a series of discrete dots (Earnshaw & Heck, 1985); alternatively it may have a helical appearance (Boy de la Tour & Laemmli, 1988). As already mentioned, electron microscopy of histone-depleted chromosomes revealed very variable images of scaffolds, no doubt in part due to swelling during the extraction procedure; electron microscope images of less swollen chromosomes suggest that scaffolds may consist of two main interconnected fibres in each chromatid (Zhao *et al.*, 1991). Although some of this variation in scaffold structure may be attributable to, for example, differences in the stage of the cell cycle, or of the type of cell examined, it seems clear that a large part of the variation must be artefactual, resulting from differences in preparation procedures.

If the structure of the scaffold remains uncertain, its composition is much more closely defined: it consists of two principal non-histone proteins, and a number of minor proteins that have not yet been characterized (Lewis & Laemmli, 1982). In addition, a number of centromeric proteins are tightly associated with the scaffold, supporting the evidence from electron microscopy that the centromeres (using the word in its broadest sense) are an integral part of the chromosome scaffold. These will be described in Chapter 12. The two principal scaffold proteins were originally named Sc I and Sc II. Protein Sc I (170 kDa) has turned out to be Topo IIα, an enzyme involved in a large number of processes that involve the untwisting of DNA (Wang, 1996). In mitotic chromosomes, Topo IIα is probably involved chiefly in the processes of condensation and segregation, although scaffold Topo IIα would not necessarily be required for these processes, as the enzyme is distributed throughout the width of the chromosome during the processes of condensation and segregation (Sumner, 1996). Nevertheless, the occurrence in the SARs of Topo IIα cleavage sites suggests that the enzyme has some physiological function in the scaffold, although its function in the scaffolds may be primarily structural.

The other main scaffold protein, Sc II (135 kDa), is a member of the SMC (Structural Maintenance of Chromosomes) family of proteins, which, like Topo IIα, are involved in chromosome segregation and condensation (Saitoh *et al.*, 1994; Heck, 1997; Hirano, 1998). Protein Sc II and similar proteins in *Xenopus* chromosomes (XCAP-C and XCAP-E) are all condensins, which are proteins involved in chromosome condensation (Heck, 1997; Hirano, 1998). Unlike Topo II, Sc II is only present in the mitotic chromosome scaffold, and is absent from the interphase nuclear matrix (Saitoh *et al.*, 1994). The mechanism by which SMC proteins produce chromosome condensation is described in Section 6.5.

6.4 Chromosome condensation – the final stages

As noted in Section 6.3, it has been proposed that the formation of chromatin loops attached to a scaffold is the final stage of chromosome condensation, and that differences in the length of chromatin loops are the primary determinant of the diameter of the chromatids. Although there is evidence that the loops could be of the correct size to produce chromatids of the diameters that have been observed, there is nevertheless good reason to suppose that there is another stage of condensation above that provided by the loops. Not only is this implied by the thickening of chromosomes from a diameter of about 0.2 μm to about 0.7 μm as they pass from prophase to metaphase (in mammals: El-Alfy & Leblond, 1989; El-Alfy *et al.*, 1994), but structural changes leading to further condensation have actually been observed. These structural changes take two forms that, if not mutually exclusive, at least seem difficult to reconcile with each other. These two modes of condensation are chromosome coiling and condensation into chromomeres.

6.4.1   Chromosome coiling

The coiling of chromosomes has been observed and studied for a very long time (for reviews see Huskins, 1941; Manton, 1950), particularly in plant chromosomes. In mammalian chromosomes, coiling can be induced by special treatment during preparation (Ohnuki, 1965), or may occur spontaneously (Fig. 6.4), but is only seen in a very small proportion of chromosomes. Coiling of chromosome scaffolds has been described (Rattner & Lin, 1985; Boy de la Tour & Laemmli, 1988), and it is estimated that such coiling would produce a ninefold packing of the chromatin (Rattner & Lin, 1985), possibly more than is actually needed for full condensation of metaphase chromosomes.

In the model described by Stack & Anderson (2001), coiling is regarded as the fundamental means of chromosome condensation. It is proposed that condensation is the result of shortening of a contractile core, which would be situated at one side of the chromatid. The contracted chromosomes would not show a hollow centre provided that the contraction was strong enough.

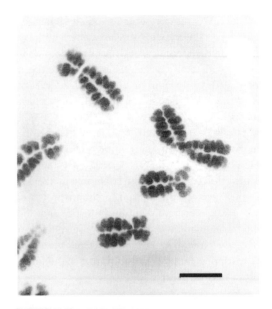

Figure 6.4 Scanning electron micrograph (backscattered mode and negative contrast) showing coiling of human chromosomes. Scale bar = 3 μm. Reproduced with permission from Sumner (1991) *Chromosoma* **100**, 410–418. © Springer-Verlag.

At centromeres the cores would remain in close contact with each other until anaphase, and so coiling could not occur until the sister centromeres had separated.

If coiling is a fundamental feature of chromosome structure, it may be asked why it is so rarely seen. It can, of course, be argued that the spirals are so closely packed together that they cannot usually be resolved, although it might be supposed that at intermediate stages of contraction more detail of the spirals should be visible. If a metaphase chromosome with a diameter of 0.7 μm were produced by the coiling of a prophase chromosome with a diameter of 0.2 μm, there would be a hole 0.3 μm in diameter along the centre of the chromatid. Such a large hole is never seen, so either the model is completely wrong, or the coiling is more irregular than supposed in the simple model just described or is combined with some other mode of condensation. Methods of inducing coiling in chromosomes often involve mechanical damage or

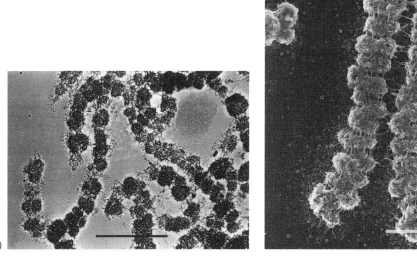

(a) (b)

Figure 6.5 Chromomeres. (a) Scanning electron micrograph of human pachytene chromosomes. Scale bar = 5 μm. Reproduced with permission from Sumner (1986) *Chromosoma* **94**, 199–204. © Springer-Verlag. (b) Scanning electron micrograph of a mouse mitotic chromosome. Scale bar = 2 μm. Reproduced with permission from Sumner (1998) *Advances in Genome Biology* **5A**, 211–262. © JAI Press.

drastic chemical treatments, which might be expected to induce all sorts of artefacts, although observations of spirals in living cells seem to be incontrovertible evidence of their reality (Manton, 1950). It could also be that organisms with large chromosomes (in which, for purely technical reasons, spirals would be more easily visible) actually need the extra condensation afforded by coiling to provide sufficient condensation, whereas organisms with small chromosomes can manage with a lower degree of condensation and might not require coiling. Organisms with larger amounts of DNA in their genomes have chromosomes that are not only longer, but are also fatter than those from organisms with small DNA amounts (Fig. 1 in Macgregor & Varley, 1988). The relationship between genome size, chromosome diameter, packing ratio, loop size and other relevant parameters has yet to be explored systematically for organisms with large and small genomes, although there is some evidence that organisms with larger C-values have larger loops (Buongiorno-Nardelli *et al.*, 1982).

6.4.2 Chromomeres

Another way in which chromosomes can undergo a final level of condensation is by the formation of chromomeres, which may be defined as aggregations of chromatin fibres that have no obvious orientation. Like coils, they are by no means always visible, yet there is one stage – pachytene of meiotic prophase (Fig. 6.5a) – at which they seem to be invariably present. During prophase and metaphase of mitosis, chromomeres are much less commonly seen but, like coils, they do occur in a very small proportion of chromosomes (Fig. 6.5b); they have also been described as part of the process of chromatin condensation in early prophase (El-Alfy & Leblond, 1989; El-Alfy *et al.*, 1994).

The characteristic pattern of distribution of chromomeres on chromosomes and the close correlation between the distribution on chromosomes of pachytene chromomeres and G-bands (Section 10.2.2) show that chromomeres are not random aggregations, but are specific structures with defined locations. However, neither the mechanism by which they form nor the factors that determine their localization is yet known. In spite of this, it is possible to propose a way in which the formation of chromomeres would be compatible with the loops-and-scaffold model. It has already been remarked that the scaffold does not always appear to be continuous, but is formed of a series of discrete dots (Earnshaw & Heck, 1985); these would form the foci for the aggregation of the loops. As the scaffold material itself aggregated and formed a continuous structure, so the chromomeres would aggregate to form a continuous cylinder. In fact, Cook (1995) has put forward such a model, in which loops radiate from transcription factories that aggregate to form chromomeres, which in turn fuse to form a cylindrical chromatid. There are, however, some problems with the details of this model: if the scaffold is based on transcription factories, these would be expected to contain RNA polymerase (Section 5.3), yet this enzyme is not a component of mitotic scaffolds. Heterochromatin (Chapter 7) is not generally transcribed, and therefore does not contain RNA polymerase. Moreover, chromosomes are divided into gene-rich and gene-poor regions (Section 10.2.2), and the latter would obviously have fewer attachments to transcription factories, and much larger loops, yet in general the loops seem to be fairly similar in size throughout most of the length of the chromosomes (Craig *et al.*, 1997). Regardless of these details, it is clear that condensation into chromomeres need not be incompatible with a loops-and-scaffold model for the chromosome, although chromomeres appear to be part of the process of condensation of loops and scaffolds, rather than an extra level of condensation beyond that provided by the loops and scaffold.

6.5 Biochemistry of condensation

At least three classes of proteins have been implicated in chromosome condensation: histones, topoisomerase II and SMC proteins. Traditionally, phosphorylation of histone H1 has been associated with chromosome condensation. In *Tetrahymena*, H1 and other linker histones are not essential, but their absence does result in reduced chromosome condensation (Shen *et al.*, 1995). On the other hand, condensation can take place without H1 phosphorylation in mouse cells (Guo *et al.*, 1995), and the role of H1 phosphorylation could be to loosen the binding of H1 to chromatin to allow access to other condensation factors (Hirano, 2000).

Phosphorylation of serine 10 in the N-terminal tail of histone H3 is strongly associated with condensation of mitotic chromosomes, both spatially and temporally (Hendzel *et al.*, 1997; Houben *et al.*, 1999; Section 4.2.6). Phosphorylation may act by reducing the affinity of the H3 tails for DNA, thus allowing access of condensation factors to the DNA (Sauvé *et al.*, 1999; Hirano, 2000).

It has been shown experimentally that Topo IIα is required for chromosome condensation: inhibition or immunodepletion of Topo IIα inhibits condensation (Giménez-Abián *et al.*, 1995). It is not yet clear, however, how Topo IIα participates in condensation (Hirano, 2000). It decatenates DNA by cutting one double-stranded (ds) DNA molecule, passing another dsDNA molecule through the gap and then resealing the gap (Fig. 6.6); this is necessary for the separation of daughter DNA molecules after replication (Section 2.3.1), and for the separation of sister chromatids at anaphase (Section 2.3.3). It is a major component of the chromosome scaffold (Section 6.3.2) but is also distributed throughout the body of the chromosome at prophase (Sumner, 1996), at the time when the chromosomes are condensing. It is possible that decatenation is necessary to allow condensation to proceed, and conversely that after the chromosome has condensed the condensation is stabilized by intramolecular recatenation (Hirano, 2000).

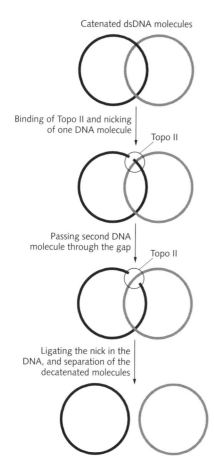

Catenated dsDNA molecules

Binding of Topo II and nicking of one DNA molecule

Topo II

Passing second DNA molecule through the gap

Topo II

Ligating the nick in the DNA, and separation of the decatenated molecules

Figure 6.6 The mechanism of action of topoisomerase II. Although the decatenation of circular DNA molecules is shown, exactly the same principles apply to linear DNA molecules constrained by attachment to a protein matrix.

The proteins whose functions in chromosome condensation are known best are the condensins. Two condensin complexes are known: 8S condensin, which in *Xenopus* consists of the SMC proteins XCAP-C and XCAP-E; and 13S condensin, which also contains XCAP-D2, -G and -H (which are not SMC proteins). Protein XCAP-E is homologous to the ScII scaffold protein (Section 6.3.2). Equivalent proteins have been identified in other organisms, including yeasts (Table 6.2). Only the 13S condensin complex functions in chromosome condensation, and it does so by inducing positive supercoils in DNA and binds to them to stabilize them (Fig.

6.7) (Hirano, 2000; Holmes & Cozzarelli, 2000). This reaction requires the hydrolysis of ATP; it is also necessary for XCAP-D2 to be phosphorylated (Kimura *et al.*, 1998).

Many details of condensin action still have to be elucidated. The supercoiling model just described has been worked out on DNA, and it is not known how well it would apply to chromatin. Nor is it known exactly what chromosomal substructure condensin would act on. The 13S condensin complex is very large, perhaps extending for $0.1\,\mu m$ (Holmes & Cozzarelli, 2000), so it could potentially act over quite large distances. There is only about one condensin complex per 10 kb of DNA, but if this were in a 30 nm chromatin fibre, it would be equivalent to about 0.6–$0.7\,\mu m$ (at 0.34 nm per base, 10 kb of DNA would occupy $34\,\mu m$, which, with a packing ratio of 50 for a 30 nm fibre, would equal $0.68\,\mu m$). With the average length of a loop being about $0.5\,\mu m$ (Table 6.1), there would be roughly one condensin complex per loop. Remember, however, that during condensation the chromosomes shorten and thicken; it remains to be shown how condensin does this, but it is evidently not simply a matter of contracting loops.

An intriguing addition to the proteins involved in chromosome structure and condensation is the giant (~2 MDa) protein titin. Titin was originally described as an elastic protein in muscle, but it is also present in chromosomes, and mutations in titin cause defects in condensation (Machado & Andrew, 2000).

6.6 The periphery of the chromosome

Unlike most cellular organelles, chromosomes have no membranes to separate them from the rest of the cell from the time the nuclear membrane breaks down at prometaphase until it is reformed at telophase. Nevertheless, chromosomes do have a distinct surface layer, which may provide protection from the surrounding cytoplasm but also has other functions. This layer has been given many names (Table 6.3), but it will be referred to here as the chromosome periph-

Table 6.2 Proteins of the 13S condensin complex.

Xenopus	Saccharomyces cerevisiae	Schizosaccharomyces pombe	Caenorhabditis elegans	Drosophila
XCAP-C*	SMC4	Cut3p	DPY-27[†]	
XCAP-D2 (pEg7)	LOC7	Cnd1		
XCAP-E*	SMC2	Cut14	MIX-1[†]	
XCAP-G	YCG1	Cnd3		
XCAP-H	BRN1	Cnd2	DPY-26	Barren

*An SMC protein.
[†]Required for dosage compensation (Section 8.4.1).
Data from Heck (1997); Hirano *et al.* (1997); Sutani *et al.* (1999).

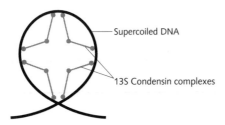

Supercoiled DNA

13S Condensin complexes

Figure 6.7 Condensation of chromosomal DNA as a result of induction of supercoiling by the 13S condensin complex.

Table 6.3 Names applied to the chromosome periphery.

Chromosome periphery
Chromosome surface
Halo surrounding the chromosomes
Outer surface of chromosomes
Pellicle
Perichromosomal layer
Perichromosomal region
Peripheral chromosomal material (PCM)
Sheath
Surface domain of chromosomes

ery, following the usage of Hernandez-Verdun & Gautier (1994).

The chromosome periphery covers almost all the chromosome, except for the centromeric region, where the surface is occupied by the kinetochores (Section 12.3), and the nucleolus organizer regions (NORs), where remnants of the nucleolus remain attached to the chromosome (Chapter 11). The chromosome periphery consists of closely packed fibrils, and dense granules 11–16 nm in width. With appropriate preparation, the periphery can be demonstrated as a dense layer surrounding the body of the chromosome (Fig. 6.8). The periphery forms during prophase, and disappears at telophase, so the chromosome periphery is a dynamic structure, probably having several functions (Hernandez-Verdun & Gautier, 1994).

The chromosome periphery consists of proteins and ribonucleoproteins (Hernandez-Verdun & Gautier, 1994), some of which are listed in Table 6.4. These proteins are not a homogeneous class, but originate from different parts of the nuclei, and are associated with the chromosomes at different stages of mitosis. It is therefore likely that the chromosome periphery has several different roles, and these are discussed below.

A role for the chromosome periphery in chromosome condensation has been proposed, based on the temporal correlation between the attachment of certain proteins to the chromosome and the time of condensation (Hernandez-Verdun & Gautier, 1994). This is improbable, given the involvement of condensins in chromosome condensation, and their localization in the chromosome scaffold (Section 6.5). Similarly, it is unlikely that the chromosome periphery has a structural role, which would be comparable to the exoskeleton of arthropods. There is more evidence that the periphery may provide protection against cytoplasmic components, because large molecules appear to be unable to penetrate into mitotic chromosomes (Yasuda & Maul, 1990).

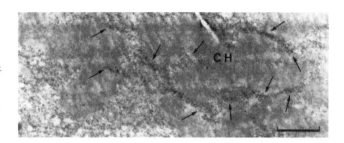

Figure 6.8 The chromosome periphery. Perichromosomal material (arrows) surrounds chromosomes (CH). Scale bar = 1 μm. Reproduced with permission from Gautier *et al.* (1992) *Chromosoma* **101**, 502–510. © Springer-Verlag.

Table 6.4 Proteins of the chromosome periphery.

Protein	Molecular weight	Origin/function	Stage of mitosis*	Refs
Fibrillarin	34 kDa	Nucleolus	P → T	1,2
INCENPs	135, 155 kDa	See text	P → M	3
Ki-67	345, 395 kDa	Nucleolus	P → T	1
Ku proteins	70 kDa	Nucleolus		4
Lamin B receptor			A → T	1
Lamins		Nuclear envelope	T	
p103		Nucleolus	P → T	1
p400+		Nucleoplasm	P → T	1
p52		Nucleolus	P → T	1
p66		Nucleolus	P → T	1
Perichromin	33 kDa	Nuclear envelope	P →	1
Perichromonucleolin		Nucleolus	P → T	1
Peripherin	27–31 kDa	? Conservation of chromosome structure	M → A	1
Protein B23	38 kDa	Nucleolus		1
Ribocharin		Nucleolus (*Xenopus*)	M → A	1
Ribosomal protein S1		Late stages of rRNA processing	M → A	1
snoRNPs†				1
snRNPs‡	28 kDa		M → A	1

*A, anaphase; M, metaphase; P, prophase; T, telophase.
†snoRNPs, small nucleolar ribonucleoproteins.
‡snRNPs, small nuclear ribonucleoproteins.
References: 1, Hernandez-Verdun & Gautier (1994); 2, Yasuda & Maul (1990); 3, Cooke *et al.* (1987); 4, Wachtler & Stahl (1993).

The overwhelming evidence is that the chromosome periphery is a means of transport for various proteins. Nucleolar proteins are bound to the surface of the chromosome throughout mitosis, and are then incorporated in newly formed nucleoli at the end of telophase. In the yeast *Saccharomyces cerevisiae*, which has a closed mitosis (i.e. the nuclear envelope never breaks down), there is no superficial layer of proteins on the chromosomes, and nucleolar components are transferred into the daughter cells by partition of the nucleolus (Hernandez-Verdun & Gautier, 1994). Other, non-nucleolar proteins may also be carried through mitosis in a similar way.

Other proteins found in the chromosome periphery are what have been termed 'chromosomal passengers' (Earnshaw & Bernat, 1991), which use the chromosomes as a means of reaching the correct position on the metaphase plate to carry out functions at anaphase and later stages of mitosis. Some of the proteins originally described as passengers are centromeric, and will be described further in Chapter 12. One group of passenger proteins, INCENPs, are at first widely distributed over the chromosomes, then become concentrated in the centromeric region at metaphase and become detached from the chromosomes in late metaphase, after which it is possible that the INCENPs determine the location of the cleavage furrow (Earnshaw & Bernat, 1991).

Although the terms 'chromosomal passenger' and 'passenger protein' were originally coined to describe proteins carried by the chromosomes to a specific part of the cell or mitotic apparatus, where they would perform their ultimate functions, it could be argued that a large proportion, if not all, of the proteins in the chromosome periphery are in some sense passengers. The nucleolar proteins, for example, evidently become attached to the chromosomes so that they do not become dispersed in the cytoplasm during mitosis, but remain available in the vicinity of the chromosomes to provide material for the formation of a new nucleolus when the time comes. Such proteins have no function on the chromosome, but are merely using it as a means of transport. They are indeed passengers and, like passengers on a bus or train, get on and off at various times and places.

6.7 Meiotic and mitotic chromosomes compared

It would seem reasonable to suppose that meiotic chromosomes are built on a similar plan to that of mitotic chromosomes, and indeed they have a core or scaffold from which loops radiate (Moens & Pearlman, 1990), they condense by forming chromomeres (Fig. 6.5a), and can form spirals (Manton, 1950; Nokkala & Nokkala, 1985). It should be noted that whereas chromomeres are characteristic of the pachytene stage of prophase, spirals are seen most clearly at metaphase I and later stages, including the second meiotic division, raising the possibility that different mechanisms of condensation are used at different stages. The metaphase chromosome is very much shorter than the pachytene chromosome, indicating that substantial additional condensation occurs between pachytene and metaphase.

The main difference between meiotic and mitotic chromosomes is in the nature of their scaffold or core structures, meiotic prophase chromosomes having the synaptonemal complex (SC) (Section 2.5.2) instead of the less complex scaffold structure found in mitotic chromosomes (Section 6.3). This difference must not be exaggerated. Meiotic metaphase chromosomes have perfectly good scaffolds that can be stained with silver and that run along the centre of the chromatid (e.g. Suja et al., 1992), just like those in mitotic chromosomes. The problem arises when comparisons are made between the SC, usually at the pachytene stage of meiotic prophase, and the scaffold of mitotic metaphase chromosomes. Much has been made of the fact that there is only a single lateral element per chromosome (i.e. per two chromatids) in meiotic prophase, rather than one for each chromatid (i.e. two per chromosome) at mitotic metaphase. In addition, the mitotic scaffold is located centrally in the chromatid, whereas the lateral element is peripheral in the meiotic prophase chromosome. On the basis of such differences, the scaffold and the SCs have often been regarded as independent structures (e.g. Heyting, 1996), although there is no real evidence for this, and certain proteins from the lateral elements of the SC are also found in meiotic metaphase cores (Stack & Anderson, 2001).

It is not valid to compare meiotic prophase chromosomes with mitotic metaphase chromosomes; instead, the comparison should be made with mitotic prophase chromosomes, which have only a single scaffold (Giménez-Abián et al., 1995; Sumner, 1996). Both the lateral elements and the scaffolds can be stained with silver and contain Topo II (Heyting, 1996), although these are not highly specific. Unfortunately, detailed comparisons of the composition of the lateral elements and the mitotic scaffolds have not been made. There is no doubt that the SC contains many proteins that do not occur in the mitotic scaffold, but this is hardly surprising because it has additional functions. Similarly, there are many types of DNA sequences associated with the SC that do not associate with the mitotic scaffold (Moens, 1994); such sequences are often involved in recombination, so again it is not surprising that they would only be present in the SC. It thus seems more likely that the lateral elements and the SC as a whole are a special adaptation of the chromosome scaffold to the needs of meiotic prophase, and not special structures that are lost in late prophase and replaced by completely new chromosome scaffolds at meiotic metaphase I.

6.8 There is still much to be learnt about chromosome structure

It is both surprising and unfortunate that we have such a poor understanding of the highest levels of chromosome organization. The organization of the scaffold and its relationship to specific DNA sequences is still poorly understood, in spite of its importance for numerous functions of chromosomes. The mechanism of chromosome condensation, one of the fundamental features of mitosis and meiosis, is only just beginning to be unravelled. Nevertheless, good progress is being made in understanding these aspects of chromosome organization, and much should be clearer in a few years' time.

On the whole, particularly in this chapter, chromosome organization has been described as if it were essentially similar throughout the chromosome. In many fundamental respects this is indeed true, but it should not blind us to the fact that there is extensive differentiation along chromosomes, such as the existence of centromeres, heterochromatin, secondary constrictions, and so on, all of which have superimposed a distinctive pattern on the basic chromosome organization. The next few chapters will be concerned with such aspects of chromosomal differentiation.

Constitutive heterochromatin

7

7.1 What is heterochromatin?

Heterochromatin is perhaps the most misused word and the least understood concept in the whole of the study of chromosomes. It has been used to describe different concepts, and our understanding and definition of heterochromatin has shifted as our knowledge has increased and as different methods have become available to study it.

Heterochromatin was first defined by Heitz in 1928 as chromatin that did not, unlike the rest of the chromatin, decondense at the end of telophase, but instead remained compact throughout interphase, and was found to be condensed even at the beginning of prophase. It is therefore characteristic of heterochromatin that it contracts much less during prophase than the remainder of the chromatin, known as euchromatin, does (Balicek *et al.*, 1977), and thus occupies a greater proportion of the chromosome length at metaphase than it does at prophase. Brown (1966) classified heterochromatin into two main classes: facultative heterochromatin, which is permanently condensed chromatin that occurs in only one of a pair of chromosomes, and thus has the same DNA composition as the chromatin of its euchromatic homologue; and constitutive heterochromatin that occurs at the same site in both homologues of a chromosome. Facultative heterochromatin is often associated with sex chromosomes and sex differentiation (Sections 8.3.7 and 8.4.3). However, facultative heterochromatin is not necessarily restricted to one of a pair of chromosomes, and is better regarded as regions that are epigenetically repressed and are heterochromatic for only part of the life cycle. Constitutive heterochromatin, on the other hand, may be regarded as a substance (or rather, a group of substances with common properties) that tends to have a DNA base composition substantially different from that of the euchromatin, and is untranscribable because of its DNA composition.

Heterochromatin characteristically shows little or no genetic activity, and there has been an undesirable tendency to refer to all condensed, genetically inactive chromatin as heterochromatin. In extreme cases, such as the entire nucleus of nucleated erythrocytes in vertebrates, and sperm heads in most organisms, this would mean that all chromatin, regardless of its composition, would be heterochromatin, and render the term heterochromatin essentially meaningless. Although heterochromatin is generally transcriptionally inactive, transcription of heterochromatin has been described in certain plants (Nagl & Schmitt, 1985) and vertebrates (Varley *et al.*, 1980; Sperling *et al.*, 1987), although the function, if any, of the RNA produced is not known. Heterochromatin is typically late replicating (John, 1988), but not all late-replicating regions of chromosomes are necessarily heterochromatin.

In fact, heterochromatin should be defined not merely by its condensation, time of replication and genetic inactivity, but also by its staining properties (C-banding and other methods, see

Section 7.3) and by the type of DNA and proteins it contains (Sections 7.3.1 and 7.3.2). Nevertheless, there are segments of chromatin that may have some, but not all, of the accepted properties of heterochromatin but are still regarded as heterochromatin. In the molecular era, it may be necessary to use rather different criteria from those based on, for example, staining properties.

7.2 Where is constitutive heterochromatin on the chromosomes?

Blocks of heterochromatin can occur in virtually every part of chromosomes. Nevertheless, they occur preferentially in certain parts of chromosomes, and are found at specific sites on specific chromosomes.

Virtually all chromosomes have blocks of heterochromatin in the centromeric region, and these blocks may vary in size between very large and very small. In the cat family (Felidae), among others, the centromeric heterochromatin is very small (Pathak & Wurster-Hill, 1977), and may correspond merely to that region of the chromatin associated with the kinetochore (Chapter 12). In other species, the centromeric (or strictly, the paracentromeric) heterochromatin can form large blocks (Fig. 7.1).

Constitutive heterochromatin is also found quite commonly in the terminal (non-centromeric) regions of chromosomes (Fig. 7.2), although a large proportion of species lack terminal heterochromatin on most or all of their chromosomes. A common mode of chromosomal evolution is the formation of heterochromatic short arms on chromosomes that are acrocentric or telocentric in related species (Section 16.3.5). Least common are interstitial blocks of heterochromatin, which are nevertheless not uncommon in, for example, insects and plants with large chromosomes (Fig. 7.2). It should be noted that different blocks of heterochromatin in the same species often differ in DNA composition (Sumner, 1990). Blocks of heterochromatin are also usually heteromorphic, that is, they often differ in size between individuals of the same

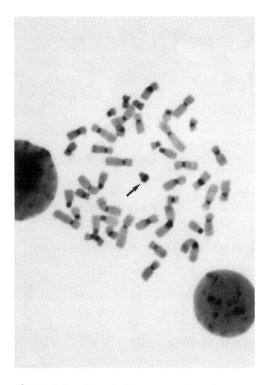

Figure 7.1 A C-banded human metaphase, showing centromeric heterochromatin on all chromosomes, plus heterochromatin on the long arm of the Y chromosome (arrow). Reproduced with permission from Sumner (1972) *Experimental Cell Research* **75**, 304–306. © Academic Press.

species, and between homologues in the same individual (Fig. 7.3).

7.3 What is constitutive heterochromatin made of?

It is not particularly straightforward to study whether or not particular segments of chromosomes fail to decondense at the end of mitosis, and it was a great advance when it became possible to study constitutive heterochromatin using a relatively simple staining technique called C-banding (Box 7.1). C-Banding (Figs 7.1 & 7.2) was shown to stain almost all segments of constitutive heterochromatin that had been identified by their failure to decondense in interphase, although there were a few exceptions (John,

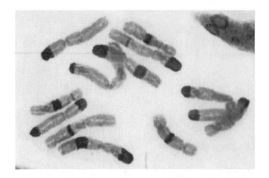

Figure 7.2 A C-banded metaphase from the plant *Scilla sibirica*, showing terminal and interstitial heterochromatin. Reproduced with permission from Vosa (1973) *Chromosoma* **43**, 269–278. © Springer-Verlag.

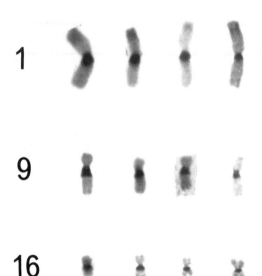

Figure 7.3 Heteromorphism of heterochromatin. Human chromosomes 1, 9 and 16, each showing a series of decreasing size of C-bands.

1988; Sumner, 1990). However, study of yeast chromosomes has shown that these have segments that have properties typical of heterochromatin (Grunstein, 1998): resistance to nucleases, late replication and induction of position effect variegation (PEV) and related phenomena (Section 7.4.5). Because of the small size of their chromosomes, which makes cytogenetical study very difficult, it would hardly be expected that C-banding could be demonstrated in yeasts. Sim-

ilarly, in *Drosophila* and probably other organisms, telomeric regions of chromosomes have properties of heterochromatin, but without showing distinctive staining (Cryderman *et al.*, 1999).

7.3.1 What sort of DNA is found in constitutive heterochromatin?

Most segments of constitutive heterochromatin on eukaryotic chromosomes contain high concentrations of highly repeated (satellite) DNA, which is found only at low levels or not at all in euchromatin. There are no common properties of base sequence or length of repeating unit in these highly repeated DNAs (Section 3.3.1.1). They can vary in composition from highly A+T-rich to highly G+C-rich, and in length from a 2 bp repeat to repeating units of hundreds or thousands of base pairs (Beridze, 1986; Sumner, 1990). The involvement of classical satellites in constitutive heterochromatin may be connected with common properties of secondary DNA structure (many satellite DNAs are bent; Martinez-Balbas *et al.*, 1990), repetition itself (the length of repeating DNA units coincides in some cases with that of the nucleosomal repeat; Strauss & Varshavsky, 1984) or simply a freedom to produce multiple copies of a sequence in heterochromatin without any deleterious effects. In any case, not all segments of heterochromatin contain highly repeated DNA (John, 1988). A number of cases have been reported in which constitutive heterochromatin appears to be made up only of middle repetitive sequences. In fact, there is increasing evidence that transposable elements may accumulate in heterochromatin (Ananiev *et al.*, 1998b; Dimitri & Junakovic, 1999; CSHL/WUGSC/PEB *Arabidopsis* Sequencing Consortium, 2000). Although accumulation of transposable elements in heterochromatin might be less damaging because heterochromatin is generally inactive (Section 7.1), such sequences may have a more positive role in chromosome structure and function (Dimitri & Junakovic, 1999). Even within a species or within a single chromosome there can be different types of DNA in constitutive heterochromatin, and it is clear that the properties of heterochromatin

Box 7.1 C-Banding of chromosomes

C-Banding is a method of staining chromosomes that is specific for most constitutive, but not facultative, heterochromatin. Chromosome preparations are treated successively with dilute acid, alkali (barium hydroxide), warm saline and stained with Giemsa dye (Sumner, 1972). The alkali hydrolyses the DNA, which has been depurinated by fixation in alcohol–acetic acid and by the dilute acid treatment, and the hydrolysed DNA is extracted by the saline incubation (Holmquist, 1979). Because of its more compact nature, the DNA is not extracted so easily from the constitutive heterochromatin, which is stained more strongly as a result (Figs 7.1 & 7.2).

Table 1 Stages of the C-banding technique.

Fixation in alcohol–acetic acid	Depurinates DNA
Treatment with dilute acid	Depurinates DNA
Barium hydroxide	Hydrolysis of depurinated DNA
Saline incubation	Extraction of hydrolysed DNA
Giemsa staining	Selective staining of constitutive heterochromatin, which is more resistant to extraction

cannot depend simply on the presence of a specific DNA sequence. In fact, there is growing evidence that repetition itself can cause the formation of heterochromatin. In *Drosophila*, vertebrates and plants, multiple copies of a transgene can form heterochromatin (Henikoff, 1998; Hsieh & Fire, 2000). How repetition, by itself, might induce the formation of heterochromatin is not known.

Although there is no consistent pattern of DNA sequence in constitutive heterochromatin, there is one feature of DNA that is commonly found in heterochromatin and appears to be important for the condensation of heterochromatin. This is cytosine methylation (Section 3.5). High levels of methylation are found in satellite DNAs of many plants and mammals (Beridze, 1986; CSHL/WUGSC/PEB *Arabidopsis* Sequencing Consortium, 2000), although many organisms, such as yeasts, *Caenorhabditis* and *Drosophila*, have little or no 5-methylcytosine in their DNA. Cytosine methylation can be demonstrated not only by chemical analysis, but also by immunolabelling chromosome preparations for 5-methylcytosine (Fig. 7.4). Demethylation of cytosine, whether occurring as a normal

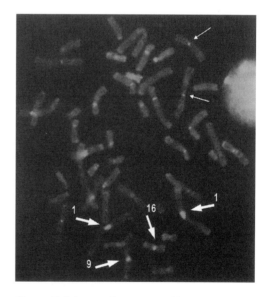

Figure 7.4 Immunofluorescence of human chromosomes showing the concentration of 5-methylcytosine in the heterochromatin. Note the large blocks of heterochromatin on chromosomes 1, 9 and 16, and smaller blocks at other centromeres (thin arrows). Micrograph kindly provided by D. Bourc'his.

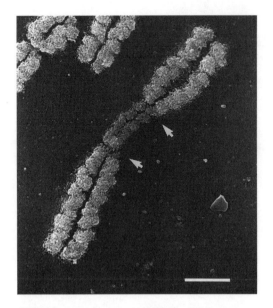

Figure 7.5 Decondensation of demethylated heterochromatin (arrows). Scanning electron micrograph of human chromosome 1 from a patient with the ICF syndrome (Section 17.7) in which the chromosomal DNA is poorly methylated. Scale bar = 2 μm.

developmental process (e.g. in spermatozoa; Martin *et al.*, 1983), as a pathological state (e.g. ICF syndrome; Miniou *et al.*, 1994), or by treatment of cells with 5-azacytidine (Schmid *et al.*, 1983a), causes decondensation of hetero-chromatin (Fig. 7.5).

7.3.2 Are there specific heterochromatin proteins?

Although blocks of constitutive heterochromatin do not owe their properties to a characteristic type of DNA, it is clear that certain proteins, protein motifs or simple modifications are found in heterochromatin from a wide variety of organisms (Table 7.1). Histone H4 is under-acetylated in heterochromatin of plants (Belyaev *et al.*, 1997), *Drosophila* (Turner *et al.*, 1992) and mammals (Jeppesen *et al.*, 1992). Underacetyla-tion is associated with transcriptional inactivity, so this finding is hardly surprising in view of the general inactivity of heterochromatin. Acetylation results in nucleosome remodelling so that the DNA is more accessible to transcription factors (Section 4.2.4); however, experimental acetyla-tion of histones in heterochromatin does not prevent C-banding (Halleck & Schlegel, 1983), a technique whose specificity appears to depend on the heterochromatin remaining compact and inaccessible (Sumner, 1990). The acetylation of histone does not, therefore, appear to be a primary determinant of heterochromatin structure.

Heterochromatin protein 1 (HP1) and similar proteins have been found in a wide variety of organisms (Eissenberg & Elgin, 2000; Table 7.2), although not all such proteins are components of heterochromatin. The amino-terminal region of these proteins contains a region known as the chromodomain (*chromo*some *o*rganization *modi*-fier *domain*), connected by a 'hinge' region to a 'chromo shadow domain' (Fig. 7.6; Eissenberg & Elgin, 2000). The chromodomain is one of the regions of HP1 responsible for binding to het-erochromatin. The HP1 does not bind directly to the DNA in heterochromatin, but instead binds to histone H3 methylated at lysine 9 (Bannister *et al.*, 2001; Lachner *et al.*, 2001). As previously described, H3 methylation is associated with transcriptional inactivity (Section 4.2.6). The chromo shadow domain appears to be required for self-association of HP1-type molecules to form dimers (Eissenberg & Elgin, 2000). The HP1-type proteins form complexes with several other proteins, including SU(VAR)3–7 and SU(VAR)3–9 in *Drosophila* and TIF1-alpha and -beta and CAF1 (chromatin assembly factor) in mouse, and, like HP1, these proteins can localize to heterochromatin. In *Drosophila* and *Xenopus*, the origin recognition complex proteins ORC1 and ORC2 interact with HP1, although the sig-nificance of this is not clear (Wallrath, 1998; Eissenberg & Elgin, 2000). Binding of HP1 to the lamin B receptor could be involved in the localization of heterochromatin and condensed chromatin next to the nuclear envelope (Wall-rath, 1998). Protein HP1 binds to the nuclear envelope through its chromodomain, and this interaction might be involved in reassembly of the nuclear envelope (Kourmouli *et al.*, 2000).

The early embryo of *Drosophila* is syncytial, and the first 13 divisions are passed through very

Table 7.1 Proteins of constitutive heterochromatin (not exhaustive).

Protein	Species	Properties	Refs
α-Protein	African green monkey	Nucleosome positioning protein. Binds alpha-satellite	1
Underacetylated histones H3, H4	*Peromyscus*	Acetylation does not prevent C-banding	2
	Human	Very low levels of acetylation in major blocks of heterochromatin	3
	Vicia faba		4
Histone H4 acetylated at Lys 12	*Drosophila*	Enriched in heterochromatin	5
HP1		See Table 7.2	
HP1-interacting proteins	*Drosophila*	Form complexes with HP1	7
cp17.3	*D. virilis*	? Histone variant associated with satellite DNA	8
cp75	*D. virilis*	? Equivalent to D1 in *D. melanogaster*	8
Suvar (3)7	*Drosophila*	Zinc-finger protein required for PEV	6, 9
Su(var) 231	*Drosophila*	Suppressor of PEV	9
Modulo	*Drosophila*	DNA-binding protein required for PEV	6, 9
GAGA factor	*Drosophila*	Transcription factor. Binds to AAGAG and AAGAGAG sequences in satellite DNA	6
Ikaros	Mouse	Location changes during cell cycle	10
MeCP2	Mouse	5-Methylcytosine-binding protein	11
HMGA1a	Mammals	Concentrated in heterochromatin	12
TIF1beta			
SP100			
Suvar39H1/2			
ATRX			
ACF			
DNMT3b			
Helios			
PcG complex (RING1; BMI1; hPc2)			
RAP1	Yeast		
SIR 2,3,4	Yeast		
Swi6	*S. pombe*		
Clr4	*S. pombe*		
Rik1	*S. pombe*		

References: 1, Strauss & Varshavsky (1984); 2, Halleck & Schlegel (1983); 3, Jeppesen *et al.* (1992); 4, Belyaev *et al.* (1997); 5, Pirotta (1997); 6, Lohe & Hilliker (1995); 7, Wallrath (1998); 8, Viglianti & Blumenfeld (1986); 9, Reuter & Spierer (1992); 10, Brown *et al.* (1997); 11, Lewis *et al.* (1993); 12, Disney *et al.* (1989).

Figure 7.6 Structure of HP1 proteins.

rapidly (Orr–Weaver, 1994). During this period, no heterochromatin is detectable, either by the criterion of condensation or by C-banding (Vlassova *et al.*, 1991), and it is interesting to note that HP1 does not become associated with the chromosomes until heterochromatin becomes visible towards the end of the syncytial stages (James *et al.*, 1989). The concentration of HP1 in heterochromatin appears to be associated with phosphorylation of this protein (Eissenberg *et al.*, 1994). In embryos in which the HP1 is non-functional, chromosomes do not condense properly and chromosome morphology and segregation are defective (Lohe & Hilliker, 1995).

In yeasts, various heterochromatin proteins

Table 7.2 HP1-like proteins.

Species	Protein	Chromosomal location
Schizosaccharomyces pombe	Swi6p	Heterochromatin
Tetrahymena thermophila	Hhp1p	Condensed chromatin of macronuleus
Caenorhabditis elegans	emb|CAB07241	?
	Gi|3702834	?
Planococcus citri	pchet1	Male-specific chromatin
	pchet2	?
Drosophila melanogaster	HP1	Heterochromatin, telomeres, some other sites
Drosophila virilis	DvHP1	?
Xenopus laevis	Xhp1alpha	?
	Xhp1gamma	?
Gallus domesticus	CHCB1	?
	CHCB2	?
Mus musculus	mHP1alpha	?
	M31 (MoMOD1)	Heterochromatin
	M32 (MoMOD2)	Euchromatin
Homo sapiens	HP1alpha	Heterochromatin
	HP1beta	Heterochromatin
	HP1gamma	Euchromatin
Mammals	HP1gamma	Euchromatin and heterochromatin (Minc et al., 2000)

Data from Eissenberg & Elgin (2000).

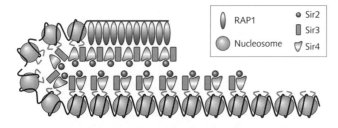

	RAP1	● Sir2
		▮ Sir3
	Nucleosome	▽ Sir4

Figure 7.7 Structure of telomeric heterochromatin proteins in budding yeast, *Saccharomyces cerevisiae*. Reproduced with permission from Strahl-Bolsinger *et al.* (1997) *Genes & Development* **11**, 83–93. © Cold Spring Harbor Laboratory Press.

have been identified, and their interactions with each other and with DNA have been analysed, so the molecular structure of the heterochromatin is quite well understood (Grunstein, 1998). In *Saccharomyces cerevisiae* there are no HP1-like proteins, and instead silent information regulator (SIR) proteins, which have no homologues in other organisms, are used to produce heterochromatin. The protein RAP1 binds to the telomeric sequence $C_{1-3}A$ and then recruits SIR3, followed by SIR4 and SIR2 (Fig. 7.7). As

the heterochromatin spreads into nucleosomal chromatin, SIR3 and SIR4 interact with the N-terminal domains of histones H3 and H4. In yeast heterochromatin, histone H4 is underacetylated on lysines 5, 8 and 16, but not on lysine 12, reflecting the situation in mammals and *Drosophila* (Table 7.1). It may be necessary for the histones to be deacetylated before binding can occur, and it must be significant that SIR2 has deacetylase activity (Khochbin *et al.*, 2001).

In fission yeast (*Schizosaccharomyces pombe*),

the formation of centromeric heterochromatin depends on underacetylation of histones (Ekwall *et al.*, 1997), and also requires the centromere-specific proteins Clr4 (a histone H3 methylase equivalent to human Suvar39H1), Rik1, Chp1 (a chromodomain protein) and Swi6 (an HP1-like protein) (Pidoux & Allshire, 2000). Both Swi6 and Chp1 require Rik1 and Clr4 to bind to the repetitive centromeric DNA, and Swi6 is also a component of the other heterochromatic domains in *S. pombe* chromosomes, the telomeres and the mating-type loci. Inheritance of the heterochromatic state depends on the histone being underacetylated (Ekwall *et al.*, 1997).

7.4 What does heterochromatin do?

It is a general presumption that constitutive heterochromatin is inactive, inert material. In *Drosophila*, in which it is possible to manipulate the genome easily, and alter the amount and position of the heterochromatin, it can be shown that, in general, such alterations have no effect on the viability of the flies (Yamamoto & Miklos, 1978). Similarly, in humans, studies of heteromorphisms in thousands of newborns failed to reveal any effects on the phenotype of the differences in the amount of heterochromatin (Bobrow, 1985; Hsu *et al.*, 1987). Certain organisms – parasitic nematodes (Müller *et al.*, 1996), copepods (Beerman, 1977), hagfish (Nakai *et al.*, 1995) and others – eliminate C-banded heterochromatin in somatic cells in early development, suggesting that the heterochromatin has no function in the soma. During polytenization of dipteran chromosomes, the heterochromatin is not replicated (Section 15.2.3), suggesting that it is of no importance. In *Drosophila* (Weiler & Wakimoto, 1995) and *Arabidopsis* (CSHL/WUGSC/PEB *Arabidopsis* Sequencing Consortium, 2000) only an extremely low number of genes map to heterochromatin, and in humans no genes have been mapped to C-bands (Bickmore & Craig, 1997). The sequences of certain short, highly repeated DNAs found in heterochromatin appear to be such that no sensible polypeptide sequence could be translated

from them. There is thus a substantial body of evidence that constitutive heterochromatin is essentially inactive material with few, if any, phenotypic effects – 'junk' DNA according to some.

In spite of all this, there is abundant and increasing evidence that constitutive heterochromatin can contain genes and other functional DNA sequences, can have important functions and effects in the germ line even though it has been eliminated from the soma, can affect chromosome segregation and is the cause of position effect variegation (PEV). Some of these (e.g. a role in segregation) may be regarded as functions of heterochromatin, in that the heterochromatin is necessary for the performance of a particular action; others are to be regarded as effects, in which the heterochromatin is not essential and the activity that it affects can take place in the absence of the heterochromatin, but with somewhat different parameters. These functions and effects will now be considered. It will be noticed that many of the examples given come from *Drosophila*. This is partly because *Drosophila* is uniquely thoroughly studied both from a genetic and a cytogenetic standpoint, so more is known about it. It nevertheless has distinctive features of its own (e.g. achiasmate male meiosis without synaptonemal complexes) so that conclusions drawn from *Drosophila* do not necessarily apply to other species.

7.4.1 Chromatin elimination and diminution

As mentioned above, chromatin elimination occurs in a variety of organisms (John, 1988), but has been studied in most detail in parasitic nematodes (Müller *et al.*, 1996). The essential features are that, at specific cell divisions in very early development, in those cells that will give rise to the soma, the chromosomes break up into a large number of smaller chromosomes and most of their heterochromatin is lost and disintegrates. In *Parascaris univalens*, between 80% and 90% of the total DNA is lost, but in *Ascaris suum* only about 25% is lost. No loss of heterochromatin occurs in those cells that give rise to the germ line. The eliminated heterochromatin consists largely of highly repeated satellite DNAs, which are then largely absent from the future somatic cells; in *Ascaris suum* some middle repetitive sequences are also eliminated. As well as the repeated sequences, three single-copy genes have been identified that are eliminated with the repeated sequences but are retained in the germ cells (Müller *et al.*, 1996). Thus although heterochromatin is eliminated in somatic cells but retained in the germ line, it might be that not only repeated sequences but also specific genes embedded in the heterochromatin are required for germ line function. In copepods the eliminated chromatin may also contain sequences that are not genetically inert (Standiford, 1989).

7.4.2   Genes and other functions in *Drosophila* heterochromatin

Although the short, highly repeated sequences of satellite DNAs in *Drosophila* heterochromatin obviously cannot act as genes, several true protein-coding genes have been identified in heterochromatin (Lohe & Hilliker, 1995; Weiler & Wakimoto, 1995). In some cases, at least, these genes cannot function properly if translocated to euchromatin, and thus the argument that such genes might represent tiny regions of euchromatin embedded in the heterochromatin does not seem to be tenable. *Drosophila melanogaster* Y chromosomes contain at least nine genes, of which six are fertility factors (Pimpinelli *et al.*, 1986; Carvalho *et al.*, 2001). The Y chromosome is typically heterochromatic in somatic cells, but it decondenses in spermatocytes and at the appropriate stage forms a lampbrush chromosome (Section 14.4), and the fertility factors form typical lampbrush loops.

As well as conventional genes, *Drosophila* heterochromatin contains highly repeated sequences that have genetic effects (Pimpinelli *et al.*, 1986; Gatti & Pimpinelli, 1992). The *Responder* locus, part of the *Segregation Distorter* system, consists of a 120 bp sequence repeated up to 2500 times (Doshi *et al.*, 1991). Another heterochromatic effect is *ABO*, that rescues the maternal-effect lethality caused by the euchromatic mutation *abo*. Interestingly, *ABO* functions in the earliest stages

of embryogenesis before transcription of euchromatic genes has begun, even if the *ABO* sequences have been introduced by the sperm. *ABO* occur on both the X and no. 2 chromosomes, but only two doses are necessary to counteract the effect of the *abo* mutation. Thus, like the nematode heterochromatin, which is eliminated in somatic cells, *Drosophila* heterochromatin contains typical protein-coding genes at a low density, but also contains other factors that interact with more typical euchromatic genes.

7.4.3   Effects of heterochromatin on pairing and meiosis

It has been proposed several times that constitutive heterochromatin has an important role in homologous pairing in meiosis, but evidence for this is weak. Although pre-existing associations of blocks of heterochromatin in chromocentres in interphase may help to keep homologues in close proximity, there is no real evidence that they play a fundamental role in pairing (John, 1988). On the other hand, heterochromatin does have some negative effects on pairing and crossing-over. Crossing-over is usually absent in heterochromatin, although there are some exceptions (John, 1990). This absence of crossing-over is often associated with a lack or delay of synaptonemal complex (SC) formation in such regions (John, 1988). There may also be differences in the structure of the SC in heterochromatic regions (John, 1990).

Effects of constitutive heterochromatin on the number and position of chiasmata at meiosis are widespread. As a general rule, the presence of a block of heterochromatin inhibits the formation of chiasmata in its vicinity (John, 1988; Sumner, 1990). However, in *Allium*, chiasmata are formed preferentially adjacent to blocks of heterochromatin, and in the absence of such blocks the chiasmata are less localized (Loidl, 1982). Heterochromatin can also affect the number of chiasmata, as well as their distribution (John, 1988; Sumner, 1990). In some cases, heterochromatin increases the number of chiasmata, but in others (e.g. the grasshopper *Atractomorpha*; Miklos & Nankivell, 1976), the number of chiasmata is reduced by an increased amount of heterochromatin. In some species heterochromatin has no significant effect on chiasma distribution or number (Attia & Lelley, 1987). Thus although heterochromatin can clearly have effects on meiotic pairing and crossing-over, there is no consistency to these effects, and the idea that heterochromatin might have a general function in controlling these aspects of meiosis cannot be sustained.

7.4.4 A role for constitutive heterochromatin in chromosome segregation

A large number of studies have suggested various functions for heterochromatin in the processes of chromosome segregation. In the fission yeast *S. pombe*, underacetylation of histone and the presence of Swi6 are necessary for the formation of heterochromatin, and hyperacetylation of histones or mutation of Swi6 leads to chromosome loss at mitosis (Ekwall *et al.*, 1997). In fact, Swi6 is needed so that the Rad21 subunit of cohesin can bind to centromeres and ensure cohesion until anaphase (Bernard *et al.*, 2001). Twelve *csp* genes are also involved in heterochromatin organization, and mutation of these also leads to defects in chromosome segregation (Pidoux & Allshire, 2000). In female meiosis in *Drosophila*, the small chromosome 4 never undergoes crossing-over, and the X chromosome fails to cross over in 5–10% of oocytes. In most organisms, failure to undergo crossing-over and formation of chiasmata inevitably leads to non-disjunction (Section 2.5.3), but the segregation of chromosomes 4 and X in female *Drosophila* is highly regular, with only about 0.1% non-disjunction. This is achieved by intimate pairing of the heterochromatin of the homologous chromosomes (Hawley & Theurkauf, 1993; Irick, 1994; Dernburg *et al.*, 1996). In these chromosomes, the euchromatin separates before metaphase I, just as it does in chiasmate chromosomes. Non-homologous chromosomes will also segregate regularly from each other in most cases, using a system known as distributive segregation; this mechanism does not involve any association of heterochromatin.

However, both the homologous and non-homologous segregation systems make use of the same protein, known as nod.

Meiosis in male *D. melanogaster* is always achiasmate in all chromosomes, and no SCs are formed. However, the mechanisms used to ensure proper pairing and segregation differ from those used in females. The XY pair do use repeated sequences to ensure proper pairing. For a long time the pairing site has been known as the collochore, but it is only recently that it has been shown to consist of multiple copies of a 240 bp sequence from the intergenic spacer (IGS) that separates the individual copies of the ribosomal DNA repeats, which are in the heterochromatin of the X and Y chromosomes (Hawley & Theurkauf, 1993; Irick, 1994; Lohe & Hilliker, 1995). In the sibling species *D. simulans* there is no ribosomal DNA on the Y chromosome, but it nevertheless has multiple copies of the 240 bp sequence to ensure pairing with the X (Lohe & Hilliker, 1995). Heterochromatin is not invariably used to ensure homologous pairing in *Drosophila* meiosis; in the second chromosome the pairing sequences are distributed throughout the euchromatin, but none are in the heterochromatin (Irick, 1994).

There is also evidence that paracentromeric heterochromatin is involved in holding together mitotic chromosomes until anaphase (Lica *et al.*, 1986; Sumner, 1991). This, with the centromere itself, is the last region of the chromosomes to separate; moreover, it contains a high concentration of topoisomerase II at this stage (Sumner, 1996), which might act on the repeated sequences in the heterochromatin to ensure their separation at this stage (Sections 2.3.1 and 12.4.1). Evidence in support of a role for repeated DNA in holding sister chromatids together comes from an experiment in which human alpha-satellite was inserted into a hamster chromosome. Although the alpha-satellite did not induce the formation of a kinetochore (Chapter 12), it did delay sister chromatid separation (Warburton & Cooke, 1997).

A further effect on chromosome segregation is claimed for paracentromeric heterochromatin (Vig, 1987). It has been reported that: different chromosomes separate their sister chromatids at slightly different times at the beginning of anaphase; the larger the block of heterochromatin adjacent to the centromere, the later the chromatids will separate; and in extreme cases this could lead to aneuploidy. It has not been established that such effects are of biological significance, however.

7.4.5 Position effect variegation (PEV)

Position effect variegation is the random, stable and clonally inherited inactivation of genes brought into the proximity of heterochromatin, resulting in mosaic expression of such genes within a tissue. The mosaicism results from the variable degree of spreading of a heterochromatinizing factor or factors from the heterochromatin into the adjacent euchromatin.

Many reviews of PEV have appeared in recent years (Karpen, 1994; Weiler & Wakimoto, 1995; Elgin, 1996; Wakimoto, 1998; Wallrath, 1998), and many of the basics of the phenomenon are now reasonably well understood. In fact, PEV can be produced in a number of different ways: by heterochromatinization spreading from heterochromatin into the adjacent euchromatin; by positioning euchromatic genes adjacent to blocks of heterochromatin in interphase nuclei; by elimination of DNA sequences; and by failure to amplify DNA sequences in polytene chromosomes (Section 15.2.3).

In *Drosophila* polytene chromosomes, regions containing genes subject to PEV look like heterochromatin, that is, they show a more condensed structure with an indistinct pattern of bands, and in some cases they are under-replicated (John, 1988; Reuter & Spierer, 1992; Karpen, 1994; Weiler & Wakimoto, 1995). The heterochromatinization is correlated with suppression of gene expression. Chromosome regions subject to PEV show greater resistance to nucleases than euchromatin and a more regular packing of their nucleosomes (Wallrath & Elgin, 1995), as well as binding of the heterochromatin protein HP1 (Belyaeva *et al.*, 1993; Fanti *et al.*, 1998). The degree of PEV is affected by some 150 genes that are either enhancers or suppressors of PEV.

Although PEV was originally described in *Drosophila*, a similar suppression of gene expression as a result of moving euchromatic genes next to heterochromatin has also been described in mouse (Wallrath, 1998). However, several situations have now been described in which PEV occurs in regions that do not form classic heterochromatin. These include the telomeric regions of *Drosophila* chromosomes, and the centromeric and telomeric regions of yeast chromosomes (Cryderman *et al.*, 1999). None of these regions show the condensed, deeply stained appearance typical of heterochromatin, although yeast chromosomes are in any case too small for satisfactory cytological studies. The ends of *Drosophila* chromosomes consist of multiple copies of certain transposons and certain other repetitive sequences, and thus have one of the characteristics of heterochromatin (Cryderman *et al.*, 1999). In yeasts, PEV at both centromeres and telomeres is associated with changes in histone acetylation and in nucleosome organization (Ekwall *et al.*, 1997), and in fission yeast (*S. pombe*) silencing of genes in regions adjacent to the centromeres is associated with the spreading of Swi6 protein into the silenced region (Partridge *et al.*, 2000). Similarly, at yeast telomeres, silencing (telomeric position effect, TPE) is associated with spreading of heterochromatin proteins into the silenced regions (Section 13.4; Fig. 7.7). Situations such as these lead to a definition of heterochromatin as material that can silence adjacent genes.

If silencing and heterochromatinization can spread along chromosomes from existing blocks of heterochromatin, what is to stop the whole chromosome from becoming heterochromatinized and inactivated? One answer is the presence of boundary elements or insulators, which prevent silencing from spreading past them. Insulators are defined as DNA sequences that act as a neutral barrier to the influence of neighbouring elements (Bell & Felsenfeld, 1999). Insulators that stop the spread of silencing by heterochromatin have been identified in *S. cerevisiae* and *Drosophila* (Bell & Felsenfeld, 1999; Bi & Broach, 2001). The available evidence indicates that the normal nucleosome structure is disrupted, thus

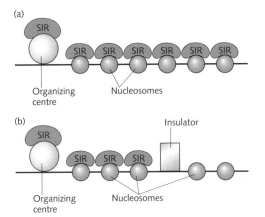

Figure 7.8 Proposed structure for an insulator in budding yeast, *S. cerevisiae*. (a) Silencing spreads from an 'organizing centre' along the chromosome, with the SIR complex binding to adjacent hypoacetylated nucleosomes. (b) In the presence of an insulator containing Rap1 protein, spreading of the SIR complex from one nucleosome to the next is blocked.

preventing propagation of the heterochromatin structure (Bi & Broach, 2001; Fig. 7.8). The *TEF2*-UAS insulator in *S. cerevisiae* contains the consensus sequence for the DNA-binding protein Rap1, which we have already seen binds to non-nucleosomal telomeric DNA (Section 7.3.2), while the *Drosophila gypsy* insulator consists of 12 binding sites for the zinc-finger protein su(Hw) and the mod(mdg4) protein.

Spreading of gene repression from adjacent heterochromatin on the same chromosome is not the only mechanism of PEV; in some cases the gene or genes subject to PEV are remote from the blocks of heterochromatin. In such cases it is thought that the interphase chromosomes become folded in such a way that the inactivated genes are brought adjacent to heterochromatin, either on the same chromosome or on another chromosome (nuclear compartmentalization: Lohe & Hilliker, 1995; Marcand *et al.*, 1996). In mice, inactivation of certain genes is correlated with an association with heterochromatin that contains the Ikaros protein (Brown *et al.*, 1997), although in this case it does not lead to variegation.

Position effects have been implicated in

various human diseases (Kleinjan & van Heyningen, 1998), where transcription of a gene has been affected by a rearrangement outside the coding sequences and the promoter region. However, the mechanisms of such position effects are at present far from clear, and it has not been established that heterochromatin is involved in such cases.

7.5 Applications of heterochromatin staining

The original uses of specific staining methods for constitutive heterochromatin were for the identification of chromosomal sites of heterochromatin (Figs 7.1 & 7.2), the study of variation of constitutive heterochromatin (Fig. 7.3), the use of heteromorphism of heterochromatin as a marker to distinguish homologues, the study of phenotypic effects of heterochromatin, the study of chromosomal evolution and, in perhaps the majority of species, as an important tool for chromosome identification.

Identification of sites of heterochromatin, their size, range of variability and their DNA composition are all essential aspects of the characterization of a species' karyotype. As we have already seen (Section 7.4), blocks of heterochromatin can vary considerably in size without any obvious effect on the whole organism, although the amount of heterochromatin can have effects at the chromosomal level (Section 7.4.3). Although variation in heterochromatin within a species seems to be relatively neutral, closely related species may nevertheless differ greatly in the amount of heterochromatin in their chromosomes (Section 16.3.5).

A very significant fraction of human reproductive loss and genetic disease is the result of aneuploidy or polyploidy: with few exceptions, fetuses with such chromosome imbalances are lost during pregnancy (Section 17.2). Heteromorphism of heterochromatin has been used to identify which parent the additional chromosome or set of chromosomes has come from. Similarly, heteromorphisms have been used to distinguish donor cells from those of the recipient in bone marrow transplants, thus indicating whether or not the transplant has been successful. Although simple staining methods have largely been supplanted in these applications by studies using DNA sequences, the staining methods still have the advantage of being able to ascertain the origin of a single cell.

Chromosomes of mammals and other higher vertebrates can be identified by methods such as G-banding that produce distinctive patterns throughout the length of the chromosomes. For practical purposes, such methods do not work for chromosomes of plants, invertebrates and lower vertebrates, so staining of heterochromatin has to be the principal method for identifying chromosomes in the majority of living species. The importance of C-banding for this purpose is not immediately obvious to those who work on, for example, mammalian chromosomes, who cannot only use a variety of banding techniques, but also chromosome painting, to identify chromosomes.

7.6   Heterochromatin today

At the beginning of this chapter, an attempt was made to define heterochromatin. It will have become clear that the idea of heterochromatin simply as blocks of highly repetitive DNA that do nothing is an oversimplification. Indeed, the presence of specific proteins, such as HP1, is likely to be more of a universal property of heterochromatin than any specific type of DNA. While some effects of heterochromatin are probably a result of the presence of a block of condensed chromosome, it is clear that in *Drosophila*

at least, there are several real functions in heterochromatin: specific genes, certain genetic factors that are clearly not conventional genes and regions that ensure proper chromosome segregation. A role for heterochromatin in segregation is probably widespread, but the detailed gene mapping studies on mouse and human have so far failed to locate any genes in the heterochromatin of these organisms. Only time, and more detailed studies, will show whether *Drosophila* is exceptional in having genetic factors in its heterochromatin.

Position effect variegation (PEV) is another phenomenon found in *Drosophila* that turns out to be widespread, possibly even universal. The paradox here is that it can apparently be caused by regions of chromatin that are not typical heterochromatin. Nevertheless, on the basis that they cause PEV, such regions are often referred to as heterochromatin. Perhaps they really do have the properties of classic heterochromatin, but are too small to distinguish with a light microscope. Specific proteins that condense and inactivate chromatin may be a better marker for heterochromatin than specific DNA sequences. As we learn more about the molecular organization of heterochromatin, our definitions of it are undoubtedly changing.

Websites

Position effect variegation
www.hhmi.org/science/genetics/henikoff.htm
Telomeric position effect in yeast
www.isrec.ch/recherche/gasser_lab.asp

Sex chromosomes and sex determination

8.1 What are sex chromosomes?

Many animals and a few plants that reproduce sexually and have separate sexes have sex chromosomes. Sex chromosomes are usually one pair of chromosomes that are the same in one sex but different in the other, and are believed to carry factors that determine the sex of the carrier (although direct evidence for this is lacking in most cases). It is clear that sex chromosomes have evolved independently many times: often differentiated sex chromosome systems are only found in the more highly evolved members of a group, while less highly evolved members have identical karyotypes in both sexes (Section 8.2). In addition, there are a number of different sex chromosome systems that could not easily have evolved from one another. In this chapter we shall describe the different sex chromosome systems that have been found, the ways in which they appear to have evolved, the mechanisms by which they determine sex and the phenomenon of dosage compensation, by which two copies of a chromosome in one sex produce the same amount of gene products as a single copy in the other sex. It is appropriate to deal with sex chromosomes immediately after a chapter on heterochromatin, because the formation of heterochromatin seems to be important in the evolution of sex chromosomes, and facultative heterochromatin is important not only in some sex determination systems but also in some cases of dosage compensation.

Perhaps the most familiar sex chromosome system is that in which males are XY and females are XX. Such systems are found in nearly all mammals, many insects and in other groups (Table 8.1), but their wide distribution and the fact that each group appears to have ancestors without differentiated sex chromosomes indicate that XX/XY sex chromosome systems have evolved independently many times. Studies of the mechanisms by which sex is determined in these organisms reinforce this conclusion. Similarly XX/XO and ZZ/ZW sex determination systems are found in very diverse groups of organisms (Table 8.1), and again must have evolved more than once.

8.2 The evolution of sex chromosomes

Many organisms do not have differentiated sex chromosomes, and environmental factors can be important in determining sex. For example, hormones or behaviour can determine sex in some lower vertebrates (e.g. Shapiro, 1994), temperature can determine sex in Chelonia, Crocodilia and other reptiles (Deeming & Ferguson, 1988), and in the marine mollusc *Crepidula fornicata* sex changes with age (Fretter & Graham, 1962). No doubt there is also a genetic component in sex determination in these organisms, because genes must specify the substrate on which the environmental factors can work, but such systems are very plastic and can produce very skewed ratios of males to females. Because of the regularity of

Table 8.1 Sex chromosome systems.

System	Males	Females	Examples
XX/XY	XY	XX	Most mammals; many insects; some plants
XX/XO	XO	XX	Grasshoppers and many other insects; nematodes
$X_1X_1X_2X_2/X_1X_2Y$	X_1X_2Y	$X_1X_1X_2X_2$	Certain mammals, insects and spiders
XX/XY_1Y_2	XY_1Y_2	XX	Certain mammals, insects and spiders
ZZ/ZW	ZZ	ZW	Birds; some reptiles; Lepidoptera
Haplodiploidy	Haploid	Diploid	Hymenoptera (bees and wasps)
Elimination of one parental set of chromosomes	Haploid	Diploid	Mealy bugs

For more detailed information on chromosomal sex-determining mechanisms in different organisms, see Bull (1983).

segregation of chromosomes, sex chromosome systems generally produce more or less equal numbers of males and females.

Sex chromosomes are believed to have evolved from situations in which sex was determined by a single gene with two alleles on an identical pair of chromosomes. One sex (the homogametic sex) would be homozygous for the gene, while the other sex (the heterogametic sex) would be heterozygous (Jablonka & Lamb, 1990; Lucchesi, 1994). It has been proposed that the first stage in the differentiation of sex chromosomes would have been suppression of crossing-over between the heterozygous chromosomes. There are various reasons why this might occur. One is that there would be strong selective forces favouring suppression of crossing-over between the allele determining the heterogametic sex and mutations in genes that benefit the heterogametic sex but harm the homogametic sex. Another is the desirability of keeping together the various genes involved in sex differentiation: for example, the mammalian Y bears not only the male-determining gene, but also genes involved in spermatogenesis. Not only is a spermatogenesis gene of no value in a female, but a male without it is obviously sterile. Once crossing-over has been suppressed, it becomes inevitable that genes are lost from the Y (or W) chromosome and that it becomes heterochromatinized and degenerate (Jablonka & Lamb, 1990). Lack of crossing-over reduces the likelihood that deleterious mutations could be repaired (Section 3.6), and thus they will accumulate.

Although the evolution of suppression of crossing-over between heterogametic sex chromosomes has not been observed directly, it is possible to follow the process of heterochromatinization. In fact, there are many species of lower vertebrates in which the sex chromosomes are morphologically identical, but one differs from the other in containing a block of heterochromatin (see Jablonka & Lamb, 1990, for references). In a few cases, there is no heterochromatin visible by staining methods, but one of the sex chromosomes has a late replicating region in one sex. Late replication, of course, tends to be associated with genetic inactivity (Sections 7.1 and 10.2.2) and heterochromatinization (Section 7.1). In snakes, a complete series can be assembled from primitive species in which there is no differentiation of sex chromosomes either by morphology or staining, through species in which the sex chromosomes are morphologically identical but differentiated by the presence of heterochromatin, to advanced species in which the chromosomes are also morphologically distinguishable (Jones & Singh, 1985).

Suppression of crossing-over might lead to heterochromatinization, or vice versa. In some cases, however, the sex chromosomes may become differentiated by structural changes. Examples are known in which the heteromorphism of sex chromosomes is due to pericentric or paracentric inversions (Jablonka & Lamb, 1990). Because crossing-over in such regions leads to duplications or deletions, which are often lethal, crossing-over tends to be suppressed in such regions. Once crossing-over has been suppressed in this way, heterochromatinization and degeneration are likely to follow.

Detailed mechanisms for degeneration of Y (or W) chromosomes have not been elucidated (Charlesworth & Charlesworth, 2000). However, once crossing-over has been suppressed, and as a result genes have become inactivated by mutation, there is no longer any selective pressure to retain them.

8.3 Sex chromosome systems and mechanisms of sex determination

8.3.1 XX/XY sex determination in mammals

In mammals, the Y chromosome is absolutely essential for the production of males. The Y chromosome carries the testis determining factor (*Tdf*), and once testis formation has been induced, other male characteristics are induced by testicular hormones. As well as the normal XY males and XX females, humans and mice with no Y chromosome, whether XO, XXX or with even more X chromosomes, invariably develop as females (though with some abnormalities, see Section 17.2.2), and those with Y chromosomes, whether XXY, XXXY, XYY or whatever, develop as males (again with some abnormalities).

Cytologically, the X chromosome in eutherian mammals is in many respects very like an autosome: it comprises about 5% of the haploid chromosome complement, and in females it can pair with its homologue and undergo crossing-over throughout its length. However, it differs from autosomes in having a lower density of genes (Deloukas *et al.*, 1998), and its set of genes is largely conserved throughout eutherian mammals (whereas autosomes have undergone extensive rearrangement in mammalian evolution). The X also contains a greater than average proportion of genes involved in sex determination and reproduction (Graves, 2001). The Y chromosome, on the other hand, is in many respects quite distinct from the autosomes, often containing a large amount of heterochromatin, very few genes and having only a small region of homology with the X chromosome.

In men, the distal part of the short arm of the Y chromosome is the pseudoautosomal region, while the *Tdf* (or *Sry*) gene is in the proximal part of the short arm. The proximal part of the long arm is euchromatic and contains spermatogenesis gene(s), while the distal part consists of a large block of heterochromatin (Fig. 8.1). The pseudoautosomal region is very small and invariably pairs with the homologous region on the X at meiosis, and always forms a single chiasma with the X; it thus behaves in the same way as autosomes. Like autosomes, the pseudoautosomal region contains a few genes that have no connection with sex determination (Graves, 1994). The mouse Y chromosome is organized quite differently, with its pseudoautosomal region at the end of the long arm and its testis determining factor near the centromere.

The sex chromosomes of marsupials and monotremes are rather different from those of eutherians, and may represent a more ancestral condition (Graves, 1996). Marsupial X chromosomes are generally smaller than those of eutherians (about 3% of the haploid chromosome complement), the Y chromosome is extremely small and there is no pseudoautosomal pairing region. The Y does not control all aspects of male

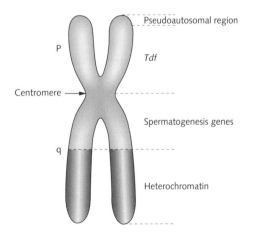

Figure 8.1 The main components of the human Y chromosome. The pseudoautosomal region forms the tip of the short arms, and pairs with the homologous region of the X at meiosis to form a chiasma. The rest of the Y is not homologous with the X, and carries genes for male sex determination (*Tdf*) and spermatogenesis.

differentiation in marsupials. Monotreme sex chromosomes, on the other hand, are large and show extensive homology with each other, and carry many genes that are autosomal in eutherians (Graves, 1996).

8.3.2   XX/XY sex determination in *Drosophila*

Unlike mammals, the Y chromosome is not required for sex determination in fruit flies (*Drosophila*), in which the sex depends solely on the ratio of X chromosomes to autosomes (Nöthiger & Steinmann-Zwicky, 1987); XO individuals are phenotypically normal males except that they are sterile, because the Y chromosome carries a number of fertility factors (Section 14.4).

Sex determination in *Drosophila* works through a key gene, *Sex-lethal* (*Sxl*), which is switched off in XY males and on in XX females. Through a cascade of control genes, either male or female differentiation genes are repressed. The critical feature of the process is the counting of the chromosomes, both autosomes and X chromosomes, but the mechanisms for this are not clear and may well differ in soma and germline (Nöthiger & Steinmann-Zwicky, 1987; Cline, 1993; Parkhurst & Meneely, 1994; Cline & Meyer, 1996).

8.3.3   XX/XO sex determination in *Caenorhabditis*

The Y chromosomes are often small (sometimes scarcely visible with a light microscope) and generally carry few genes, so it is not surprising that many organisms can manage without one. Although a few species of mammals have been reported without a Y chromosome, it is among grasshoppers and nematodes that the XX/XO sex determination system is most widespread. In the nematode *Caenorhabditis*, where this system has been analysed in most detail, it is clear that, as in the *Drosophila* XX/XY system, sex determination depends on the ratio of X chromosomes to autosomes (Meyer, 2000).

Although in *C. elegans* XX individuals are her-maphrodites, it seems clear that this is simply a modification of the XX female system found in the vast majority of nematodes and indeed in other species of the genus *Caenorhabditis* (Hodgkin, 1987). Genes involved in the X chromosome counting mechanism and in sex determination have been identified (Nicoll *et al.*, 1997; Carmi *et al.*, 1998). Interestingly, one of these genes, *SEX-1*, is distantly related to the mammalian *Dax1*, a gene on the X chromosome that, when duplicated, causes sex reversal in XY males (Ramkissoon & Goodfellow, 1996). Thus sex-determining mechanisms, even in organisms as far apart as mammals and nematodes, may have features in common, and the distinction between mechanisms depending on the presence of a Y chromosome and those that assess the ratio of X chromosomes to autosomes may not be as hard and fast as had been supposed.

8.3.4   ZZ/ZW sex determination systems

Where the male is the homogametic sex, and the female heterogametic, the sex chromosomes are known as ZZ and ZW, respectively, although in principle there is no difference from XX/XY systems. Like XX/XY sex determination systems, ZZ/ZW systems have evolved more than once, being found principally in the Lepidoptera, in many reptiles and in birds (Table 8.1). As with XX/XY systems, the W chromosome tends to be small and largely heterochromatic, though 'less advanced' species tend to have W chromosomes that are more similar in size to the Z chromosome (Traut & Marec, 1997; Ogawa *et al.*, 1998). Although it has been proposed that Y chromosomes (and W chromosomes) evolved by degeneration of a chromosome that was originally homologous to the X (or Z), there is evidence that primitive Lepidoptera had a ZZ/ZO sex determination system, and that the W was formed by fusion of an autosome to a Z chromosome, followed by degeneration; in some cases this W chromosome appears to have been lost, with secondary formation of a ZZ/ZO constitution (Traut & Marec, 1987).

The avian Z and W chromosomes clearly differentiated from a pair of autosomes, and have

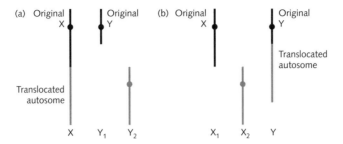

Figure 8.2 Multiple sex chromosome systems. (a) Components of an XX/XY_1Y_2 sex chromosome system. An autosome has been translocated to the original X; the homologue of the autosome ('Y_2') must therefore always segregate with the original Y ('Y_1') to maintain a balanced karyotype. (b) An $X_1X_1X_2X_2/X_1X_2Y$ system. The same considerations apply as in (a), but this time the autosome has been translocated on to a Y.

no homology with the X and Y of mammals (Fridolfsson *et al.*, 1998). The Z and W chromosomes pair with each other at meiosis and, as in mammals, there appears to be a pseudoautosomal region in which crossing-over occurs (Chandra, 1994). The precise mechanism of sex determination in birds is not yet clear, but may be based on the ratio of Z chromosomes to autosomes rather than being due to specific sex-determining genes (Chandra, 1994). In any case, there is some degree of plasticity in sex determination in birds: in female birds, only the left gonad differentiates into an ovary, and if the ovary is removed, the right gonad differentiates into a testis.

8.3.5 Multiple sex chromosome systems

Several organisms have developed multiple sex chromosome systems, with two or more 'X' or 'Y' chromosomes. Such 'extra' sex chromosomes are in fact the result of fusions between authentic sex chromosomes and autosomes: the homologous autosome not translocated to the sex chromosome must nevertheless be segregated as if it were a sex chromosome to maintain a balanced karyotype (Fig. 8.2). Similar multiple sex chromosome systems have also been reported in groups with ZZ/ZW sex determination systems (e.g. Traut & Marec, 1997).

8.3.6 Haplodiploidy

So far, the chromosomal sex-determining mechanisms described have involved differentiation

of a pair of chromosomes. In haplodiploidy, however, sex is determined by the number of complete sets of chromosomes: haploid individuals are male, and diploid individuals are female. Thus in most cases males develop from unfertilized eggs, and females from fertilized eggs. Such a sex-determining mechanism is found in various groups of arthropods, including mites (Acari), thrips (Thysanoptera) and bees, ants and wasps (Hymenoptera) (Beukeboom, 1995).

Mechanisms of sex determination by haplodiploidy have been studied almost entirely in the Hymenoptera, and it is clear that there must be a number of different mechanisms (Poirié *et al.*, 1993; Beukeboom, 1995). The simplest is the one-locus multi-allele model, in which heterozygotes are always female, but hemizygotes or diploid homozygotes are male; this mechanism occurs in several species. Many other models, however, have failed to attract any experimental support. Recently it has been proposed that imprinting (Chapter 9) could be responsible for sex determination in some Hymenoptera (Beukeboom, 1995), and such a mechanism has now been found in the parasitic wasp *Nasonia vitripennis* (Dobson & Tanouye, 1998). In imprinting, chromosomes from one parent are marked in some way so that they can be distinguished from their homologues in the zygote or embryo. In *N. vitripennis*, sex determination depends on the presence in the embryo of correctly imprinted paternal chromosomes. Some males of this species also have a supernumerary chromosome, PSR (paternal sex ratio), that specifically eliminates all paternal chromosomes from the fertilized egg (Beukeboom, 1995).

8.3.7 Scale insects: imprinting and facultative heterochromatin

Among the scale insects (including mealy bugs, Pseudococcidae), there is one of the more curious manifestations of chromosomal differentiation associated with sex. There are no sex chromosomes, and both males and females develop from fertilized eggs. In embryos that develop into females, both parental sets of chromosomes remain euchromatic, but in male embryos one set of chromosomes either becomes facultative heterochromatin (see Section 7.1 for a definition) or is eliminated (Brown & Nur, 1964; Nur, 1990). The heterochromatinized set of chromosomes is always derived from the male parent, and this is therefore an example of genomic imprinting (Chapter 9).

Strictly speaking, this seems to be a case of sex differentiation rather than sex determination, and also seems to have features in common with X chromosome inactivation in mammals (Section 8.4.3). However, although methylation of DNA occurs in scale insects, differences in methylation between maternal and paternal genomes have not been detected. Paternal DNA does, however, contain a nuclease-resistant fraction that is associated with the nuclear matrix and consists partly of middle-repetitive sequences; the same DNA sequences are not nuclease resistant in females (Khosla et al., 1999). Histone H4 is also hypoacetylated in the paternally derived genome, typical of inactive, condensed chromatin (Ferraro et al., 2001). There is an interesting parallel with the Y chromosome of Drosophila (Section 8.3.2), because an intact set of paternally derived, heterochromatic chromosomes appears to be necessary for male fertility (Brown & Nur, 1964).

8.3.8 Sex chromosomes in plants

Most plants do not have differentiated sex chromosomes, even if they are dioecious, and in many cases sex chromosomes seem to have evolved recently, and occur sporadically in unrelated groups. Species such as Melandrium album (Vyskot

et al., 1993) and Silene latifolia (Filatov et al., 2000) have XX/XY systems, with the Y chromosome playing an important role in determining maleness. Repetitive DNA sequences have accumulated on the Y chromosome both in S. latifolia (Filatov et al., 2000) and in Rumex acetosa (Shibata et al., 2000), which has an XX/XY_1Y_2 sex chromosome system. There is some evidence for dosage compensation by methylation of one X chromosome in females of M. album (Vyskot et al., 1993), as in mammals (Section 8.4.3).

8.4 Dosage compensation: coping with different numbers of X chromosomes in the two sexes

A consequence of having two X chromosomes in females but only one in males is that females would be expected to produce twice as much of the gene products coded by the X chromosome as males would. Although the precise dose of some gene products is not critical, there would probably be enough genes whose dosage was critical for this to be a problem. Accordingly, it is not surprising that mechanisms have evolved to equalize the amounts of gene products produced by the X chromosomes in both sexes. The mechanism used to achieve this dosage compensation differ from one organism to another (Meller, 2000): in female mammals one of the two X chromosomes is switched off, while in Drosophila males the single X has to work twice as hard as each of the two X chromosomes in females. In Caenorhabditis elegans, transcription from the two X chromosomes in hermaphrodites is down-regulated so that it equals that from the single X in males. Although dosage compensation is so widespread in XX/XY and XX/XO systems, evidence for it in the ZZ/ZW system of birds remains tentative, and possible mechanisms are still under discussion (Ellegren, 2002).

8.4.1 *Caenorhabditis*: down-regulation of both X chromosomes in hermaphrodites

In *Caenorhabditis*, dosage compensation is achieved using a pathway that has some components in common with the sex-determining mechanism (Dawes *et al.*, 1999; Kuroda & Kelley, 1999). The dose of X chromosomes is indicated by X-signal elements that control the expression of the gene *xol-1* (XO lethal). Expression of *xol-1* produces a male phenotype, and its inactivity results in a hermaphrodite (Meyer, 2000). In hermaphrodites, *xol-1* is inactive, and this allows the sex determination and dosage compensation genes *sdc-1*, *-2* and *-3* to be active. As a result, their gene products SDC-2 and -3, as well as another protein, DPY-30, bind to the two X chromosomes of hermaphrodites. In the presence of SDC-2 and -3, the proteins DPY-26, –27, -28 and MIX-1 also bind to the X; of these proteins DPY-27 and MIX-1 are SMC proteins, which are required for chromosome condensation (Section 6.5). As a result of this condensation, the level of transcription from both X chromosomes in the hermaphrodites is reduced (Marin *et al.*, 2000; Meyer, 2000) (Fig. 8.3). Failure of dosage compensation in XX worms is lethal (Meyer, 2000). In male *C. elegans*, *xol-1* is active, and negatively regulates the sex determination and dosage compensation genes *sdc-1*, *-2*, and *-3*. There are thus no SDC-2 and -3 proteins to bind to the chromosome, and so the series of events that lead to X chromosome condensation in hermaphrodites cannot occur (Fig. 8.3).

8.4.2 *Drosophila*: making the X work harder

In *Drosophila* males, the X chromosome has a more diffuse structure, produces twice as much RNA as each of the X chromosomes in females and failure of dosage compensation in either sex is lethal (Baker *et al.*, 1994; Gorman & Baker, 1994; Lucchesi, 1998). Dosage compensation of genes on the male X appears to be controlled individually, and certain genes are not subject to compensation: for example, genes whose expres-

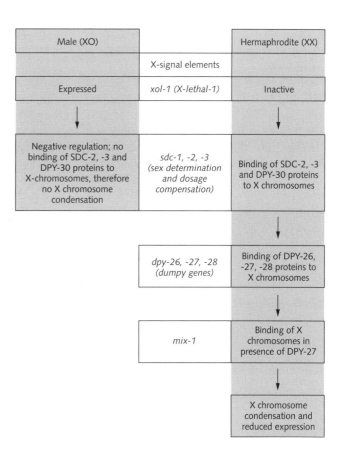

Figure 8.3 Dosage compensation in *Caenorhabditis elegans*. See text for further explanation.

sion is restricted to one sex, such as the yolk-protein genes, and genes that are present on both X and Y (e.g. *bobbed*) (Baker *et al.*, 1994). Dosage compensation is controlled by the same gene, *Sxl* (*sex-lethal*), that controls sex determination. A group of five male-specific lethal genes, *maleless (mle), male-specific lethal (msl) -1, -2* and *-3* and *males absent on the first (mof)*, forms a complex (Msl) that binds strongly to the male X chromosome. The MOF protein is a histone acetyltransferase that specifically acetylates lysine 16 on histone H4 (Akhtar *et al.*, 2000), thereby altering the chromatin structure of the X (Franke & Baker, 2000); as noted earlier, histone acetylation is associated with a more open chromatin structure and active transcription (Section 4.2.4). Two non-coding RNAs, *rox1* and *rox2*, also bind to the male X chromosome at the same sites as the MSL protein, and the MOF histone acetyltransferase requires *rox2* to bind to the X through its chromodomain (Akhtar *et al.*, 2000; Franke & Baker, 2000) (Fig. 8.4). In females, *Sxl* inhibits transcription of *rox* genes and translation of *msl-2*, which inhibits the other MSL proteins from binding to the female X chromosome (Franke & Baker, 2000).

Of the hundreds of sites on the X chromosome that bind the Msl complex, 30–40 appear to be particularly strong sites, and two of these sites correspond to the *rox1* and *rox2* genes. It is postulated that the Msl complex binds first to these sites, and then spreads to the remaining sites on the X whose dosage is compensated (Kelley *et al.*, 1999). Thus dosage compensation may not occur autonomously for each gene on the X chromosome, but rather appears to spread over small groups of genes.

8.4.3   Mammals: switching off all X chromosomes except one

Mammals achieve dosage compensation in yet another way, by switching off all X chromosomes except one, so that both males and females have only one active X chromosome. That superfluous X chromosomes are switched off, rather than just one X, is shown in human females with extra X chromosomes, who have only one active X regardless of whether they have three, four or even more X chromosomes; similarly, males with Klinefelter's syndrome, who have more than one X as well as a Y (usually XXY), have only one active X.

The inactive X is late replicating, is largely inactive transcriptionally and its histones are hypoacetylated (Jeppesen, 1997). In interphase it forms a compact mass of facultative heterochromatin against the nuclear envelope, which is called the Barr body (see Fig. 5.2b). Once an X chromosome has been inactivated, it and its descendants normally remain inactivated through many cell generations throughout the life of the individual. In eutherian mammals, X chromosome inactivation is random in the embryo, so that the body of females is a mosaic of tissues with one or the other X chromosome active. This is demonstrated clearly in the coat coloration of certain mammals, for example, the tortoiseshell or calico cat, which has patches of orange and black fur. The colours are produced by different alleles of a gene on the X chromosome, and therefore two X chromosomes are needed to produce tortoiseshell cats. Tortoiseshell cats are thus normally XX females; the rare males have an XXY sex chromosome complement. In marsupials and in

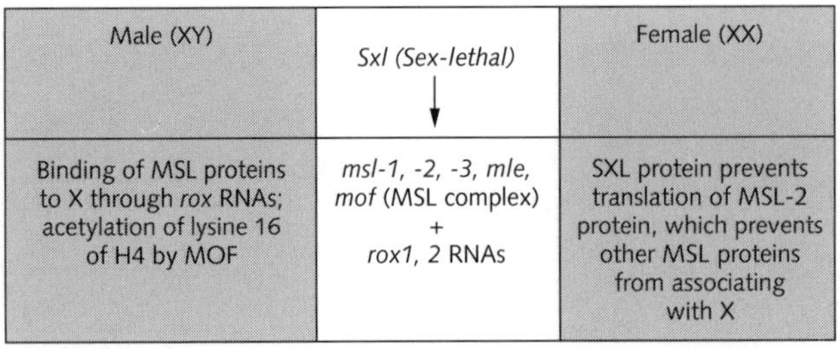

Male (XY)	*Sxl (Sex-lethal)*	Female (XX)
Binding of MSL proteins to X through *rox* RNAs; acetylation of lysine 16 of H4 by MOF	*msl-1, -2, -3, mle, mof* (MSL complex) + *rox1, 2* RNAs	SXL protein prevents translation of MSL-2 protein, which prevents other MSL proteins from associating with X

Figure 8.4 Dosage compensation in *Drosophila*. See text for further explanation.

Table 8.2 Stages of mammalian X chromosome inactivation.

Process	Mechanisms
Choice of chromosome	? Autosomal factor blocking inactivation of a single X
Initiation of inactivation	Stabilization of *Xist* transcripts
Spread of inactivation	Coating inactive X with *Xist* RNA
Maintenance of inactivation	DNA methylation; hypoacetylation of histones; late replication; heterochromatinization

See text for further explanation.

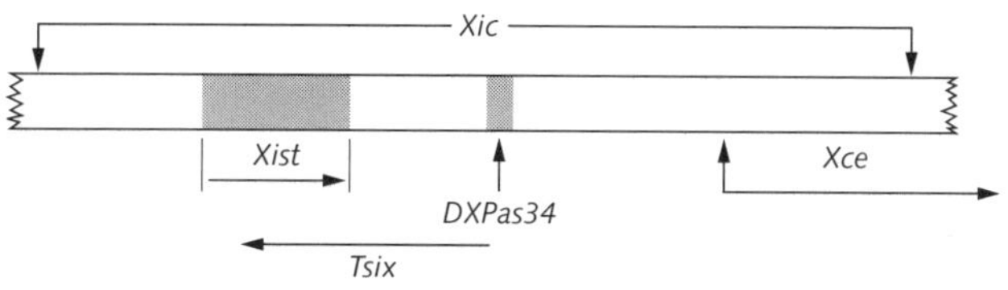

Figure 8.5 The mammalian X chromosome inactivation centre. Arrows indicate the direction of transcription.

placental tissues of eutherians, however, it is normally the paternal X chromosome that is inactivated (Graves, 1996). Because both X chromosomes are active in early embryos, the paternal X chromosome must be imprinted in these situations (Chapter 9). The X chromosome inactivation is less complete and less stable in marsupials than in eutherians (Graves, 1996), and although the marsupial X is late-replicating, it is not always condensed in interphase.

How does the mammalian dosage compensation system work? It seems to involve at least four stages: choice of chromosomes, initiation of inactivation, spread of inactivation and maintenance of inactivation (Table 8.2). The mechanism by which X chromosomes are chosen for inactivation or activation is not at all clear, although it has been postulated that there is an autosomal factor present in a limited amount that would be sufficient to block inactivation of only one X (Panning & Jaenisch, 1998). It is consistent with such a mechanism that in triploid female embryos (69,XXX) either one or two X chromosomes remain active, and in tetraploids (92,XXXX), two X chromosomes are inactivated and two remain active.

For initiation of inactivation, a region known as the X inactivation centre (Xic) is required, which has been defined by studying chromosome translocations and deletions as a single spe-

cific region on the X chromosomes of mice and humans, without which the chromosome cannot be inactivated (Lee & Jaenisch, 1997). This region contains two important sequences: *Xce*, the X-controlling element; and the *Xist* (X-inactive specific transcript) gene (Fig. 8.5). Different alleles of *Xce* affect the susceptibility to inactivation of the chromosome that carries them, possibly through differences in DNA methylation (Avner *et al.*, 1998). The *Xist* gene has been studied much more intensively and is clearly implicated in initiation and spreading of inactivation. It was identified as a non-coding RNA transcribed only from the inactive X and specifically coats the inactive X. In early embryonic cells, *Xist* is transcribed from both X chromosomes, but the transcripts are unstable; when inactivation is initiated, transcripts from the future inactive X become stabilized and coat the chromosome, whereas transcription from the future active X ceases (Panning & Jaenisch, 1998). The *Xist* gene appears to be controlled, at least to some extent, by a sequence known as *Tsix*, which is synthesized from the strand opposite to *Xist* and completely overlaps it, and may act as an antisense RNA to control *Xist*. The *Tsix* sequence appears to determine which X chromosome(s) will be silenced, without affecting counting of X chromosomes or their silencing (Mlynarczyk & Panning, 2000).

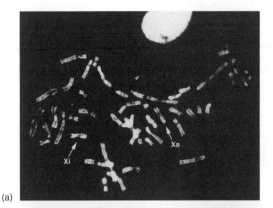

(a)

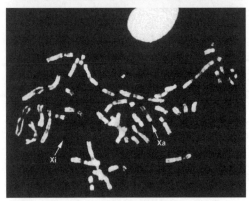

(b)

Figure 8.6 Underacetylation of histone H4 on the inactive X in a human cell: (a) DNA; (b) acetylated histone H4; labelling is absent from the inactive X. Xa, active X chromosome; Xi, inactive X chromosome. Reproduced with permission from Jeppesen & Turner (1993) *Cell* **74**, 281–289. © Cell Press.

The *Xist* gene may not be sufficient to establish inactivation (Clemson *et al.*, 1998), and it is certainly not required to maintain inactivation, although transcription of *Xist* from the inactive X continues throughout life (Clemson *et al.*, 1996). In fact, coating of the inactive X with *Xist* RNA is the first visible evidence of X chromosome inactivation, followed by silencing of X-linked genes and late replication. Formation of facultative heterochromatin, involving hypoacetylation of histones (Fig. 8.6), enrichment in macroH2A proteins (variants of histone H2A; Chadwick & Willard, 2001b) and methylation of CpG islands develop later, and may be responsible for maintenance of X chromosome inactivation rather than its establishment (Avner

& Heard, 2001), although methylation may not be required in marsupials (Graves, 1996).

Little is known about the mechanism of spreading of X inactivation, but it was proposed by Mary Lyon that LINE repeated sequences (Section 3.2.2) could be involved. In support of this, it has been found that the human X is enriched in the L1 class of LINEs, that L1 sequences are fewer in regions that escape inactivation and that L1 sequences may serve to propagate inactivation along the X chromosome (Bailey *et al.*, 2000).

The X chromosome inactivation in mammals is not complete: apart from the pseudoautosomal region (Sections 8.3.1 and 8.5), a few specific regions contain active genes (Disteche, 1995, 1999). These regions lack methylation of CpG islands, their histones are acetylated and they replicate their DNA early. Up to 20% of all genes on the human X may escape inactivation, although a much lower proportion of genes may be active on the mouse inactive X (Carrel *et al.*, 1999; Disteche, 1999). Escape from inactivation may be a secondary phenomenon, as such genes may be silent during development (Lingenfelter *et al.*, 1998). On the other hand, in X;autosome translocations, inactivation can spread into the autosomal segment, producing position effect variegation (Russell, 1983).

8.5   Sex chromosomes at meiosis and gametogenesis

Differentiated sex chromosomes have special problems at meiosis that have been solved in various ways. The X and Y (or Z and W) chromosomes often have very limited regions that can pair at meiosis; sometimes there is no pairing or, if there is, no chiasma is formed. In species with XO sex determination, there is no other chromosome for the X to pair with. Normally unpaired chromosomes cause delay or breakdown of meiosis (Section 2.5.2), but obviously special mechanisms have been developed to deal with such cases. Many examples are described by John (1990), but in most cases the molecular mechanisms involved are unknown. In the achi-

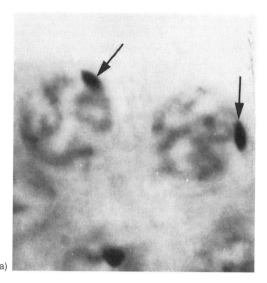

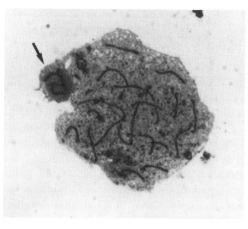

(a)

(b)

Figure 8.7 Sex chromosome bodies in male meiotic prophase. (a) Condensed X chromosome (arrows) in pachytene spermatocytes of the cricket, *Acheta domesticus* (XO male). (b) The XY sex chromosome body (arrow) in a human pachytene spermatocyte, prepared so as to show the synaptonemal complexes. Micrograph kindly provided by R.M. Speed.

asmate males of *Drosophila melanogaster*, pairing of the sex chromosomes is ensured by the close approximation of the spacers of the ribosomal DNA repeats, which occur on both chromosomes (McKee, 1996). Other *Drosophila* chromosomes use heterochromatin for pairing (Section 7.4.4), and perhaps many other species use a similar system. In many of the species described by John (1990) the sex chromosomes are heteropycnotic (i.e. condensed), and in meiotic prophase they form a dense sex body (Fig. 8.7a). Condensation of the sex chromosomes in heterogametic males is very widespread, and also occurs in mammals (Fig. 8.7b).

Most mammals solve the pairing problem by having a small pseudoautosomal region at one end of their X and Y chromosomes; these regions are homologous, pair at meiosis, form a synaptonemal complex (Fig. 8.8) and have a single obligatory chiasma. In the mouse, the proteins M31 (equivalent to HP1beta) and histone macroH2A1.2 are localized to the pseudoautosomal region until anaphase I, and may help to prevent premature desynapsis (Turner *et al.*, 2001). Marsupials in general do not have a pseudoautosomal region on their sex chromo-

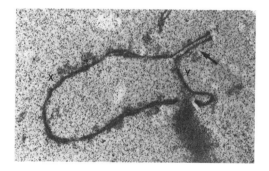

Figure 8.8 Synaptonemal complex preparation of a human male spermatocyte, showing pairing of the short arms of the X and Y to form an SC (arrow), while the greater parts of both the X and Y chromosomes (X, Y) are unpaired. Micrograph kindly provided by R.M. Speed.

somes, and although the X and Y may form end-to-end attachments, there is no synapsis or crossing-over (Graves *et al.*, 1998). It may be that pseudoautosomal regions are relatively recent translocations of autosomal material on to the sex chromosomes (Graves *et al.*, 1998), and that absence of pairing and crossing-over between the X and the Y is the norm.

The XY body (formerly, and incorrectly,

known as the sex vesicle) in male mammalian meiosis is condensed, late replicating and transcriptionally inactive (Borsani & Ballabio, 1993) and contains specific proteins (Kralewski & Benavente, 1997; Turner *et al.*, 2000). Most interestingly, this is the only situation in males in which *Xist* is expressed (Migeon, 1994; Ayoub *et al.*, 1997). Conversely, in female meiosis in mammals, both X chromosomes are active and uncondensed, and no *Xist* is transcribed. However, there is no methylation of the X in either male or female germ cells (Lyon, 1993), unlike the methylation of CpG islands in the inactive X of somatic cells.

8.6   Sex chromosomes: different means, the same ends

It has not been possible in this chapter to cover the complete range of sex-determining mechanisms, not even those that involve chromosomal sex determination. Although many organisms manage well with a pair of differentiated sex chromosomes, the degree of differentiation is highly variable. Other organisms manage without sex chromosomes, or use multiple sex chromosome systems or even more bizarre arrangements

(Fredga, 1994). Even among organisms with straightforward sex chromosome pairs, there are various ways of dealing with the dosage problem that arises from having two X or Z chromosomes in one sex but only one in the other sex. Birds apparently ignore the problem and get away with it, but in every group in which dosage compensation and chromosome inactivation has been studied in detail, different mechanisms are used. This is, of course, consistent with the evidence that chromosomal sex-determining mechanisms have evolved independently numerous times. The study of these phenomena is valuable in its own right, of course, but additionally throws light on mechanisms of gene regulation and imprinting (Chapter 9).

Websites

www.ultranet.com/~jkimball/Biology/Pages/S/SexChromosomes.html

www.molbio.mu-luebeck.de/biology/research/ephestia.htm

www.rrz.uni-hamburg.de/biologie/b_online/e11/11a/htm

Dosage compensation in Drosophila
sdb.bio.purdue.edu/fly/polycomb/msl2-7.htm

Imprinting

9.1 What is imprinting?

Imprinting is a process whereby modifications can be made to chromosomes or genomes in the parental generation, most probably in the gametes, so that there are functional differences between paternal and maternal genes or chromosomes in the offspring (Barlow, 1994; Reik & Walter, 1998). Imprinting is an epigenetic phenomenon, that is, a stable change in the course of development. It is established afresh in the germ line in each generation, is stably inherited throughout somatic cell divisions and is the cause of parent-of-origin-specific expression of certain genes. The phenomenon has been referred to as gametic, genomic, genetic, gene, germinal, chromosomal or parental imprinting, of which the first two have become most popular; in the context of genetics and chromosomology, the word imprinting without any qualification is commonly used, a practice followed in this book. Whereas classical Mendelian genetics presupposes that the parental genomes are essentially equivalent, the existence of imprinting shows that they are not. Some examples of imprinting were given in the previous chapter, in relation to sex determination and differentiation: the heterochromatinization and sometimes the loss of paternally derived chromosomes in scale insects (Section 8.3.7); and the preferential inactivation of the paternal X in female marsupials and in the extraembryonic tissues of female rodents. However, imprinting is also found in other situations, although it must be pointed out that a number of other phenomena that do not involve differential gene expression according to parental origin have also been referred to as imprinting (Barlow, 1994); such phenomena are not considered here.

In this chapter, the phyletic distribution of imprinting, and the form it takes in different organisms, will be summarized, the mechanisms of imprinting will be described and the functions (if any) of imprinting will be discussed.

9.2 Which organisms show imprinting?

Imprinting has been found in flowering plants (angiosperms), some insects and in mammals and a few other organisms (Morison *et al.*, 2001; http://www.geneimprint.com; http://cancer.otago.ac.nz:80/IGC/Web/home.html). So far it has not been found in the well-studied nematode *Caenorhabditis* (Reik & Walter, 1998).

9.2.1 Imprinting in plants

In flowering plants, imprinting has to occur in the gametophytes, the haploid generation resulting from meiosis that only undergoes limited growth and division before differentiating into germ cells. Androgenetic or gynogenetic embryos are more or less normal, as are haploid plants, so that effects of imprinting seem to be less in plants than in animals (Messing & Grossniklaus, 1999). However, in the endosperm, which results from a separate fertilization event, both genomes are required. Endosperm with the incorrect ratio of paternally and maternally derived genomes fails to grow properly, both in maize and in *Arabidopsis*, with the consequent death of the embryo (Martienssen, 1998). Other imprinted genes affect endosperm pigment and storage protein synthesis (Messing & Grossniklaus, 1999). In *Arabidopsis* the imprinted *Medea* gene controls seed development (Messing & Grossniklaus, 1999).

9.2.2 Imprinting in insects

It was in insects that imprinting was first recognized and defined (Crouse, 1960). In the fly *Sciara coprophila*, there are two sets of autosomes in the zygote (one from each parent), one maternal X chromosome and two paternal X chromosomes. In the germ line and the female soma, one paternal X is lost during embryogenesis, while both paternal X chromosomes are lost from the male soma. During spermatogenesis, the paternal set of autosomes is also lost, at the first meiotic division. These observations indicated that the chromosomes had become labelled according to the parent from which they were derived. The somewhat similar situation in scale insects, in which the paternal set of chromosomes becomes heterochromatinized and often lost, has already been described in Section 8.3.7.

In *Drosophila*, mutations have been reported that cause the loss of either the maternal or the paternal chromosome set in zygotes (Golic *et al.*, 1998). In addition, a situation has been reported in which the level of expression of the *white* eye-colour gene depends on whether it is inherited maternally or paternally. Although none of these situations is entirely normal, they do demonstrate that *Drosophila* has the potential for gametic imprinting.

9.2.3 Imprinting in mammals

It is among mammals that imprinting is most widely distributed and has been most intensively studied. Evidence for imprinting in mammals originally came from the rare cases in which

both genomes are derived from the same parent: androgenetic if derived from the father, and gynogenetic if derived from the mother. Neither androgenetic nor gynogenetic fetuses develop normally, but even more interestingly they develop differently. In humans, androgenetic zygotes form hydatidiform moles, in which the embryo itself dies and the placenta grows excessively. Conversely, gynogenetic zygotes form ovarian teratomas, in which the placenta fails to grow properly and the embryo itself is poorly differentiated. Similar observations have been made in mice, which have the added advantage of being experimentally tractable, so that embryos can be created by transplanting maternal or paternal pronuclei into zygotes to produce not only straightforward androgenetic or gynogenetic embryos, but also, by using nuclei with chromosome translocations, embryos with uniparental disomy (i.e. pairs of chromosomes or parts of chromosomes derived from the same parent). Studies using such embryos indicate that specific parts of chromosomes are involved in imprinting (Peterson & Sapienza, 1993). Imprinted chromosome regions have been mapped in detail in the mouse (Beechey *et al.*, 2001; Fig. 9.1), and lists of imprinted genes have been compiled for humans (Morison *et al.*, 2001; http://cancer.otago.ac.nz:80/IGC/Web/home.html; http://www.geneimprint.com). In general, homologous chromosome regions are imprinted in mouse and humans, although not every gene that is imprinted in the one species is imprinted in the other (Surani, 1994); for example, *IGF2R* is imprinted in mouse but not in humans (Morison *et al.*, 2001). Molecular studies have confirmed that imprinted genes are generally clustered (Reik & Walter, 2001a), and that imprinting is therefore a chromosomal phenomenon rather than a characteristic of individual genes. It has been estimated that 100–200 genes (Horsthemke *et al.*, 1997) are imprinted, of which about 50 have been identified in mouse and man (http://cancer.otago.ac.nz:80/IGC/Web/home.html).

Imprinting probably occurs in all eutherian mammals. In marsupials (Section 8.4.3) it is always the paternal X chromosome that is in-

Mouse imprinted genes, regions and phenotypes

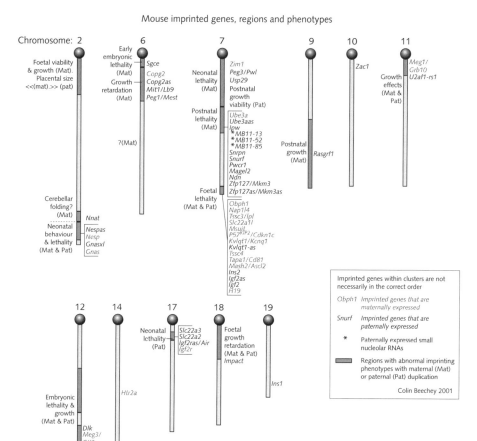

Figure 9.1 The mouse imprinting map, showing imprinted chromosomal regions (▬) and imprinted genes that are paternally or maternally expressed. Reproduced with permission from Beechey *et al.* (2001) MRC Mammalian Genetics Unit, Harwell, Oxfordshire. World Wide Web Site – Genetic and physical imprinting map of the mouse. http://www.mgu.har.mrc.ac.uk/imprinting/all_impmaps.html

activated, so this must also be imprinted. Many genes that are imprinted in eutherians are also imprinted in marsupials (John & Surani, 2000); however, certain genes that are imprinted in other mammals do not appear to be imprinted in monotremes. Conversely, it is a reasonable assumption that in groups in which partheno-genesis can occur (as it does very occasionally in most other groups of vertebrates), imprinting either is absent or does not involve essential genes. Even in mammals, knock-outs and uni-parental disomy of certain imprinted genes do not appear to have particularly marked effects (Hurst & McVean, 1998).

A number of human genetic diseases result from alterations to imprinted genes. Prader–Willi syndrome (PWS) and Angelman syndrome (AS) are the result of deficiencies in gene expression from the paternal chromosome region 15q11–q13 (PWS) or from the corresponding maternal region (AS) (Jiang *et al.*, 1998), and each occurs with a frequency of about 1 in 15000 births (Nicholls *et al.*, 1998). Both diseases can occur as a result of deletion of an approximately 4 Mb region in 15q11–q13 (Fig. 9.2), or in some cases as a result of uniparental disomy (i.e. the same chromosomal region on both homologues is derived from the same parent). About 5% of

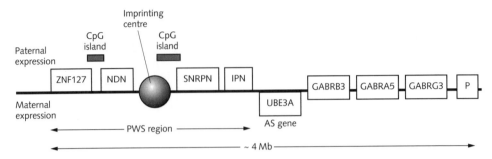

Figure 9.2 Structure of the chromosomal region (15q11–q13) deleted or altered in Prader–Willi and Angelman's syndromes.

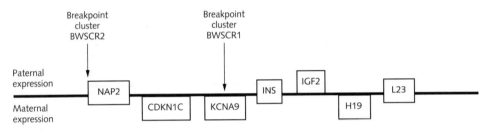

Figure 9.3 Structure of the imprinted chromosomal region (11p15.5) in Beckwith–Wiedemann syndrome.

PWS and AS patients have abnormal methylation in the imprinted region. A few AS patients have loss-of-function mutations in the *UBE3A* gene (Jiang *et al.*, 1998; Nicholls *et al.*, 1998). Beckwith–Wiedemann syndrome (BWS) is a result of loss of imprinting in the chromosomal region 11p15.5 (Reik & Maher, 1997) (Fig. 9.3); it occurs most commonly as a result of biallelic expression of *IGF2*, and also by paternal disomy, or by silencing (with or without methylation) of the *H19* gene (Reik & Maher, 1997). Wilms' tumour shows a loss of imprinting in the same region (Feinberg, 1994). Such diseases are not only of clinical significance (Section 17.6), but their investigation has helped to throw light on the mechanism of imprinting.

9.3   How does imprinting work?

In mammals (Reik & Walter, 1998; Feil & Khosla, 1999; Reik & Walter, 2001a), and probably in plants (Martienssen, 1998; Messing & Grossniklaus, 1999), the DNA of imprinted regions is methylated, forming differentially methylated regions (DMRs). In general, it is the silent allele that is methylated, but there are several imprinted genes where the active allele is methylated (Jiang *et al.*, 1998; Reik & Walter, 1998; Feil & Khosla, 1999; Reik & Walter, 2001a) (Fig. 9.4). The importance of methylation is emphasized by the fact that loss of methylation, either induced experimentally in methyltransferase-deficient mice (Reik & Walter, 1998) or naturally occurring in pathological states such as PWS and AS (Horsthemke *et al.*, 1997), disrupts imprinting and has significant phenotypic effects. As well as DNA methylation, paternal and maternal alleles of imprinted genes differ in their chromatin structure and time of replication. Active alleles have a more open chromatin structure that is hypersensitive to nuclease digestion, while inactive alleles have compact chromatin that is not accessible to nucleases (Feil & Khosla, 1999). Inactive alleles are replicated later than the active ones (Efstratiadis, 1994; Horsthemke *et al.*, 1997; Feil & Kelsey, 1997). All these differences, in fact, also distinguish tran-

	Maternal methylation	Paternal methylation	No methylation
Maternal repression	Peg1/Mest Peg3 Nnat Snrprn Znf127* U2afbp-rsl Impact Ndn	Rasgrf1	Igf2*
Paternal repression	Kcnqt1/Kvlgt1* Igf2r* Nesp*	H19	Ube3a* Copg2*

Figure 9.4 Correlation between repression and methylation of imprinted genes. For certain genes, methylation is not correlated with repression, and instead repression is caused by antisense RNA. Data from Reik & Walter (2001b).

* Repression by antisense RNA

(a)   (b)   (c)

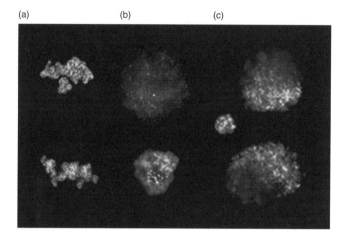

Figure 9.5 Differential demethylation of chromatin in the early mouse embryo. Nuclei are immunolabelled to show sites of 5-methylcytosine (5-MeC; light grey). (a) A zygote, 3 h after fertilization, showing a high level of 5-MeC in both pronuclei. (b) Pronuclei 8 h after fertilization: the male pronucleus (upper) has become demethylated. (c) A two-cell embryo 32 h after fertilization, showing the paternal chromatin (unmethylated; dark grey) segregated from the maternal chromatin (still methylated). Reproduced with permission from Mayer et al. (2000) *Nature* **403**, 501–502. © Macmillan Magazines Ltd.

scriptionally active and inactive chromosome regions in which genes are not imprinted (Sections 4.2 and 10.2.2). There is also evidence of somatic pairing of imprinted chromosomal regions during S phase, which might be important for the maintenance of imprinting (Riesselmann & Haaf, 1999).

How is methylation established and maintained, and what determines that specific regions are imprinted? It is self-evident that the methylation required for imprinting must be acquired in the germ line, as this is the only stage at which maternal and paternal genomes are separate, but in fact the process of imprinting is much more complicated than simply establishing a pattern of methylation in the germ line that is then maintained throughout life. Methylation of DNA is completely eliminated in the germ line, both from imprinted and non-imprinted genes, but then a sex-specific pattern of methylation is established, so that for imprinted genes the pattern differs between oocytes and sperm (Reik & Walter, 1998). Surprisingly, fertilization is followed by demethylation during the cleavage stages of the embryo; the paternal genome is demethylated shortly after fertilization, before the first cleavage, but the maternal genome is not demethylated until after several cleavage divisions have taken place (Mayer et al., 2000; Reik & Walter, 2001a; Reik et al., 2001) (Fig. 9.5). This is followed by *de novo* methylation, which is often maintained throughout the rest of development and into adult life (Jaenisch, 1997), but in some cases changes during development (Reik & Walter, 2001a). A specific DNA methyl-transferase, Dnmt1o, has been identified that methylates imprinted genes in the eight-cell embryo (Dean & Ferguson-Smith, 2001).

Throughout these processes, the differential methylation of imprinted genes must be maintained, during both the demethylation and remethylation processes. This implies that the methylation associated with imprinted genes must lie within a different type of chromatin structure from that of other genes, whether methylated or not, but information on this point is lacking. In fact, it has been found that the crucial methylation is in CpG islands known as 'imprinting boxes'. The imprinting box of the *Igfr2* gene is in an intron, and that of the *H19* gene is upstream of its promoter. Methylation of these (and presumably other) imprinting boxes is resistant to demethylation during the cleavage stages, and the non-methylated boxes of their alleles are resistant to the methylation that occurs after implantation (Jaenisch, 1997). It remains unclear how the specific methylation of imprinting boxes is established during gametogenesis. There is no evidence for common DNA sequences in imprinting boxes (Tilghman, 1999), although it is possible that they might show common higher order structure. A number of imprinted genes do show large numbers of direct DNA repeats, which show allele-specific methy-lation and might show distinctive secondary structure (Neumann & Barlow, 1996; Constância *et al.*, 1998). Two imprinting centres have been shown to act as silencers in *Drosophila*, indicating that they can bind specific chromatin factors (Reik & Walter, 1998); such factors could mark them for germ line methylation (in mammals), and protect them against unwanted demethylation and methylation later in development. How methylation might spread from an imprinting centre to the rest of the imprinted region is not yet known (Reik & Walter, 2001a).

Several mechanisms seem to be used to control transcription of imprinted genes (Reik and Walter, 2001a). Inactivation of promoters by methylation, associated with underacetylation of histone, is a familiar mechanism (Sections 3.5 and 4.2.4) that is used in some cases. For imprinted genes that are active in spite of being methylated, it has been suggested that they contain silencers that are inactivated by methylation; this is the mechanism with DMR1 (differentially methy-lated region 1) of the *Igf2* gene, for example.

In the case of *H19* and *Igf2* an insulator or boundary element is used (Schmidt *et al.*, 1999) (Fig. 9.6). In the maternally derived chromo-some, the imprinting control region (ICR) upstream of *H19*, as well as *H19* itself, are unmethylated, so the CTCF protein (which also acts to repress many non-imprinted genes) can bind to the ICR (Reik & Murrell, 2000). The CTCF bound to the ICR acts as an insulator, so that the enhancers that control both *H19* and *Igf2* cannot gain access to the latter gene. In the

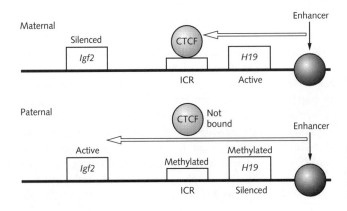

Figure 9.6 The role of a boundary element (insulator) in regulating the differential expression of the imprinted *H19* and *Igf2* genes.

paternally derived chromosome, both *H19* and the ICR are methylated, CTCF is prevented from binding and the enhancers have no difficulty in gaining access to *Igf2*.

Several imprinted genes are associated with antisense RNA transcripts; all these transcripts are themselves imprinted, are paternally expressed (Fig. 9.4) (unlike *Tsix*, the antisense RNA to *Xist* in X-chromosome inactivation; Section 8.4.3) and mostly repress paternally derived genes (Reik & Walter, 2001a,b). The antisense RNAs are, in some cases at least, transcribed from introns of the genes they repress. The mechanisms by which the antisense RNAs repress their genes have not yet been established.

The discussion above on imprinting mechanisms is based on imprinting in eutherian mammals, and on only a limited number of genes that have been studied intensively. It must not be assumed that imprinting mechanisms are necessarily the same in all organisms that show imprinting. It has already been mentioned that imprinting can occur in *Drosophila*, in which there is negligible methylation. Another, perhaps more significant, example is X chromosome inactivation in marsupials (Section 8.4.3). In marsupials, it is always the paternal X that is inactivated, and it must therefore be imprinted. However, methylation does not appear to be involved in X chromosome inactivation in marsupials and, perhaps because of this, inactivation is less stable than in Eutheria. Instead, histone acetylation appears to be the essential factor, not only in the marsupial X, but also in the inactive X in mouse extraembryonic tissues (Wakefield *et al.*, 1997; John & Surani, 2000). Even in eutherians, some imprinted genes are not methylated (Tilghman, 1999).

In plants, imprinting appears to be a two-stage process: DNA methylation occurs first, and the methylated DNA is thought to attract specific proteins that would lead to a change in chromatin structure (Messing & Grossniklaus, 1999). Interestingly, the *Medea* locus in *Arabidopsis*, which is itself imprinted, encodes a polycomb-group protein (Goodrich, 1998) that is believed to affect chromatin structure and may be involved in imprinting.

9.4   What is imprinting for?

It is generally supposed that diploidy is advantageous, one reason being that deleterious mutations on one chromosome can be masked by a normal allele on the homologous chromosome. Why, therefore, should diverse organisms revert to a situation that is essentially haploidy in parts of their genomes? Numerous hypotheses have been put forward, and some seem more plausible (or at least less implausible) than others, although none is yet wholly convincing (Jaenisch, 1997; Hurst & McVean, 1998; Spencer, 2000). Considering that imprinting must have arisen independently several times, it is probably not realistic to suppose that there could be only a single explanation. In many insects (Section 9.2.2) imprinting seems to be intimately involved in sex determination, and in marsupials it is a factor in X chromosome dosage compensation (Section 8.4.3). Most discussion about the role of imprinting has focused on eutherian mammals, however. The most popular, but nevertheless controversial, explanation for imprinting in Eutheria is the conflict hypothesis (also referred to as 'parental conflict' or 'genetic conflict'). The basis of this hypothesis is that there is a 'conflict' between the maternal and paternal genomes in the case of multiple paternity (Jaenisch, 1997; Haig, 1999; Spencer, 2000). Because mammalian offspring are uniquely dependent on their mother for a limited amount of nourishment, both *in utero* and after birth, it is in the mother's interest to restrict the growth of the offspring uniformly, so that as many as possible survive. If, however, there are multiple fathers, it is in the interest of each father to maximize the growth of his own offspring at the expense of offspring of other fathers. Imprinting would be favoured if maternal and paternal requirements favoured different levels of gene products from specific loci (Haig, 1999). At first sight the apparent lack of imprinting in the egg-laying monotremes might appear to support this hypothesis, but in fact the monotreme egg develops *in utero* for several weeks and is nourished by the mother (John & Surani, 2000); the same considerations should therefore apply as in other mammals.

One prediction of the hypothesis is that many imprinted genes should affect growth, and that paternally expressed genes should enhance growth while those expressed from the maternally derived chromosomes should inhibit growth. In general this seems to be true, but other observations do not clearly support the conflict hypothesis (Hurst & McVean, 1998). In plants there are imprinted genes that do not affect morphogenesis and are not subject to parental conflict, although there are others, such as *Medea* in *Arabidopsis*, that control growth during seed development and appear to be more akin to mammalian imprinting genes in their behaviour (Messing & Grossniklaus, 1999). Only the maternal allele of *Medea* is expressed in the endosperm, and this restricts growth, consistent with the parental conflict hypothesis (Mora-Garcia & Goodrich, 2000).

An alternative proposal is that imprinting in mammals might be related to brain development (Tilghman, 1999; John & Surani, 2000). A large number of imprinted genes are expressed in the brain, and correlations have been made between certain aspects of behaviour and imprinting. Nevertheless, a coherent hypothesis linking imprinting to the brain and behaviour has not yet been formulated.

The existence of imprinting is a practical problem in cloning of mammals by nuclear transfer. Only a very small proportion of embryos produced in this way result in live births, and even those that do have some abnormalities (Humphreys *et al.*, 2001; Rideout *et al.*, 2001).This is most probably due to inadequate genetic reprogramming of the imprints in the donor nuclei; embryos derived from embryonic stem cells, which possibly need less reprogramming, seem to do better than embryos derived from somatic cells. Even those cloned animals that survive to adulthood may have quite widespread dysregulation of transcription, indicating that mammalian development may in fact be quite tolerant of abnormalities in imprinting.

Imprinting is a fascinating and, until recently, wholly unexpected phenomenon. There is now a considerable, though as yet far from complete, understanding of its mechanisms in eutherian mammals, although the reasons for its existence remain uncertain. It is, however, clear that imprinting must have evolved independently several times, as it is produced by different mechanisms in different groups of organisms, and almost certainly has different selective advantages in different groups. Comparative studies will be essential for a complete understanding of imprinting phenomena.

Websites

Imprinting in the mouse
http://www.mgu.har.mrc.ac.uk/imprinting/imprinting.html

Imprinting in other organisms, particularly mammals
www.geneimprint.com
http://cancer.otago.ac.nz:80/IGC/Web/home.html

Euchromatin and the longitudinal differentiation of chromosomes

10

10.1 What is euchromatin?

When Heitz recognized heterochromatin (Chapter 7) in 1928, he distinguished it from euchromatin (the 'true' chromatin), which showed the 'normal' behaviour of decondensing at the end of mitosis and becoming diffuse in the interphase nucleus. The genes were believed to be in the euchromatin, and absent from heterochromatin, although the latter is not always true (Section 7.4.2). Equally, euchromatin is not uniformly packed with genes, but contains regions of relatively high and relatively low gene density. It is also convenient to exclude from a consideration of euchromatin the various specialized regions of chromosomes that are described in the following chapters: the nucleolar organizer regions (Chapter 11), which themselves are euchromatic, but are usually embedded in heterochromatin; the centromeres (Chapter 12), which are heterochromatic; and the telomeres (Chapter 13), which, although they lack the distinctive staining properties of heterochromatin, nevertheless have some of its characteristics (Section 7.4.5). If this makes it sound as if euchromatin is what is left when everything that can be clearly defined is taken away, there is perhaps a grain of truth in this, as euchromatin can be divided up into a number of different categories with a wide range of properties, and is not a single substance. The different categories of euchromatin are distributed in a characteristic pattern along the chromosomes to produce the

longitudinal differentiation of chromosomes that is the subject of this chapter.

10.2 Euchromatin and chromosome banding in mammals

When mammalian chromosome preparations are treated and stained in a variety of ways, reproducible patterns of transverse bands are produced. These are the chromosome bands (Fig. 10.1), and the essential point is that the patterns are the same regardless of the method used to produce them, although they do vary with the degree of contraction of the chromosomes (Fig. 10.2). Practical details of banding techniques are not relevant here: they are summarized in Boxes 10.1–10.3.

10.2.1 A note on nomenclature of bands

There is some confusion about what to call individual bands. The Standing Committee on Human Cytogenetic Nomenclature defined a band as 'part of a chromosome which is clearly darker or lighter with one or more banding techniques' (ISCN, 1995). It is implicit in this definition that a chromosome treated with a G-banding method will consist of G-positive and G-negative (or G-dark and G-light) bands, and similarly for Q-banding, R-banding and all the other types of banding. A consequence of this is that there is no way of referring to a band independently of the technique used to produce it,

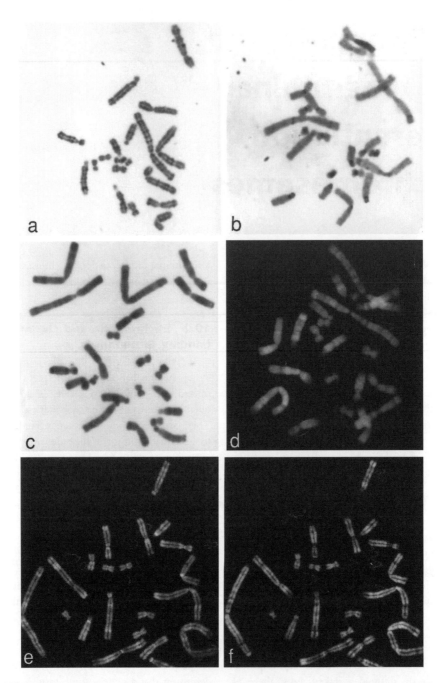

Figure 10.1 Chromosome bands produced on metaphase Chinese hamster ovary (CHO) chromosomes using different methods. G-Banding (a), Q-banding (d) and DAPI (f) produce essentially the same patterns, while R-banding (c) and mithramycin (e) produce the opposite pattern. This can be seen most clearly by comparing mithramycin (e) and DAPI (f) banding on the same set of chromosomes: each pattern is the reciprocal of the other. The C-banding pattern (b) is restricted to small regions and is not related to the pattern of bands throughout the length of the chromosomes that is seen with the other methods. Reproduced with permission from Sumner (1994) *European Journal of Histochemistry* **38**, 91–109. © Società Italiana di Istochimica.

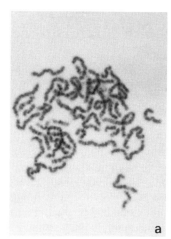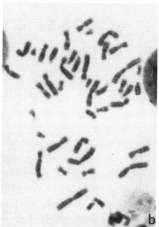

Figure 10.2 G-Banded human chromosomes at (a) prophase and (b) metaphase. Although in fact the overall patterns of banding are similar, there are many more bands visible in the prophase chromosomes, and the pale bands occupy a greater proportion of the chromosome (see also Fig. 10.6). Reproduced from Sumner (1976) *Kew Chromosome Conference*, pp. 17–22, published by North-Holland.

Box 10.1   G-Banding

G-Banding is by far the most widely used method for staining euchromatic chromosome bands, because it is easy and reliable to perform, the staining is permanent and it does not require a fluorescence microscope. For these reasons, it is the principal method in clinical cytogenetics, as well as in many other cytogenetical studies in mammals. G-Banding provides the standard karyotype of humans and other mammals of cytogenetic importance and the framework for a standardized method of describing the locations of genes, breakpoints and other features on chromosomes (ISCN, 1995). G-Banding is performed simply by treating chromosome preparations with warm 2 × SSC or dilute trypsin, then staining with Giemsa (Gosden, 1994; Barch *et al.*, 1997; Sumner & Leitch, 1999; Czepulkowski, 2001; Rooney, 2001) (Figs 10.1a & 10.2).

Box 10.2   R-Banding

Among the other methods that can be used for demonstrating euchromatic bands, R-(reverse) banding (Fig. 10.1c) is of particular interest, because the pattern is complementary to that produced by G-banding. As a result, the terminal regions of chromosomes are generally stained, which makes it easier to determine the limits of the chromosomes, unlike G-banding where the chromosome ends are often indistinct. A set of R-bands that is especially resistant to treatment and that retains its staining when other R-bands have lost theirs is known as T-bands (terminal bands). Use of both G- and R-banding successively on the same chromosome preparations can help to localize the sites of chromosome breaks more precisely than by using either method on its own.

R-Banding is usually obtained by incubating chromosome preparations in phosphate buffer at 85–90°C, followed by Giemsa staining (Gosden, 1994; Barch *et al.*, 1997; Sumner & Leitch, 1999; Czepulkowski, 2001; Rooney, 2001).

Box 10.3 Banding with fluorochromes

Methods of banding using fluorochromes that are specific for particular DNA bases are historically important (the first modern banding technique used the fluorochrome quinacrine mustard; Sumner, 1990) but not now generally used in routine cytogenetics, as they do not provide extra information compared with G- or R-banding, the banded preparations are not permanent and tend to fade when illuminated and a special (and expensive!) fluorescence microscope is needed. However, some fluorochromes provide useful counterstains for fluorescence *in situ* hybridization (see Box 5.1); DAPI and chromomycin A₃ are most widely used as counterstains.

Staining methods are very simple: the chromosome preparation is simply immersed in a dilute solution of the fluorochrome for a short time, and then mounted with a suitable mountant designed to retard fading (Gosden, 1994; Barch *et al.*, 1997; Sumner & Leitch, 1999; Czepulkowski, 2001; Rooney, 2001). Examples are shown in Fig. 10.1d (Q-banding, using quinacrine) and Fig. 10.1f (DAPI), both of which produce patterns similar to G-banding, and Fig. 10.1e, which shows banding with mithramycin (similar to chromomycin A₃) and gives a pattern similar to R-banding.

Table 10.1 Characteristics of euchromatic bands in mammalian chromosomes.

G-Bands	R-Bands
Positive G-bands	Negative G-bands
Positive Q-bands	Negative Q-bands
Negative R-bands	Positive R-bands
A+T-rich DNA	G+C-rich DNA
Late replicating DNA	Early replicating DNA
Early condensation	Late condensation
Pachytene chromomeres	Interchromomeric regions
Little recombination	Meiotic pairing and recombination
Nuclease insensitive	Nuclease hypersensitive
Low concentration of genes	High concentration of genes
Low level of histone acetylation	High level of histone acetylation
High level of H1 subtypes	Low level of H1 subtypes
HMGA1a present	HMGA1a absent
Rich in LINEs (long intermediate repetitive DNA sequences)	Rich in SINEs (short intermediate repetitive DNA sequences)
Low level of chromosome breakage	High level of chromosome breakage

After Sumner (1998b).

which makes it difficult to describe a particular class of bands succinctly. To avoid this problem, a convention has grown up to refer to darkly staining, positive G-bands, or the corresponding bands produced by any other technique, simply as G-bands, while darkly staining, positive R-bands, which are more or less equivalent to the weakly staining G-bands, are referred to as R-bands. This convention will be used here except when it

might cause ambiguity, in which case a full description will be given. The equivalence between different types of bands is given in Table 10.1.

To identify individual bands, a numbering system has been devised (ISCN, 1995). In the example shown in Fig. 10.3, from the human karyotype, each chromosome is numbered according to its size (1 being the largest) and

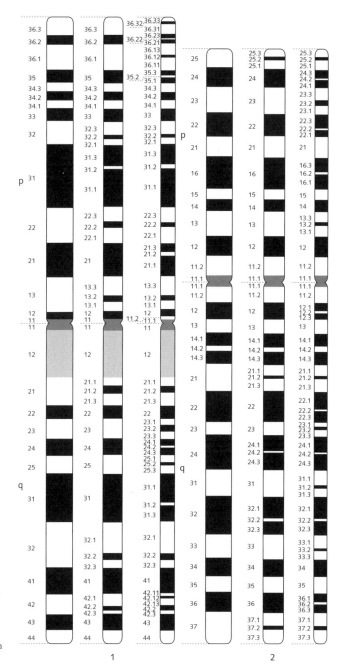

Figure 10.3 Diagram of G-banding patterns of human chromosomes 1 and 2, showing the numbering system used for the bands at different degrees of chromosome contraction. Left: chromosome 1; right: chromosome 2. For each, the banding pattern is shown that corresponds to a total of 350, 550 and 850 bands in the haploid genome. Reproduced from ISCN (1981), © March of Dimes Birth Defects Foundation.

the arms are designated by letters: p (*petit*) for the short arm, and q (because it follows p in the alphabet) for the long arm. Each arm is then divided up by 'landmarks' – conspicuous bands that are selected to divide up the arm into smaller segments and are numbered (1, 2, 3, etc.) outwards from the centromere. Within each of these regions, less conspicuous bands are visible, and these are numbered from 1 upwards within each region, again counting from the centromeric direction; G- and R-bands are numbered in the same series. Thus starting from the centromere on the short arm of chromosome 1 the first band is 1p11, and the terminal band is designated as 1p36 – the sixth band in the third landmark region (Fig. 10.3).

This system was devised for metaphase chromosomes showing about 350 bands in a haploid chromosome set. When it was found that more elongated chromosomes, either from naturally occurring prophase cells or induced by various treatments, displayed a larger number of bands (up to 1250 for routine purposes, although a maximum of 2000 was claimed; Yunis, 1981), it was necessary to modify the system to allow for the extra bands. Because the prophase bands fuse together to form the metaphase ones, a simple system of describing subdivisions of each band was added, by using additional figures after a decimal point. To take our example of band 1p36, at the level of 550 bands in a haploid chromosome set, three sub-bands can be identified in this region and are designated 1p36.1, 1p36.2 and 1p36.3. With even more elongated chromosomes showing about 850 bands, sub-band 1p36.3 can be divided further, the subdivisions being numbered 1p36.31, 1p36.32 and 1p36.33.

Similar systems have been used for numbering the bands on the chromosomes of most mammals whose chromosomes have been studied intensively (Sumner, 1990, p. 16), and among birds in the domestic fowl (chicken). The only variant on the system is in the mouse, in which letters have been used to designate the landmark regions of the chromosomes (Evans, 1989); because all the chromosomes in this species are telocentric, there is also no need to distinguish p and q arms.

10.2.2 G- and R-bands compared

A large number of banding methods, particularly those using fluorochromes, produce patterns that depend on the base composition of the chromosomal DNA. Fluorochromes with a preference for A+T-rich DNA, such as quinacrine, DAPI or Hoechst 33258, produce patterns similar to G-banding, and those with a preference for G+C-rich DNA, such as chromomycin A_3

or mithramycin, produce patterns similar to R-banding. Similar results are obtained using antibodies against specific nucleotides, and R-banding itself may depend on DNA base composition (Sumner, 1990, pp. 115–118). These observations therefore clearly indicate that there is a difference in base composition between G- and R-bands, a conclusion reinforced by observations to be described below. The average difference in base composition between G- and R-band DNA was believed to be quite small, only a few per cent; for example, Holmquist *et al.* (1982) stated that G-band DNA is 3.2% richer in A+T than R-band DNA. Now that the human genome has been almost completely sequenced, it is clear that there is much greater variation than this, with extremes of 33.1% and 59.3% G+C content (IHGSC, 2001). This degree of variation is on a much finer scale than that detectable by chromosome banding, but even at a scale of >3.9 Mb, comparable with visible bands (Drouin *et al.*, 1994), G+C contents as low as 36% and as high as 50% have been found. These differences are clearly ample to produce base-specific banding, for which in any case there is unlikely to be a linear relationship between fluorescence intensity and base composition (Sumner, 1990). There is no evidence that G-banding methods depend in any way on DNA base composition, and it seems more likely that these methods rely on some difference in chromatin structure that has not yet been defined.

Replication banding (Box 10.4) produces patterns resembling G- and R-banding (Fig. 10.4), although there are some small differences (Drouin *et al.*, 1994), and the detailed pattern is highly dependent on timing. R-Bands replicate during the early part of S phase, and G-bands during the late part; heterochromatin (C-band material) is usually the last to be replicated. Many workers have claimed that there is a sharp discontinuity in the middle of the S phase at a point when the early replicating bands have completed their DNA synthesis, and the late replicating bands have not yet begun to synthesize their DNA, consistent with the existence of a mid-S phase checkpoint (Section 2.2.2.1); others have failed to detect such a break, and have found that

Box 10.4 Replication banding

Deoxyribonucleic acid is replicated during the S phase of the cell cycle, but not all the DNA is replicated simultaneously. Replication banding (Fig. 1) shows the pattern of early and late DNA replication on chromosomes. It is produced by culturing cells in the presence of bromodeoxyuridine (BrdU), which is incorporated into DNA instead of thymidine. After fixation and spreading of the chromosomes on a slide in the usual way, the BrdU can be detected either by photolysis followed by Giemsa staining (Fig. 10.5), or by labelling with anti-BrdU (Fig. 10.4). If the BrdU is present during the early part of the S phase, the DNA that is replicated first will be labelled; if it is present during the later part of the S phase, late replicating DNA will be labelled. With careful experimental design, DNA that is replicating during intervals as short as a few minutes can be identified. Patterns of early and late replication banding tend to correspond to R- or G-banding in mammals (Section 10.2.2), with very late replicating regions tending to be heterochromatic (Section 7.1). Interestingly, replication patterns

can be produced in a wide variety of organisms that do not show G- or R-bands on their chromosomes (Sumner 1998b).

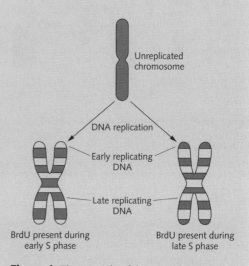

Figure 1 The principle of detecting chromosome replication using bromodeoxyuridine (BrdU) labelling (replication banding).

Chromosome 1

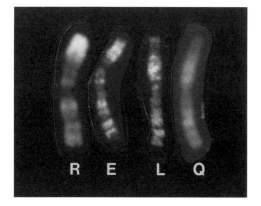

Figure 10.4 Replication bands compared with R- and Q-bands on human chromosome 1. The early replication pattern (E) corresponds to that of the R-bands (R), while the late replication pattern (L) corresponds to that of the Q-bands (Q). Replication patterns revealed by immunofluorescence following bromodeoxyuridine (BrdU) incorporation; R-bands demonstrated by acridine orange staining.

synthesis is continuous throughout S phase (Sumner, 1990, p. 243; Drouin et al., 1994). Detailed analysis of replication timing shows that each band replicates at a specific time (Bickmore & Craig, 1997) and can be placed in one of a large number of distinct time periods. The time of replication of a chromosome segment is to a large extent an inherent property of that segment; in general, specific bands replicate at the same time even when translocated to another chromosome. In the yeast *Saccharomyces cerevisiae* it is quite clear that some replication origins initiate early, and others initiate late (Brewer *et al.*, 1993; Bickmore & Craig, 1997). Nevertheless, replication times of chromosome bands can be changed, for example in cases of position effect variegation (Section 7.4.5) when euchromatic regions of chromosomes are heterochromatinized by being placed next to a block of heterochromatin, or next to yeast telomeres (Bickmore & Craig, 1997). Similarly, in the inac-

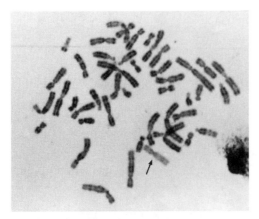

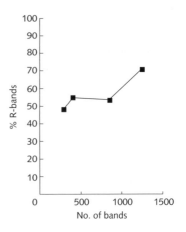

Figure 10.5 A human female metaphase spread, stained to show replication patterns. Early replicating regions are dark, and late replicating regions pale, giving an R-banding pattern; the late-replicating inactive X chromosome (arrow) is entirely pale. Reproduced from Sumner (1983) *Science Progress* **68**, 543–564, published by Blackwell Scientific Publications.

Figure 10.6 Graph showing the increase in the proportion of the chromosomes that is G-banded as the chromosomes contract. Data from Bickmore & Craig (1997).

tive X chromosome of female mammals, replication is generally later than in the active X, but the R-bands still replicate before the G-bands (Drouin *et al.*, 1990) (Fig. 10.5). Changes in replication patterns have also been noted at the site of a specific translocation in cancer (Karube & Watanabe, 1988). The segregation of chromatin into early and late replicating domains is consistent with the differences in DNA base composition between G- and R-bands, because it was established many years ago that early replicating DNA is relatively G+C-rich, while late replicating DNA is relatively A+T-rich (Sumner, 1990).

G-Bands, which are late replicating, are the first to condense in prophase, and the early replicating R-bands are the last to condense (Drouin *et al.*, 1994). This is obvious in meiosis, where the pachytene chromosomes are condensed into chromomeres, separated by less condensed interchromomeric regions (see Fig. 6.5a). The pattern of chromomeres resembles that of G-bands on the same chromosome. The process of chromosome condensation is quite complicated. The number of bands in a human haploid metaphase set of chromosomes is about 350, but the maximum detectable in prophase chromosomes is generally about 1200–1300 (Drouin *et al.*,

1994). Individual bands do not therefore simply contract, but fuse with each other. This fusion is not random, but follows a fixed sequence. Narrow bands are swallowed up in adjacent pairs of larger bands, either a narrow G-band disappearing as two R-bands merge to form one larger band, or vice versa. As the chromosomes condense, the G-bands come to form an increasing fraction of the chromosome length, while the proportion that forms R-bands decreases (Bickmore & Craig, 1997) (Fig. 10.6).

G-Bands are relatively poor in genes, and R-bands relatively rich in genes (Table 10.2). (As most of the DNA in the mammalian genome is not made up of genes and their associated sequences, such as promoters, genes form a small minority of the DNA in both G- and R-bands.) DNase sensitivity is an indirect indication of gene activity in R-bands. The CpG islands are associated with about 60% of human genes and 50% of mouse genes (Bickmore & Craig, 1997), and CpG islands are concentrated in R-bands (Fig. 10.7). Most conclusively, the human gene mapping project has shown that G+C-rich regions of the genome (i.e. R-bands) are rich in genes, while A+T-rich regions (i.e. G-bands) are gene-poor (Dunham *et al.*, 1999; Hattori *et al.*, 2000; IHGSC, 2001) (Fig. 10.8).

The other feature of DNA that differs

Table 10.2 Evidence for the differential distribution of genes in G- and R-bands.

Human trisomies compatible with live birth involve chromosomes rich in G-bands (Sumner 1990)
mRNA sequences concentrated in R-bands (Yunis *et al.*, 1977; Sumner, 1990)
DNase sensitivity of R-bands (Sumner, 1990)
Acetylation of histones in R-bands (Jeppesen, 1997; Breneman *et al.*, 1996)
CpG islands concentrated in R-bands (Bickmore & Craig, 1997)
Genes concentrated in DNA fractions with highest G+C content (Sumner, 1990; Bickmore & Craig, 1997)
Direct localization of genes (Gardiner, 1996; Bickmore & Craig, 1997; White *et al.*, 1999)
Whole genome sequencing (Human Genome Mapping Project) (Dunham *et al.*, 1999; Hattori *et al.*, 2000; IHGSC, 2001)

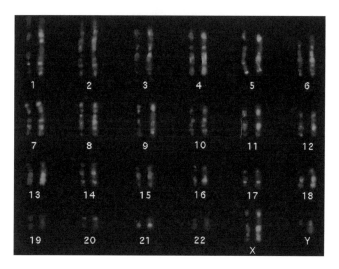

Figure 10.7 The distribution of CpG islands on human chromosomes shown by *in situ* hybridization. For each chromosome are shown: (left) hybridization with DNA in which CpG islands are close together (<100 kb apart); (right) the early replication pattern (corresponding to R-bands) demonstrated by bromodeoxyuridine (BrdU) substitution. Reproduced with permission from Craig & Bickmore (1994) *Nature Genetics* **7**, 376–382. © Nature America.

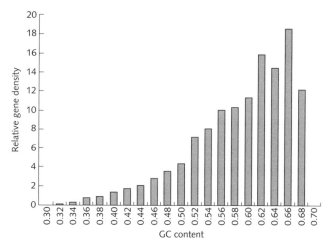

Figure 10.8 Relationship between gene density and G+C content in the human genome. Reproduced with permission from IHGSC (2001) *Nature* **409**, 860–921. © Macmillan Magazines Limited.

between G- and R-bands is the presence of different types of repeated sequences (Table 10.1). Long intermediate nuclear elements (LINEs, of which the principal family in humans is L1) are concentrated in G-bands, while SINEs (short intermediate nuclear elements, of which *Alu* is the principal family in humans, and B1 and B2 the principal types in mice; Boyle *et al.*, 1990)

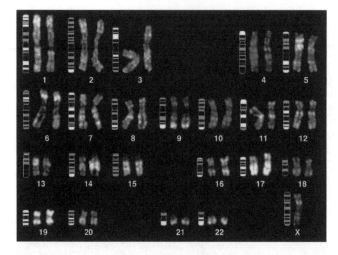

Figure 10.9 The pattern of histone H4 acetylation in human chromosomes. For each chromosome, an ideogram of the R-banding pattern is shown at the left, and the histone acetylation pattern is illustrated for each pair of chromosomes. The acetylation pattern is essentially the same as the R-banding pattern. Reproduced by permission of Wiley-Liss, Inc., a subsidiary of John Wiley & Sons, Inc., from Jeppesen (1997) *Bioessays* **19**, 67–74. © John Wiley.

tend to be concentrated in R-bands. Although this is true as a general statement, it is now known that different *Alu* sequences occur in chromosomal regions that have different base compositions (IHGSC, 2001). Older *Alu* sequences occur preferentially in G+C-rich DNA, but the youngest ones prefer A+T-rich DNA. The L1 sequences are relatively A+T-rich, with only 42% G+C, while *Alu*, B1 and B2 sequences are relatively G+C-rich at 56% G+C, thus conforming in base composition to the bands in which they are found, and possibly having a significant influence on the composition of these bands (Craig & Bickmore, 1993). Because LINEs and SINEs are mobile elements that can move around the genome by retrotransposition (Section 3.3.2), there must be factors that ensure that they can only retrotranspose into the appropriate regions of the chromosome.

Much less is known about differences in chromosomal proteins between G- and R-bands. Differences in histone acetylation have been mentioned above. Hyperacetylation of histones H3 and H4 is associated with transcriptional activity (Section 4.2.4), and patterns of histone acetylation correspond to R-banding patterns (Fig. 10.9). In particular, immunofluorescence of chromosomes with antibodies against histone H4 acetylated at lysines 5, 8 or 12 produces clear R-banding patterns (Jeppesen, 1997). Differences in histone H1 subtypes between G- and R-bands

have also been described; these may be concerned with chromatin condensation, and their concentration in G-bands is therefore not unexpected (Breneman *et al.*, 1993). Protein HMGA1a is also concentrated in G-bands, which does not seem surprising as it is reported to bind preferentially to A+T-rich DNA (Johnson *et al.*, 1988). However, HMGA1a seems to be associated with active genes, so it might have been supposed that it would be found in R-bands.

At meiosis, synapsis of homologous chromosomes begins in R-bands (Ashley, 1990), and meiotic recombination is largely restricted to R-bands (Chandley, 1986; Holmquist, 1992). This could be because specific DNA sequences required for recombination are concentrated in specific parts of the chromosomes, or simply that R-bands (i.e. the interchromomeric regions of pachytene chromosomes) have a more open structure necessary for crossing-over to occur. Differences in chromatin structure might also be responsible for the concentration of chromosome breaks in R-bands, whether induced by radiation or chemicals, or naturally occurring as in cancers or in chromosomal evolution.

10.2.3   Bands, isochores and chromatin flavours

The properties of G- and R-bands have been described above (Section 10.2.2) as if they were simply exact opposites of each other. Although this is adequate as a generalization, there are often subtle discrepancies between different types of banding: for example, the patterns produced by G- and R-banding methods are not exactly complementary, and each differs in some details from early and late replication patterns, respectively (Sumner, 1990; Drouin *et al.*, 1994). Although pachytene chromomeres are mostly A+T-rich, some terminal ones are G+C-rich (Ambros & Sumner, 1987). Not only are there such discrepancies, but there are also differences in staining intensity within the categories of both G- and R-bands (Francke, 1994). Systematic differences between bands in both the broader categories of G- and R-bands have been described as 'chromatin flavours' (Holmquist, 1992), which differ from each other not only in staining intensity, but also in many of the properties described in the previous section (Section 10.2.2). There is also quite a good correlation between chromatin flavours and isochores, long homogeneous DNA segments that differ in average base composition, gene density, etc. (Bernardi, 1989, 1993a, b). Originally G-bands were regarded as a single 'flavour', and R-bands were divided into four separate flavours on the basis of their G+C-richness and their content of *Alu* sequences. It is, however, possible to recognize subclasses of G-bands, and C-bands (Section 7.3) might be regarded as yet another flavour. A list of different flavours, and some of their properties, is given in Table 10.3. The different flavours of G-bands have not been analysed in much detail. The G_{cond} flavours are the bands that appear to be fully condensed even in early prophase, and these have the lowest concentration of genes of any of the flavours listed in the table (Bickmore & Craig, 1997). Drouin *et al.* (1994) recognized what appears to be an even more extreme set of very late replicating G-bands, which only fuse with each other during chromosome condensation to a very small extent or not at all, and to which only one gene had been mapped at that date. In any case, it must be recognized that there are probably no hard-and-fast boundaries separating the various flavours, but that there is evidently some degree of arbitrariness about the delineation of the different categories. The G_{dark} flavours are those G-bands that appear to be darkest, but do not differ radically in many respects from G-bands in general.

The 'mundane R-bands' form the greatest part of the R-bands, are neither particularly G+C-rich or *Alu*-rich and have roughly an average concentration of genes. The vAlu+ vGC− flavour is generally similar apart from having a higher than expected concentration of *Alu* sequences. The really remarkable flavours are the two very G+C-rich ones, which together occupy about 15% of the genome, but contain perhaps as many as 65% of all genes; the concentration of genes in these regions can be more than four times the average. These flavours are localized to the T-bands − a subset of R-bands that are particularly G+C-rich, resistant to the banding treatment and are often, but by no means always, at the ends of chromosomes. Thus there is something of a gradient along human chromosomes, with the most G+C-rich regions containing the highest gene concentrations towards the ends, while the more proximal regions are more A+T-rich, are gene-poor and are more likely to be G-banded (Fig. 10.1). However, as a result of rearrangement during evolution, the chromosome regions in mice that correspond to T-bands have become interstitial, although it is not known if they retain their T-banding (Bickmore & Craig, 1997, pp. 23–24). Chromosome breakage, from whatever cause, is more likely to be found in the most G+C-rich flavours (Holmquist, 1992), and T-bands are also the regions of greatest meiotic recombination (Holmquist, 1992; Bickmore & Craig, 1997, pp. 118–119). Curiously, because in general the A+T-rich G-bands tend to form meiotic chromomeres, and the G+C-rich R-bands tend to form the less condensed interchromomeric regions, many of the terminal very G+C-rich regions of chromosomes form chromomeres (Ambros & Sumner, 1987), although a detailed analysis has not been carried out to see if these actually correspond to T-bands.

10.2.4   Some details of isochores

Isochores were defined as segments of DNA of homogeneous base composition, and are from

Table 10.3 'Chromatin flavours' and their properties.

Name	Chromosomal location	Per cent of euchromatin	Base composition	Concentration of Alu sequences	Gene concentration (observed/expected)	Average separation of genes
G_{cond}	G-bands	8.3	A+T-rich		0.29	
G_{dark}	G-bands	22.6	A+T-rich		0.40	
G_{all}	G-bands	45.5	A+T-rich		0.24–0.45	71 kb
Mundane R	R-bands		G+C-rich	Low	0.61–1.0	32 kb
vAlu+vGC−	R-bands		G+C-rich	High	0.80–1.0	
vAlu–vGC+ ⎱	T-bands	15	Very G+C-rich	⎰ Low	2.26–3.44 ⎱	14 kb
vAlu+vGC+ ⎰				⎱ High	2.63–4.31 ⎰	

Data from Holmquist (1992) and Bickmore & Craig (1997).

Table 10.4 Properties of mammalian isochores.

Isochore	Base composition (% G+C)*	Per cent of genome	Gene concentration (observed/expected)	Distance between genes	Chromosomal location
L1	39 ⎫	62	0.55	64 kb	G-bands
L2	41 ⎭				
H1	45	22 ⎫	1.2	29 kb	R-bands
H2	49	9 ⎭			
H3	53	3	9.3	4 kb	T-bands

*Approximate figures for human DNA.
Data from Craig & Bickmore (1993).

>300 kb to 1 Mb in length (Bernardi, 1993b). They are, therefore, larger than individual chromatin loops, which average 63 kb (Section 6.3), but much smaller than chromosome bands, which even at a resolution of 1250 bands per haploid genome are about 2500 kb long on average (Drouin et al., 1994). Excluding highly repetitive satellite DNAs, mammalian genomes can be fractionated into five classes of isochores (Table 10.4). These can be seen to correspond very roughly with different types of bands and chromatin flavours. The isochores L1 and L2 are relatively A+T-rich, have a relatively low concentration of genes and occur mainly in G-bands. At the other extreme, isochore H3 is G+C-rich, has a very high concentration of genes and is generally located in T-bands. Nevertheless, there is not a complete correlation between isochores, and bands or chromatin flavours: the fraction of the genome occupied by L1 and L2 isochores is much greater than the fraction occupied by G-bands, and isochore H3 comprises only 3% of the genome compared with 15% for T-bands. Indeed, while G-bands appear to consist essentially of the most A+T-rich isochores, a variety of different isochores can be found in a single R- or T-band (Gardiner et al., 1990). Within each isochore, interspersed repeated sequences (SINEs and LINEs) have the same G+C-richness as unique sequences, and even viral sequences will integrate into isochores that match their own base composition. Isochores therefore appear to be a fundamental subdivision of mammalian genomes, and in some way determine the composition of genes and other sequences that they contain.

The isochore model of chromosome structure, although very valuable, has not been confirmed in every detail by human genome sequencing (IHGSC, 2001). As mentioned in Section 10.2.2, the variation in base composition along chromosomes is greater than previously suspected, and can occur over quite small distances. The idea of isochores as segments of homogeneous base composition is therefore an oversimplification. Nevertheless, at the level of both isolated DNA and in chromosomes, it is clear that most genes are in the more G+C-rich regions. The base composition of genes is correlated with that of the isochores in which they lie (Aïssani et al., 1991), but there are other differences between genes that are correlated with the base composition of different parts of chromosomes (IHGSC, 2001): in A+T-rich regions individual genes are spread out over much greater lengths, as a result of having large introns, while genes in G+C-rich regions are more compact, with smaller introns. Although human genome sequencing has confirmed that Alu sequences (SINEs) are concentrated in G+C-rich DNA, and LI sequences (LINEs) in A+T-rich DNA (IHGSC, 2001), it has not yet confirmed the correlation between base composition of the LINEs or SINEs and the region of DNA that they lie in. However, because SINEs, LINEs and other transposable sequences comprise such a large proportion of the genome (Section 3.3.2), they must have a strong influence on the composition of the regions they reside in.

10.3 Longitudinal differentiation of chromosomes in non-mammals

So far, the description of banding and longitudinal differentiation of chromosomes has been concerned largely with the situation in mammals. One reason for this is that the situation in mammals has been studied in the greatest detail, and therefore it is easiest to give a coherent account of it. However, it is clear that mammals are not representative of all eukaryotes in the way they organize their genomes, but in fact are an exception. Many of the features of longitudinal differentiation found in mammals, particularly banding with base-specific fluorochromes, and the presence of isochores of widely differing base composition, are absent in most lower vertebrates, invertebrates and plants (Table 10.5). A few features, such as pachytene chromomeres, and differentiation into early- and late-replicating segments, have invariably been found where they have been sought, suggesting that they may be universal features of eukaryotic chromosomes. On the other hand, G+C-rich isochores and banding with base-specific fluorochromes are largely confined to birds and mammals. Reptiles do not have well differentiated G+C-rich isochores, and their chromosomes do not show good banding with base-specific fluorochromes. Certain fish (eels and some thermophilic species) have G+C-rich isochores, but at best have only poor fluorochrome banding. Monocotyledonous plants also have G+C-rich isochores, but with one exception lack base-specific fluorochrome banding. Unfortunately none of these groups has been studied in the same detail as mammals have, so satisfactory correlations between the presence or absence of isochores and base-specific banding cannot yet be deduced. G-Bands are more widely distributed than bands revealed by base-specific fluorochromes; good quality G-bands can be produced in reptiles, birds and mammals, in some fish and amphibia and in a few plants. Whatever G-banding may be showing, it seems to be phyletically more widespread than longitudinal differentiation based on DNA base composition. (Failure to produce G-bands in many organisms has often been attributed to purely technical factors, although this explanation seems increasingly unlikely with the passing of the years. Whether this is so or not, a distinction must be made between patterns produced by 'traditional' G-banding methods, which, as mentioned above (Section 10.2.2), may be related to differences in chromatin conformation, and replication bands or bands produced by, for example, restriction endonuclease digestion, which, although generally similar to the pattern of traditional G-bands in mammals, are produced by fundamentally different mechanisms and are therefore demonstrating different aspects of chromosome organization.)

If mammals and other higher vertebrates are exceptional in the organization of their chromosomes by having them divided up into compartments of differing base composition, with the genes concentrated in the most G+C-rich compartments, how are genes distributed in other organisms in which the base composition is more uniform throughout the chromosomes? Although information for organisms other than mammals is sporadic, it does seem that in general genes are not uniformly distributed on eukaryotic chromosomes. This is particularly true of monocotyledonous plants such as wheat, in which mapping of CpG islands (Moore *et al.*, 1993) and direct mapping of genes (Gill *et al.*, 1993) both show that genes are concentrated towards the ends of chromosomes. In amphibia (Herrero *et al.*, 1995) and in insects (de la Torre *et al.*, 1996; Palomeque *et al.*, 1998) the regions of nuclease sensitivity – an indirect marker of sites of active genes – are concentrated towards the ends of chromosomes. A somewhat different situation may exist in birds, in which the karyotype consists of a small number of macrochromosomes and a larger number of microchromosomes, which are too small to show any significant longitudinal differentiation (apart from the centromeric heterochromatin). The microchromosomes are more G+C-rich, have a higher density of CpG islands, a higher level of histone acetylation and twice the density of genes when compared with the macrochromosomes (Smith *et al.*, 2000). Although it is tempting to suppose that avian microchromosomes

Table 10.5 Distribution of different types of banding and of G+C-rich isochores in different groups of eukaryotes.

	Pachytene chromomeres	Replication bands	DNase hypersensitivity	G-Bands	Base-specific fluorochromes		G+C-rich isochores
					Q-Bands (A+T-rich DNA)	Chromomycin R-bands (G+C-rich DNA)	
Mammals	Yes	Yes	Yes	Good	Good	Good	Yes
Birds	Yes	Yes	–	Good	Moderate	Moderate	Yes
Reptiles	Yes	Yes	–	Good	Poor	Poor or absent	Poor
Amphibia	Yes	Yes	Yes	No	No	No	No
Xenopus	–	Yes	–	Good	–	–	No
Fish	Yes	Yes	–	A few spp.	No	No	No
Eels (Anguilla)	–	–	–	Good	Poor	No	Yes
Thermophilic spp.	–	–	–	–	–	–	Yes
Insects	Yes	–	Yes	No	No	No	–
Drosophila	–	–	–	No	No	No	No
Spiders	Yes	–	–	Yes	–	–	–
Molluscs	Yes	–	–	–	–	–	–
Plants							
Monocotyledons	Yes	Yes	–	A few spp.	No	No	Yes
Lilium	Yes	–	–	–	Yes	–	–
Dicotyledons	Yes	Yes	–	–	No	No	No
Vicia	–	–	–	Yes	–	–	–

After Sumner (1998b).

have eliminated most or all of their G-band material and now consist (apart from their centromeres) only of R-band-like material, sequencing is required to establish this.

In organisms other than humans whose genomes have been sequenced, there is no compelling evidence that their genomes are compartmentalized into gene-rich and gene-poor regions with the associated properties that have been found in mammalian chromosomes. In budding yeast, *S. cerevisiae*, there is variation in G+C content along the chromosomes, but this is not correlated with variations in gene distribution (e.g. Jacq *et al.*, 1997). However, yeast chromosomes are extremely small, and may be atypical of eukaryotes for this reason. In the nematode *Caenorhabditis elegans* both G+C content and gene distribution are fairly constant along the chromosomes (*C. elegans* Sequencing Consortium, 1998), although there tends to be more recombination towards the ends of the chromosome arms than in the middle of the chromosomes. Nematode chromosomes are holocentric (Section 12.5) and thus there is no localized centromere, which might be a factor influencing chromosome organization.

10.4 The how and why of longitudinal differentiation

Some form of longitudinal differentiation, both structural and functional, is a widespread attribute of eukaryotic chromosomes. Why should this be so? What function does it serve? At present, answers to these questions are largely speculative. The segregation of genomes into gene-rich and gene-poor regions may result from a requirement for specific positioning of genes in interphase nuclei (Section 5.2), although there is little compelling evidence for this. The chromosomes are clearly carrying around with them far more DNA than they need for their purely genetic functions, and it may be necessary to segregate such DNA in chromosome segments that are essentially inactive, and therefore are condensed and late-replicating. But if such DNA is not really required, why not get rid of it? Perhaps it does have a function, but a non-genic one. Cavalier-Smith (1978) proposed that much of the DNA in nuclei had a 'skeletal' function and was concerned with maintaining nuclear size, which in turn would have all sorts of consequences for cell physiology.

Why have isochores evolved, and with them chromosome bands of distinctive base composition? Unlike chromomeres and replication bands, which are present in virtually all eukaryotes and therefore might have evolved only once, isochores and differences in base composition along chromosomes have evolved independently at least three times: in monocotyledonous plants, in birds and in mammals. Although it has been proposed that isochores are the result of selection, they may simply be the result of mutational bias (Eyre-Walker & Hurst, 2001), suggesting that they are of no adaptive significance.

Whatever the reasons, the euchromatin of eukaryotic chromosomes is divided up at several levels into subunits of structure and function. The coarsest of these levels is the chromosome bands, which can be seen with a light microscope, but each of these is subdivided, at least in higher vertebrates, into a number of isochores, and replicons are a still smaller subdivision. There is certainly some heterogeneity among the isochores within a single band, and possibly some differences in replication timing between the replicons in a band, but such heterogeneity is sufficiently minor for the chromosome bands to appear as units of uniform composition and behaviour, clearly distinguishable from adjacent bands made up of subsets of isochores and replicons having different properties.

The nucleolus and the nucleolus organizer regions (NORs)

11

11.1 The importance of nucleoli and NORs

The nucleolus is the largest and most conspicuous nuclear organelle – so conspicuous that it was recognized over 200 years ago, by Fontana in 1781 (see Schwarzacher & Wachtler, 1983, and Wachtler & Stahl, 1993, for historical reviews of studies on nucleoli). Similarly, the nucleolus organizer regions (NORs) form a conspicuous chromosomal structure – a secondary constriction – that can be stained differentially with silver (Section 11.3), as well as being easily identified by *in situ* hybridization, and thus they form the only gene that could be identified by light microscopy on metaphase chromosomes before the development of fluorescence *in situ* hybridization (FISH). The reason for the nucleolus and NORs being so prominent is, of course, that they produce and process the ribosomal RNAs that are necessary for all protein synthesis in the cell, and which are therefore required in large quantities. The ribosomal genes are present in multiple copies, even in organisms with very small genomes, such as yeasts, and the RNA they produce forms about 80% of all the RNA in the cell.

The specialization of NORs and nucleoli for the high rate of production of ribosomes is remarkable enough, but in the oocytes of some organisms it is not sufficient, and the ribosomal genes (rDNA) themselves are amplified to an enormous degree to provide enough ribosomal material to carry the embryo through the early stages of development. Although nucleoli are rightly thought of as factories for the production of ribosomes, it has been discovered in the last few years that they can be involved in various unrelated nuclear functions. All these topics form the subject of this chapter.

11.2 The ribosomal genes

The ribosomal genes (that is, the genes for ribosomal RNA – rRNA genes or rDNA) consist of a basic repeating unit made up of a non-transcribed spacer (NTS), better called the intergenic spacer (IGS) as there is evidence that it is sometimes transcribed, and the actual ribosomal genes, separated by internal transcribed spacers (ITS) (Fig. 11.1). The whole repeating unit is often G+C-rich (Miller, 1981). The intergenic spacer is usually by far the largest component of the repeating unit. In most organisms, the genes are, in order, those for 18S, 5.8S and 28S ribosomal RNA, which are transcribed as a single unit of 45S rRNA, which is then processed into the individual components. The actual size of the rRNA genes varies between species; the figures just quoted refer to vertebrates. Although the 5.8S gene always has the same size, the other genes are often smaller, for example 17S and 25S in *Tetrahymena*, or 18S and 26S in plants. In spite of these variations in size, there is considerable homology between the ribosomal genes in different organisms, but the spacers between genes are much more variable.

The number of copies of ribosomal genes is highly variable: some organisms with small genomes (e.g. protists, fungi, some insects) have less than 100 copies of rRNA genes, while at the other extreme some plants and amphibia have more than 10 000 copies (Table 11.1). Humans have about 200 rRNA genes, a figure typical of mammals. Polymorphism in the number of copies is normal, both between homologues in the same individual and between individuals. In some species, the rRNA genes are confined to a single site on a pair of homologous chromosomes, but quite often they are spread over several chromosomes (Fig. 11.2); for example, in the mouse they can be found on any of up to six pairs of chromosomes (though not on all in one individual mouse), and in humans five pairs

Figure 11.1 Diagram of the human rRNA gene repeating unit. The non-transcribed spacer (NTS) occupies 31 kb out of the total length of 44 kb, whereas the 18S, 5.8S and 28S genes are located together in the remaining 13 kb.

Table 11.1 Numbers of rRNA and 5S RNA genes in different organisms.

Species	rRNA genes	5S RNA genes
Algae		
Acetabularia mediterranea	1900	
Chlamydomonas reinhardii	150	
Euglena gracilis	800–1000	
Yeast		
Saccharomyces cerevisiae	140	150
Slime moulds		
Dictyostelium, Physarum	~100	
Angiosperms		
Allium cepa (onion)	6950	
Phaseolus coccineus (runner bean)	2000	
Pisum sativum (pea)	3900	
Triticum aestivum (wheat)	6350	
Protozoa		
Tetrahymena pyriformis	200–290	330–780
Nematoda		
Caenorhabditis elegans	55	
Insects		
Acheta domesticus (cricket)	170	
Bombyx mori (silk moth)	240	
Drosophila melanogaster	100–240	100–200
Vertebrates		
Salmo salar (salmon)	710	
Plethodon spp. (salamanders)	2000–4300	
Triturus spp. (newts)	3900–5490	
Xenopus laevis	450–760	9000–24 000
Gallus domesticus (chicken)	190–200	
Cricetulus griseus (Chinese hamster)	250	
Mus musculus (mouse)	100	
Rattus norvegicus (brown rat)	150–170	830
Homo sapiens (man)	50–280	2000

Data from Long & Dawid (1980) and Busch & Rothblum (1982).

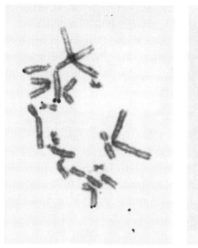

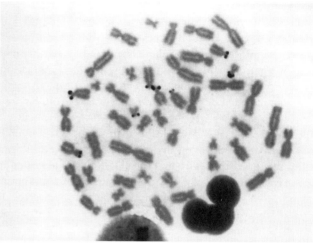

(a)

(b)

Figure 11.2 Silver-stained NORs on chromosomes of (a) CHO (Chinese hamster ovary) cells and (b) human cells. Notice that the NORs are found on several pairs of chromosomes, although in human cells all five pairs (13, 14, 15, 21 and 22) are not usually active and are therefore not stained; silver only stains active NORs. Figure 11.2 (a) reproduced with permission from Sumner & Leitch (1999) in *Light Microscopy in Biology: a Practical Approach* (ed. A.J. Lacey), pp. 151–184. © Oxford University Press.

of chromosomes carry NORs (see Long & Dawid, 1980, and Howell, 1982, for listings of the numbers and sites of NORs in different organisms). Nucleolus organizer regions can occur in a variety of locations on chromosomes: often they are near the ends, as in humans and many other species, but interstitial sites also occur.

The 5.8S, 18S and 28S RNA coded by the rRNA genes in the NORs are not the only ribosomal RNA; 5S RNA is coded for by genes that are normally at sites distinct from the NORs, and may be on one or more pairs of chromosomes. The 5S RNA genes are in clusters of hundreds or thousands of copies, and the number may be similar to, or much greater than, the number of rRNA genes (Table 11.1). The 5S genes do not form any distinctive chromosome structure, such as a constriction. The yeast *Saccharomyces cerevisiae* and the slime mould *Dictyostelium discoideum* are exceptional in having their 5S genes incorporated into the same repeating unit as the rRNA genes (Adams *et al.*, 1992). Even in other eukaryotes the 5S RNA is processed in the nucleolus, as it must be incorporated into the mature ribosome (Pederson & Politz, 2000).

Two features of the metaphase NOR need to be discussed here: the appearance of the NOR as a secondary constriction, and the significance of silver staining of the NOR. The presence of a secondary constriction could be due to a difference in structure from the rest of the chromosome, or it could be caused by a failure to condense. The possibility that length of chromatin loops is a factor determining the diameter of the chromatid has already been discussed (Section 6.3). In humans, the length of the repeating unit of rRNA genes, including the intergenic spacer, is 44.7 kb (Bickmore & Oghene, 1996). Origins of replication occur preferentially within a section of the intergenic spacer upstream from the 18S gene (Fig. 11.1), and it is such regions that are preferentially attached to the chromosome scaffold, whereas the coding sequences are preferentially found in the loops away from the scaffold. If each unit of the rRNA repeated gene represents a loop (which is far from certain), then the total length of a DNA loop in the NOR constriction would be 44.7 kb, although if more than one rRNA repeating unit should form a single loop the length would be a multiple of this. This compares with estimated loop sizes of 30–90 kb for chromosomes as a whole, with an average in the region of 63 kb (see Table 6.1). Thus if loop size were the main determinant of the highest level of chromosome structure, NORs could reasonably be expected to show up as a constriction. However, as pointed out in Section 6.3, it is far from certain that loop size determines chromosome morphology, and it would be quite reasonable to attribute the secondary constriction at NORs to delayed condensation. The NORs are very active transcriptionally, and continue transcribing RNA into prophase, and they could therefore be expected to condense later than the bulk of the chromosomes.

11.3 Silver staining of NORs and nucleoli – what does it mean?

Silver staining, under properly controlled conditions, is a highly selective method for staining interphase nucleoli and NORs on mitotic and meiotic chromosomes, and is a principal method for identifying sites of NORs on chromosomes (Sumner, 1990), although FISH is more specific. It is a characteristic of silver staining that, in species with multiple NORs, not all the NORs are usually stained; for example, in humans, no more than 7–8 out of the total of 10 are normally stained. In fact, all the available evidence indicates that silver stained NORs are sites that were transcriptionally active, or potentially so, during the preceding interphase (Sumner, 1990; Wachtler & Stahl, 1993). During spermatogenesis and oogenesis, the changes in silver staining are correlated well with known changes in rRNA synthesis (Section 11.5.2), and in *Xenopus laevis*, silver staining of NORs only appears at the stage in embryonic development at which rRNA synthesis begins. Perhaps some of the clearest evidence for a connection between rRNA transcription and silver staining comes from hybrids. Nucleolar dominance is the suppression of NOR activity of one parental set of chromosomes in hybrids, and is widespread in both plants and animals (Pikaard, 2000). In such cases, only the active NORs are stained with silver. In some plant hybrids the active NORs are less methylated than the inactive ones, but this is not true of *Xenopus* hybrids, and cannot be true of *Drosophila* hybrids (because they have almost no DNA methylation; Section 3.5). In some plants, NORs are heavily methylated even though they are active. Histone deacetylation is involved in repression of rDNA transcription in some species, and may be the immediate cause of repression in those species in which methylation is not involved. In humans, active NORs are sensitive to DNase digestion and are hypomethylated, whereas inactive NORs are less sensitive to DNase and are more highly methylated (Ferraro & Prantera, 1988), thus showing the same correlation between gene activity, methylation and nuclease sensitivity reported for other genes (Section 3.5).

Silver staining of nucleoli also occurs in interphase, and has, indeed, been known for a very long time (Derenzini *et al.*, 1994). In general, metabolically more active nuclei have more silver staining than resting nuclei; for example, phytohaemagglutinin (PHA)-stimulated lymphocytes have more nucleoli than unstimulated nucleoli (Wachtler & Stahl, 1993). It is in tumour cells that silver staining of nucleoli has become of particular interest; tumour cells tend to have more silver staining than non-tumorous cells (Derenzini *et al.*, 1994, 2000; Trerè, 2000). The relationship is in fact between the amount of silver staining and the rate of cell proliferation; the shorter the cell-cycle time, the greater the amount of silver staining in the nucleoli, which is generally measured simply as the area of silver-stained material in the nucleoli. Thus rapidly proliferating tumours show a lot of silver, but cells from slow-growing tumours may not show any differences from normal cells. In those tumours in which the amount of nucleolar silver staining is increased, this parameter has diagnostic and prognostic value. In patients with the same cancer at the same stage, those with less silver staining in their nucleoli tend to survive longer.

Considerable effort has gone into identifying the silver-staining material of NORs and nucleoli, and it turns out that several proteins in specific parts of these organelles are involved (Roussel & Hernandez-Verdun, 1994). On mitotic chromosomes, the silver-staining material forms on the outside of the chromatids at the secondary constriction, and does not form part of the chromatin itself. No more than 10% of the nucleolar proteins that stain with silver during interphase are retained on mitotic chromosomes, the rest dispersing into the cytoplasm. Six major silver-staining proteins are retained on the chromosomes, and these include one or more subunits of RNA polymerase I, and UBF, an RNA polymerase I transcription factor. This would explain why silver staining of NORs during mitosis is a good marker for active ribosomal genes. In interphase, silver staining is largely confined to regions of the nucleolus known as the fibrillar centres (Section 11.4), and here the main silver-staining proteins are nucleolin and protein B23, neither of which is directly involved in the transcription of ribosomal genes; RNA polymerase I forms a much smaller proportion of the silver-stained proteins than in mitosis.

11.4 The nucleolus in interphase

The nucleolus consists of three main components: the fibrillar centres (FCs), the dense fibrillar component (DFC) and the granular component (GC). In addition, it appears to have a skeletal component that contains a specific

protein and forms a network round the cortex of the nucleolus (Kneissel *et al.*, 2001). Some nucleoli are roughly spherical, and the different components are arranged concentrically, with the FC in the middle and the GC on the outside; nucleoli in other types of cells have more complicated shapes and structures (Schwarzacher & Wachtler, 1983; Wachtler & Stahl, 1993) (Fig. 11.3). The fibrillar centres are areas of low electron density that contain rDNA and RNA polymerase I. The DFC is usually a narrow dense zone that surrounds the FC. The GC forms the outer layers of the nucleolus, and consists of pre-ribosomal particles about 15 nm in diameter. Nucleoli are often surrounded by a layer of heterochromatin, and occasionally pieces of chromatin are seen in the interior of the nucleolus. The latter are presumably interdigitations from the exterior of the nucleolus, and not detached pieces completely surrounded by nucleolar material.

The structure of nucleoli is easily described, but it has proved much more difficult to relate the structure to function. Although transcription of the rRNA is known to occur in the nucleoli, the exact site where it occurs has not been identified (Raška *et al.*, 1995; Scheer & Hock, 1999; Medina *et al.*, 2000). Because the fibrillar centres contain both rRNA genes and the RNA polymerase I needed to transcribe them, it might be supposed that the FC must be the site of transcription. On the other hand, a body of evidence indicates that the DFC is the site of rRNA syn-

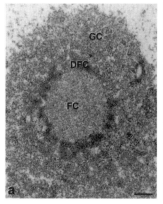 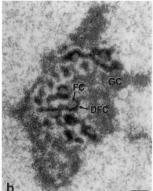

Figure 11.3 Electron micrographs of mammalian nucleoli showing contrasting structures. (a) Nucleolus from a mouse Ehrlich ascites tumour cell, showing a concentric arrangement of the fibrillar centre (FC), dense fibrillar component (DFC) and granular component (GC). (b) Nucleolus from a rat RV cell, which has a reticulated structure with a ribbon-like DFC running throughout the nucleolus. Scale bars = 0.2 μm. Reproduced with permission from Scheer & Hock (1999) *Current Opinion in Cell Biology* **11**, 385–390. © Elsevier Science.

thesis. A 'compromise' view is that the FC and the DFC form a functional continuum, and that transcription occurs in the parts of the FC that are closest to the DFC, the DFC being formed by the nascent transcripts (Raška et al., 1995). Although this seems to be the most plausible interpretation (Scheer et al., 1997; Scheer & Hock, 1999; Medina et al., 2000), the question of where this fundamental process takes place in the nucleolus has not yet been resolved.

11.4.1   Nucleolar proteins

A large number of processes occur in the nucleolus, starting from the transcription of 45S rRNA followed by cleavage into its 5.8S, 18S and 28S components and modifications of specific sites on the RNA, and packing the RNA into pre-ribosomal particles, which are then exported to the cytoplasm. Many nucleolar proteins have been identified that are involved in these processes (Olson et al., 2000), but there are many other nucleolar proteins that have not yet been adequately characterized. Some of the better-characterized nucleolar proteins are listed in Table 11.2. There are proteins such as RNA polymerase I and the transcription factor UBF that are required for the transcription of rRNA and are found in the FCs and DFC, not surprisingly as these are believed to be the sites of transcription (see above). In plants, a variant of histone H1 has been reported in nucleoli (Tanaka et al., 1999a), from which normal H1 is absent. It is plausible that this variant, p35, can modulate a specific chromatin structure required for rRNA transcription (Section 4.2.3). Other proteins are involved in processing the newly synthesized RNA, and are found in the same regions: nucleolin appears to be involved in cleavage of pre-rRNA (Ginisty et al., 1999), while fibrillarin and NAP57/dyskerin are components of small nucleolar ribonucleoproteins (snoRNPs), which are needed for three modifications to the newly synthesized rRNA: cleavage; conversion of certain uridines to pseudouridines; and methylation of ribose moieties (Maxwell & Fournier, 1995; Smith & Steitz, 1997). Ribosomal proteins such as S1, or proteins

that are involved in ribosome assembly, such as B23, are found mainly in the GC of the nucleolus, where these later stages of processing of rRNA into ribosomes are believed to occur. Nucleolar proteins, like other proteins, are synthesized in the cytoplasm, and mechanisms are needed to direct them to the nucleolus. The protein Nopp140, for example, has been found to guide fibrillarin and NAP57 to the nucleolus. Some functions of nucleolar proteins are more enigmatic. As well as its role in processing rRNA, nucleolin has been implicated in nucleocytoplasmic transport (Ginisty et al., 1999). The microtubule-associated protein (MAP) Tau has been found in the FCs (Thurston et al., 1996), where it seems likely to have some function unconnected with microtubules. Other proteins are involved in functions not traditionally associated with the nucleolus (Section 11.6).

11.5   What happens to the nucleolus during cell division?

During cell division the nucleolus breaks down, and most of it disperses although, as already mentioned (Sections 11.1 and 11.2), the chromosomal site of the nucleolus, the NOR, generally remains visible as a constriction. At the end of mitosis or meiosis, new nucleoli are formed at the NORs.

11.5.1   Mitosis

In mitotic prophase, the nucleolus is usually still visible, but disappears at prometaphase, the time of nuclear envelope breakdown. At the same time, rRNA transcription ceases, apparently due to phosphorylation of transcription factor SL1 (Scheer & Hock, 1999; Medina et al., 2000). Some nucleolar components remain at the NORs throughout metaphase and anaphase (Table 11.2), in particular RNA polymerase I and the transcription factors UBF and SL1, which include the major silver-staining proteins of the NOR. Other nucleolar components, including nucleolin, fibrillarin and No55, move to the chromosome periphery, a layer of material

Table 11.2 Some nucleolar proteins.

Protein	Molecular weight (kDa)	Site in interphase*	Site in metaphase	Function	Ref.
B23 (No38) Ribocharin	37	DFC, GC; GC	Chromosome periphery	Chaperone in ribosome assembly; B23 variant in *Xenopus*	Olson *et al.*, 2000
Perichromonucleolin	30–36	DFC	Chromosome periphery		
Nucleolin (C23)	44.5–77.1	[FC], DFC, [GC]	Chromosome periphery	Regulation of rRNA transcription; cleavage of pre-rRNA; nucleocytoplasmic transport	Ginisty *et al.*, 1999
Fibrillarin (B36)	34	FC, DFC	Chromosome periphery	snoRNP component	
Gar1 No55	55	GC	Chromosome periphery	snoRNP component	Smith & Steitz, 1997
ASE-1	55	FC	NORs		
Ki-67	345–395	DFC	Chromosome periphery		Whitehead *et al.*, 1997
RNA polymerase I S1		FC, DFC; GC	?NORs; Chromosome periphery	Transcription of rRNA; Ribosomal protein	
Topoisomerase I		DFC	NORs		
UBF (upstream binding factor)		FC, DFC	NORs	Transcription factor for rRNA	Thurston *et al.*, 1996
SL1			NORs	Transcription factor for rRNA	
Tau		FC	NORs	Microtubule-associated protein (MAP)	
Nopp 140				Guides fibrillarin and NAP57 to nucleolus	
NAP57/dyskerin				snoRNP component	
p35	28–37			Histone H1 variant in plants	Tanaka *et al.*, 1999a
NO145	111	Nucleolar cortex	NORs	Nucleolar cortical skeletal protein	Kneissel *et al.*, 2001

*FC, fibrillar centres; DFC, dense fibrillar component; GC, granular component.

Data from Wachtler & Stahl (1993), Hernandez-Verdun & Gautier (1994) and Scheer & Hock (1999) unless otherwise stated. See also Olson *et al.* (2000) for additional data.

covering the chromosome arms during mitosis (Section 6.6). Some partly processed pre-rRNAs have been found during mitosis with snoRNPs and nucleolar proteins in bodies known as nucleolus-derived foci, suggesting that elements of the rRNA processing machinery may be kept together throughout mitosis (Scheer & Hock, 1999; Dundr *et al.*, 2000).

At telophase, re-formation of nucleoli seems to be a consequence of rRNA synthesis beginning again at the NORs (Dundr *et al.*, 2000; Medina *et al.*, 2000). Nucleolar material is released from the chromosome periphery, and associates to form prenucleolar bodies, which also fuse with nucleolus-derived foci. Neither the prenucleolar bodies nor the nucleolus-derived foci contain any transcriptional machinery, but migrate to the NORs where they fuse with each other and the newly reactivated NORs and form the DFC of the new nucleolus. New FCs appear first, followed by the DFC, and finally the GC. This tends to confirm that newly synthesized rRNA is processed to pre-ribosomal particles by passing through the FC, DFC and GC in sequence. The formation of new nucleoli does not necessarily require new protein synthesis.

11.5.2   Meiosis

Meiosis (Section 2.5) is a more complicated process than mitosis, and nucleolar behaviour during meiosis is more complex than during mitosis. In both oogenesis and spermatogenesis of chordates, including several mammals, there is vigorous nucleolar activity during prophase, with a maximum at pachytene (Schmid *et al.*, 1982, 1983b; Wachtler & Stahl, 1993), which may involve the formation of new nucleoli. Ribosomal RNA synthesis ceases after pachytene in spermatogenesis, but continues into diplotene in oogenesis; in both cases synthesis is stopped entirely from metaphase I until completion of the second meiotic division. Then rRNA synthesis is resumed in the haploid spermatids, and continues almost until mature spermatozoa are formed. The reason for this post-meiotic rRNA synthesis is not clear, but it could be required for the translation of the messenger RNA that

is transcribed during the early development of spermatids (Schmid *et al.*, 1982, 1983b).

11.5.3   Amplification of nucleoli

A much more extraordinary thing happens during prophase in the oocytes of certain organisms, particularly fish, amphibia and some insects. These nuclei contain thousands of small nucleoli, and have a greater DNA content than that of a typical 4C nucleus (which is the DNA amount that would be expected in meiotic prophase). What has happened is that the ribosomal genes, with their spacers, have been amplified to a very high degree, and these amplified genes form supernumerary nucleoli. The degree of amplification is enormous: in *Xenopus laevis* the chromosomes of the oocyte nucleus contain 12.8 pg of DNA, but the quantity of amplified rDNA is no less than 30 pg. In *Triturus*, the degree of amplification is not quite so staggering, but is still very large: the amount of chromosomal DNA in the oocytes is 88 pg, and again the quantity of amplified rDNA is 30 pg. In spite of the large amount of non-chromosomal rDNA produced, the amplification is under strict control, and a fixed amount of rDNA is produced in each nucleus.

A rolling circle mechanism is used to amplify the rDNA (Fig. 11.4). The outer strand of a double-stranded circular DNA is nicked and peeled off the inner strand, and a replication fork is formed at the nick. One new strand is formed in continuity with the outer strand, using the inner strand of the circle as a template, and the other new strand is synthesized using the original outer strand as a template. Synthesis can continue for a variable number of repeats, after which the new DNA is cut off and its ends ligated. Thus a large number of circles of different sizes, each containing an integral number of rDNA repeats, is formed. It is still not clear how the original circular DNA molecules are formed from the linear rDNA in the chromosomes. A consequence of the rolling circle mechanism of amplification is that the resulting nucleoli appear as 'beaded necklaces' of different sizes (Fig. 11.5).

The amplified rDNA often appears as a mass

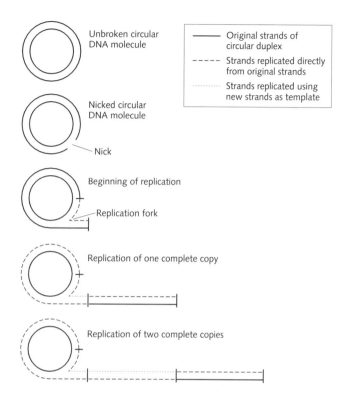

Unbroken circular
DNA molecule

——— Original strands of
circular duplex

----- Strands replicated directly
from original strands

·········· Strands replicated using
new strands as template

Nicked circular
DNA molecule

Nick

Beginning of replication

Replication fork

Replication of one complete copy

Figure 11.4 The rolling circle
mechanism of rDNA amplification.
Reproduced from Bostock & Sumner
(1978) *The Eukaryotic Chromosome*,
published by North-Holland.

Replication of two complete copies

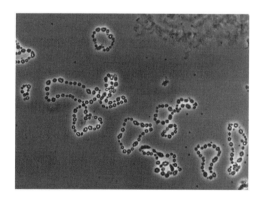

Figure 11.5 'Beaded necklace' amplified nucleoli
from a salamander oocyte. Reproduced with permission
from Macgregor (1993) *An Introduction to Animal
Cytogenetics*. © Kluwer Academic Publishers.

at one side of the nucleus, separate from the
chromosomes. This DNA and associated RNA
and proteins are released at metaphase into the
cytoplasm, where they are degraded. The function of rDNA amplification is no doubt to
provide a large stock of rRNA to form the

ribosomes that will be needed until the growing
embryo starts to synthesize its own ribosomes,
which does not occur until gastrulation in
Xenopus, for example.

11.6 What else does the nucleolus do?

Although the primary function of the nucleolus
is clearly the synthesis of rRNA and its processing into pre-ribosomes, it has long been
supposed that such a large and conspicuous
organelle would have other functions and activities. Several activities of nucleoli and NORs
during meiosis have been described, although
the proposal that the nucleolus is a site for the
formation of synaptonemal complex components
(John, 1990, pp. 91, 132) remains to be established. Nucleoli may also, by their bulk, inhibit
pairing and synapsis and synaptonemal complex
formation (John, 1990, p. 61). The use of ribosomal genes to ensure correct pairing and segrega-

tion in achiasmate *Drosophila* males has already been mentioned (Section 7.4.4), but this is clearly a highly specialized adaptation of a pre-existing chromosomal structure.

Recently it has been found that the nucleolus has important roles in cell-cycle regulation (Visintin & Amon, 2000). In the budding yeast *Saccharomyces cerevisiae*, the protein phosphatase Cdc14, which regulates exit from mitosis (Section 2.3), is sequestered in the nucleolus for most of the cell cycle and is only released during anaphase and telophase (Bachant & Elledge, 1999; Cockell & Gasser, 1999b). The Cdc14 interacts with various other nucleolar proteins, notably Net1 (or Cfi1), which remains in the nucleolus throughout the cell cycle and forms the 'REgulator of Nucleolar silencing and Telophase exit' or RENT complex. While bound to this complex, Cdc14 phosphatase activity is inhibited, but once it is released from the complex it can dephosphorylate and thereby activate the anaphase-promoting complex (APC) and the cyclin/Cdk1 complex that are required for this stage of cell-cycle progression.

Another protein whose activity is regulated in a similar way is the mammalian tumour-suppressor protein p53. The tumour-suppressor protein p19[Arf] activates p53 by sequestering a p53 inhibitor, Mdm2, in the nucleolus (Weber *et al.*, 1999). In yeast meiosis, the protein Pch2, which

is required for repression of rDNA recombination and is involved in the pachytene checkpoint that monitors proper synaptonemal complex assembly, is also located in the nucleolus (Cockell & Gasser, 1999b). Thus a picture of the nucleolus is emerging in which it is not merely a ribosome factory, but also a very convenient place to keep a variety of proteins involved in cell-cycle checkpoints and cell-cycle progression until the stage when they are needed. The precise reason for segregating such proteins in the nucleolus is not yet clear, and in any case may not always be the same, although it has been suggested that if these proteins can act on substrates in both nucleus and cytoplasm, the nucleolus may be the only place where they do not function (Bachant & Elledge, 1999). It may also be that these proteins have additional functions in the nucleolus (Visintin & Amon, 2000): in fact, it appears that Cdc14 is involved in nucleolar segregation at mitosis in yeasts, and that Pch2 prevents recombination in rDNA. The nucleolus also appears to be a site for processing and modification of various small RNAs, including the signal recognition particle RNA (Olson *et al.*, 2000; Pederson & Politz, 2000). While there is no doubt that the principal function of the nucleolus is synthesis of rRNA and processing it to produce pre-ribosomes, it is becoming clear that it is a multifunctional organelle.

Centromeres, kinetochores and the segregation of chromosomes

12

12.1 What are centromeres and kinetochores?

The centromere is the primary constriction of the chromosome, a region where the sister chromatids are held together until anaphase even after the chromosome arms have separated, and where the chromosome becomes attached to the spindle (Fig. 12.1). Attachment of the chromosome to the spindle is usually through a pair of organelles, the kinetochores, one per sister chromatid on each side of the centromere. In recent years, many DNA sequences and proteins have been identified that are associated with centromeres and kinetochores, and we are beginning to understand how they function in chromosome segregation, which of course is the essential function of the condensation of metaphase chromosomes from the interphase nucleus. Some aspects of the control of chromosome segregation have already been described in Section 2.3.3; here the emphasis will be on the function of individual components of the centromeres and kinetochores in segregation.

12.2 How are centromeres constructed?

At metaphase a centromere typically appears as a constriction in the chromosome that appears undivided while the chromosome arms are split into two chromatids (Fig. 12.1b). Three questions will be addressed in this section: why does the centromere appear as a constriction; when does the centromere divide; and are there any features of DNA that are characteristic of centromeres?

12.2.1 Why is the centromere a constriction?

We have seen in the previous chapter that there are two possible explanations, not mutually exclusive, of why nucleolus organizer regions (NORs) appear as a constriction (Section 11.2): small loop size of the DNA, and delayed condensation. The same explanations can be applied to centromeres. It has been shown that centromeric DNA in lampbrush chromosomes (Section 14.2), in meiotic chromosomes (Moens & Pearlman, 1990) and in mitotic chromosomes (Bickmore & Oghene, 1996) forms compact structures close to the chromosomal axis, even after treatment to disperse the chromatin. Strissel *et al.* (1996) have in fact shown that the human centromeric alpha-satellite DNA forms much smaller loops, with more frequent attachments to the scaffold, than the DNA of the chromosome arms. Thus a centromeric constriction could be produced simply as a result of the way in which the centromeric DNA is organized.

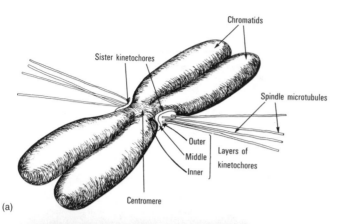

(a)

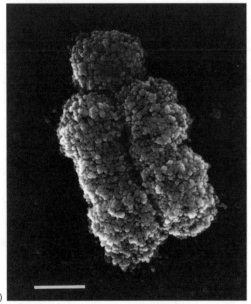

(b)

Figure 12.1 The structure of a chromosome, showing the centromere as a constriction. (a) Drawing of a chromosome, showing that it is divided into separate chromatids except at the centromere (primary constriction). The chromosome is attached to the spindle microtubules through the kinetochores, which are located at the centromere. Reproduced from Bostock & Sumner (1978) *The Eukaryotic Chromosome*. North-Holland. (b) Scanning electron micrograph of a mouse chromosome with the arms divided into chromatids, but the centromeres not split. Scale bar = 1 μm. Reproduced with permission from Sumner (1991) *Chromosoma* **100**, 410–418. © Springer-Verlag.

Nevertheless, it seems unlikely that this is the whole story. Centromeric constrictions are much less obvious on prophase chromosomes, both in mitosis and meiosis. It could be, therefore, that the chromosome arms condense and fatten as the cell proceeds towards metaphase, while the centromere does not. There are various reasons why the centromere might not condense. It could simply be that, for mechanical reasons, it cannot condense until it has divided completely. If condensation is produced by coiling, and an individual centromere became coiled, it would produce an insurmountable obstacle to separating sister centromeres (Sumner, 1991). It might

also need to remain extended to provide sufficient area for the formation of the kinetochores and for maintenance of adequate connections between sister centromeres (Rattner, 1991); in other words, the mechanical strength required of the system would need the centromeres to be extended, with a large surface area, rather than contracted.

12.2.2 When does the centromere divide?

As shown in Fig. 12.1, the centromere appears morphologically undivided, but is this really so?

Have the sister centromeres already separated by metaphase, only to be held together by a few strands of DNA that can easily be broken at the start of anaphase, or does the centromere remain as a single structure that has to be completely unravelled at the beginning of anaphase? Certainly gross morphological appearances suggest the latter, but in fact the weight of evidence seems to show that the DNA of centromeres is largely divided by metaphase, and the chromatids are only held together in very restricted regions. This has been known for many years in large plant and insect chromosomes, and has been demonstrated more recently for mammalian chromosomes (Sumner, 1998c) (Fig. 12.2). In both *Drosophila* (Carmena *et al.*, 1993) and mammalian (Bickmore & Oghene, 1996; Shelby *et al.*, 1996) chromosomes, thin strands of DNA can be

seen connecting sister centromeres at metaphase. In rare cases, whole mount transmission electron micrographs of chromosomes show two distinct, but intimately linked, sister chromatids at the centromeres (Rattner & Lin, 1987). In fact, a set of proteins known as the cohesin complex holds sister chromatids together (Nasmyth, 2001). This cohesion is established at or shortly after DNA replication, and is lost from the chromosome arms in prophase in many organisms (though not in budding yeast, *Saccharomyces cerevisiae*; Tanaka *et al.*, 1999b), so that at metaphase only the centromeric regions are still held together by cohesins (Sections 2.3.3 and 12.4.2). In fission yeast, *Schizosaccharomyces pombe*, centromeric cohesion is the result of cohesion between blocks of heterochromatin flanking the centromeres themselves, and the same may be true of mammals (Section 7.4.4). In *Drosophila* also, there appears to be a distinction between kinetochore function and centromeric cohesion (Lopez *et al.*, 2000).

If the centromere has largely divided by metaphase, when does this division occur? Unfortunately, although the literature on sister-chromatid cohesion is increasing rapidly, little of it refers specifically to centromeres (Biggins & Murray, 1999). In fission and budding yeasts, specific centromere cohesion genes have been identified (see Section 12.4.2), but at least in budding yeast a separate gene that acts during replication is required to establish sister-chromatid cohesion. Whether this includes cohesion of centromeres is not known. Only a small proportion of mammalian metaphase chromosomes show the centromere split into two, suggesting that this may be a relatively late event in metaphase.

12.2.3   What kind of DNA is needed for a centromere?

Centromeric DNAs have been identified in a variety of organisms, and some examples are listed in Table 12.1. It is important to distinguish between DNA sequences that truly belong to the centromere, and those that are merely found in the centromeric or paracentromeric regions but may not have any function related to the

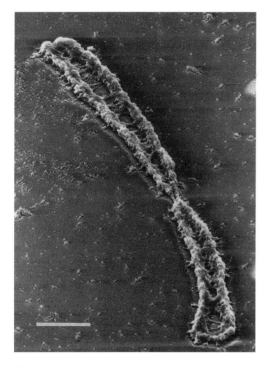

Figure 12.2 Scanning electron micrograph of a Chinese hamster ovary (CHO) chromosome, showing splitting of the centromeres, while the adjacent regions of the chromatids remain united. Scale bar = 2 μm. Reproduced with permission from Sumner (1998c) *Cell Biology International* **22**, 127–130. © Academic Press.

Table 12.1 Some centromeric DNAs.

Species	Name of DNA	Unit length	Total length	Comments	Refs
Saccharomyces cerevisiae	CDEI	8 bp	8 bp		1
	CDEII	78–86 bp	78–86 bp		
	CDEIII	26 bp	26 bp		
Schizosaccharomyces pombe	Central core	4–7 kb }	40–100 kb		2
	Repeated elements	~5 kb			
Arabidopsis thaliana		~180 bp	1000 kb		17
Triticeae (wheat, barley, rye)	CCS1	(?) 260 bp			3
Zea mays (maize)	MCS1a (CentC)	156 bp	10–24 kb	Knob-associated	4
		180 bp	15 000 copies	(neocentromeres)	5
Oryza sativa (rice)	RCS2	168 bp	~6000 copies		6
Vicia faba				No tandem repeats	7
Chironomus pallidivittatus		155 bp	1300 copies	A+T-rich with CENP-B box	8
		375 bp	100 copies	A+T-rich, chromosome 3 only	9
Drosophila melanogaster	AATAT	5 bp	Up to 150 kb	Also retrotransposons and unique sequences	10
	AAGAG	5 bp	Up to 150 kb		
Sus scrofa (pig)	Mc1	14 bp	? 10^5 copies	Metacentric chromosomes	11, 12
	Ac2	340 bp		Acrocentric chromosomes	
Pan troglodytes (chimpanzee)	Alpha-satellite		800–3500 kb per chromosome		13
Homo sapiens (man)	Alpha-satellite	171 bp	200–9000 kb	Contains CENP-B box	14
	Gamma-satellite	220 bp	0.015% of genome	X-specific; chromosome 8; G+C-rich	15
Mus musculus (mouse)	Minor satellite				16

References: 1, Hegemann & Fleig (1993); 2, Baum *et al.* (1994); 3, Aragón-Alcaide *et al.* (1996); 4, Ananiev *et al.* (1998a); 5, Birchler (1997); 6, Dong *et al.* (1998); 7, Fuchs *et al.* (1998); 8, López & Edström (1998); 9, He, H. *et al.* (1998); 10, Sun *et al.* (1997); 11, Jantsch *et al.* (1990); 12, Miller *et al.* (1993); 13, Haaf & Willard (1997); 14, Murphy & Karpen (1998); 15, Lee *et al.* (1995); 16, Wong & Rattner (1988); 17, Round *et al.* (1997).

centromere. As well as the strictly centromeric DNA, most chromosomes have blocks of heterochromatin adjacent to their centromeric regions that contain highly repetitive DNAs (Section 7.3.1) that are generally different from the centromeric sequences.

The species whose centromeric DNA sequence is understood best is the budding yeast *Saccharomyces cerevisiae* (Clarke, 1998). Its centromere consists of three centromeric DNA elements (CDEs) flanked by other sequences resistant to nuclease attack, making a total length of 220–250 bp (Fig. 12.3): CDEI is a conserved sequence, PuTCACPuTG; CDEII is a very A+T-rich sequence of 78–86 bp; and CDEIII is a highly conserved 26 bp sequence. Sequence CDEIII appears to be absolutely essential for centromeric function in both mitosis and meiosis, but changes to CDEI and CDEII, although they greatly reduce the efficiency of chromosome transmission, do not abolish the process completely (Hegemann & Fleig, 1993). The centromeres of other budding yeasts, such as *Kluyveromyces* spp., *Candida* spp. and *Yarrowia lipolytica*, are also very short, non-repeated DNA sequences, sometimes containing sequences corresponding to some or all of CDEI, -II and -III found in *S. cerevisiae*, although centromeres of one species will not work in a different species.

The centromeres of budding yeasts are not typical of the majority of eukaryotes, whose centromeres contain highly repeated DNA sequences (Table 12.1). These highly repeated centromeric sequences often contain retrotransposons, as described in human centromeres (Prades *et al.*, 1996), in maize (Ananiev *et al.*, 1998a), in *Drosophila* (Sun *et al.*, 1997) and in other organisms. However, there does not appear to be any

consensus centromeric sequence that is found in centromeric DNAs throughout eukaryotes. Although, for example, a number of primates have members of the alpha-satellite family at their centromeres, and different species of mice have minor satellite, such similarities are not seen when larger groupings are examined. Evidence that alpha-satellite might be sufficient for centromere formation was provided by experiments in which such sequences are introduced into artificial chromosomes or into abnormal sites on chromosomes of another species and produce a functional centromere (Willard, 1998). On the other hand, there are many examples of blocks of alpha-satellite that do not form centromeres, particularly in Robertsonian fusions in which only one of the two centromeres is active, although both contain alpha-satellite (Sullivan & Schwartz, 1995; Murphy & Karpen, 1998; Wiens & Sorger, 1998). To complicate matters further, mice have been found in which there are large blocks of minor satellite, but only a small specific region of these blocks forms a centromere (Mitchell *et al.*, 1993); in this case, it appears that DNA methylation might block centromeric activity in most of the minor satellite (Mitchell *et al.*, 1996). Similarly, in *Drosophila* the repeated DNAs at the centromere do not differ from the satellite DNAs found in non-centromeric heterochromatin (Wiens & Sorger, 1998). Thus no specific sequence seems to be required to form a centromere, and a centromere can be restricted to a small segment of what are apparently more-or-less identical DNA sequences. There are a number of possible explanations for these anomalies. One is that all eukaryotes do contain a small specific centromere sequence, perhaps something like those found in budding yeasts, but that this has not yet been

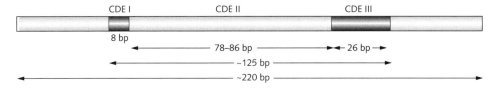

Figure 12.3 The structure of the *S. cerevisiae* centromere. Redrawn with permission from Clarke (1998) *Current Opinion in Genetics and Development* **8**, 212–218. © Elsevier Science.

found. Intensive studies have failed to discover such a sequence, and even in budding yeasts the centromeric sequences differ between species and do not function when transferred to another species, so this explanation now seems extremely unlikely. Another possibility is that sequence might not, in itself, be important, but that the features of a centromere might be produced by some aspect of higher order structure (Koch, 2000). In yeasts, centromeric DNAs show curvature, and there is also evidence for unusual DNA structures in the centromeres of higher eukaryotes (Bechert et al., 1999). However, there is, as yet, no direct evidence that a unique higher order DNA structure is essential to produce a centromere.

Even if there is no overall resemblance between centromeric DNAs in different organisms, it might be possible that there would be a short conserved sequence that is generally found in centromeric DNAs. Such a sequence, the CENP-B box, which binds to the centromeric protein CENP-B (Section 12.4.1), has actually been found. The CENP-B box has the sequence CTTCGTTGGAAACGGGA in human alpha-satellite (Masumoto et al., 1989). Similar CENP-B box sequences have since been found in centromeric DNAs from primates, mouse species (minor satellite) and Indian muntjac (Sunkel & Coelho, 1995), in the Dipteran fly *Chironomus pallidivittatus* (López & Edström, 1998) and in a number of plant species (Birchler, 1997). Nevertheless, many organisms have centromeric DNAs that lack CENP-B boxes (Goldberg et al., 1996; Kipling & Warburton, 1997); these include the alpha-satellites of African green monkey and the human Y chromosome.

In spite of this uncertainty about what features of DNA might be required for centromeric function, it might seem that some sort of highly repetitive DNA is essential to produce a centromere, even though not all regions of repetitive DNA produce centromeres. It was therefore a surprise to discover that centromeres can form in regions without any significant amount of repetitive DNA, and indeed with no obvious distinguishing characteristics at all. In humans, a number of stable marker chromosomes have been reported in which centromeric activity does not

involve alpha-satellite sequences (Choo, 1997), and detailed analysis of one such 'neocentromere' shows that it contains a wide variety of sequences, and consists largely of 'ordinary' DNA (Barry et al., 1999). Similarly, *Drosophila* mini-chromosomes have been generated with neo-centromeres that contain no recognizable centromeric DNA sequences (Williams et al., 1998). It should perhaps not have been so unexpected that centromeres can form without repetitive DNA, because the normal centromeres of the bean *Vicia faba* appear to lack any significant amount of repetitive DNA (Fuchs et al., 1998). A consequence of such observations as these is that centromere formation is now regarded as an epigenetic phenomenon (Choo, 2000), although it must nevertheless favour sites containing high concentrations of repetitive centromeric DNAs, otherwise centromeres would tend to form at random anywhere on the chromosome. In some species, indeed, centromere position is quite plastic, and can result from activation of latent or neocentromeres rather than from chromosomal rearrangements (e.g. Montefalcone et al., 1999).

Methylation has already been mentioned as a mechanism for restricting centromere action to a limited region of otherwise identical repetitive DNA, but the problem with neocentromeres is the opposite one of specifically marking a region that is to show centromere activity. One possible marker might be late replication (Csink & Henikoff, 1998); because regions containing repeated DNA tend to replicate late, they would be favoured as sites of centromeres. *Schizosaccharomyces pombe* centromeres are also underacetylated, and this is necessary for their functioning, as hyperacetylation causes chromosome loss at mitosis and disrupts the localization of centromeric proteins (Ekwall et al., 1997). However, centromeres are not necessarily the latest-replicating regions of the genome, and are not the only regions to be underacetylated. Thus although factors such as methylation, replication time and histone deacetylation could be important factors in determining that a specific region of DNA is to act as a centromere, they are signals that are generally used to modulate the activity state of chromatin (see Sections 3.5 and 4.2.4), and thus the question of what determines that a region should be a centromere, rather than what might maintain it as a centromere, remains unsolved.

12.3 How are kinetochores made?

In general, centromeres are connected to the spindle microtubules through a distinct structure known as the kinetochore, although in some organisms the spindle microtubules appear to be inserted directly into the chromatin without any specific structure. Direct insertion has been reported in chromosomes of various protozoa (Bostock & Sumner, 1978), and in yeasts. There are two types of kinetochore structure: the ball-and-cup type, which is found in higher plants and some insects (Orthoptera); and the trilaminar type, which is found in lower plants and most animals. Ball-and-cup kinetochores appear as an irregular mass (the ball), about 0.8 μm in diameter, sitting in a depression (the cup) at the centromere. The ball has lower electron density than the adjacent chromatin, but otherwise lacks distinctive features. The spindle microtubules appear to be attached on all the free surfaces of the ball. The composition of ball-and-cup kinetochores has not been studied, and therefore they will not be discussed further here.

Trilaminar kinetochores consist of a dense layer on the surface of the centromeric chromatin, an electron-lucent layer and an outer dense plate, beyond which is another electron-lucent layer known as the corona (Figs 12.4 & 12.5). The microtubules are attached mainly to the outer plate, although a few are reported to pass through it and attach to the inner kinetochore plate. About 1–120 microtubules may be attached to a kinetochore, the number of microtubules being related, approximately, to the size of the genome (Bloom, 1993; Table 12.2). The mature trilaminar kinetochore appears at late prophase or metaphase (Ris & Witt, 1981), and during prophase the kinetochore appears as an amorphous mass. Nevertheless, some components of the kinetochore are present throughout the cell cycle, regardless of its

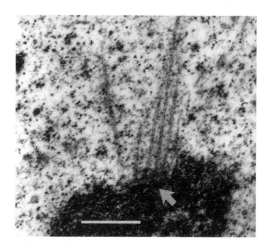

Figure 12.4 Transmission electron micrograph of trilaminar kinetochore (arrow) on a CHO chromosome, attached to the spindle microtubules. Scale bar = 0.5 μm. Reproduced with permission from Sumner (1998b) *Advances in Genome Biology* **5A**, 211–261. © JAI Press.

structural arrangement, as certain kinetochore proteins can be detected immunocytochemically even in interphase nuclei.

Trilaminar kinetochores consist essentially of protein (Section 12.4), although ribonucleoprotein components have also been claimed. No details of the latter are available. Claims have also been made that DNA is an important component of kinetochores. There can be no doubt that DNA is intimately associated with the inner kinetochore plate, because the inner plate is in close contact with the centromeric chromatin, if not actually part of it. Fibres 30 nm long, resembling chromatin fibres, have been described as components of the outer kinetochore plate, which has been reported to contain DNA, but the most recent studies do not support the view that there is any significant quantity of DNA in the outer plate (Cooke *et al.*, 1993).

12.4 Proteins of the centromere and kinetochore

Numerous proteins have now been found in the centromeric regions of chromosomes, and several have been localized to specific centromeric or

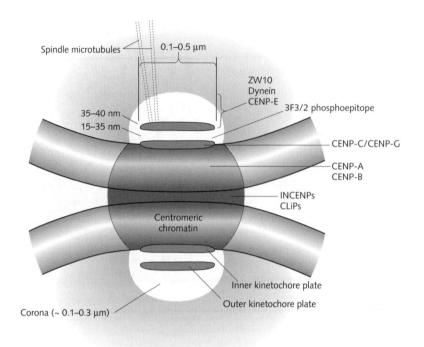

Figure 12.5 Diagram of a trilaminar kinetochore, showing the location of proteins.

Table 12.2 Numbers of microtubules attached to kinetochores in different organisms.

Species	C-Value (DNA bp)	Microtubules/ chromosome	DNA (bp)/microtubule
Chlamydomonas reinhardtii	1.09×10^8	1	5.7×10^6
Saccharomyces cerevisiae	1.4×10^7	1	0.87×10^6
Kluyveromyces lactis	1.4×10^7	1	2.3×10^6
Schizosaccharomyces pombe	1.4×10^7	2–4	1.5×10^6
Drosophila melanogaster	1.65×10^8	6–21	4.1×10^6
Locusta migratoria	6.5×10^9	18–23	2.8×10^7
Homo sapiens	3.9×10^9	20–30	6.7×10^6
Haemanthus katharinae	1.06×10^{11}	120	4.9×10^7

Data from Bloom (1993).

kinetochore substructures, analysed biochemically and their functions determined, at least in part (Rieder & Salmon, 1998; Dobie et al., 1999).

12.4.1 Mammalian centromeric proteins (Table 12.3)

Several proteins at mammalian centromeres are known as CENPs, which simply means CEN-

tromere Proteins; they do not have structural or functional features in common, but most have been identified using sera from patients with autoimmune diseases. CENP-A is a centromere-specific variant of histone H3 (Section 4.2) that is targeted to centromeres by its histone-fold domain (Shelby et al., 1997). It is present at active normal centromeres and at neocentromeres, but is absent from inactive centromeres (Willard, 1998).

Table 12.3 Proteins of mammalian centromeres and kinetochores.

Protein	Size	Location	Function
CENP-A	17 kDa	Inner kinetochore plate (Vafa & Sullivan, 1997)	Histone H3 variant
CENP-B	80 kDa	Centromere	Binds DNA
CENP-C	140 kDa	Inner kinetochore plate	Functional centromeres only
CENP-D	47 kDa		Equivalent to RCC1
CENP-E	312 kDa	Corona and outer kinetochore plate	Kinesin-like motor protein
CENP-F	367 kDa		Kinetochore assembly
CENP-G	95 kDa	Inner kinetochore plate	
CENP-H		Inner kinetochore plate	Binds CENP-C to kinetochore
BUB1		Outer kinetochore plate	Kinase complex with CENP-E; kinetochore-attachment checkpoint
BUBR1			
BUB3			Checkpoint control
Dynein			
INCENP	135 kDa	Between sister centromeres	Cytokinesis
INCENP	?150 kDa	Between sister centromeres	Cytokinesis
CLiPs		Between sister centromeres	
MCAK (mitotic centromere-associated kinesin)			Kinesin-related; spindle formation and maintenance
Arp1			Microtubule capture
p150^{Glued}			Microtubule capture
CLIP 170			Microtubule capture
Dynein		Corona/outer kinetochore plate	? Attachment to spindle
Dynactin			
Erk1			Metaphase–anaphase transition
3F3/2 phosphoepitope		Interzone	Control of metaphase–anaphase transition
ZW10		Corona/outer kinetochore plate	Metaphase–anaphase checkpoint
Mad			Kinetochore-attachment checkpoint
Topo II	170/180 kDa	Centromere	Decatenation of DNA
Poly (ADP-ribose) polymerase			Earle et al. (2000)
SUV39H1			Chromatin organization at centromeres (Aagaard et al., 2000); histone H3 methylase
Nuf2p			Wigge & Kilmartin (2001)
HEC			Human homologue of Ndc80p (Wigge & Kilmartin, 2001)

For references, see text, and Saffery et al. (2000).

CENP-A null mice die *in utero*, and show numerous problems in mitosis (Howman *et al.*, 2000). CENP-A appears to be essential for organizing centromeric chromatin.

CENP-B is a characteristic protein of many mammalian centromeres (Kipling & Warburton, 1997), and is bound to various centromeric DNAs through their CENP-B box sequences (Section 12.2.3). It can occur at inactive centromeres and is not present at all active centromeres, for example those of the human and mouse Y chromosomes, and therefore is not essential for centromere function; instead, it has been suggested that, because of its similarities to transposases, it might promote nicking and recombination of DNA and thereby promote homogenization of alpha-satellite (Kipling & Warburton, 1997). Both mitosis and meiosis proceed normally in CENP-B null mice, although such mice have lower body weight and reduced sperm production (Hudson *et al.*, 1998).

CENP-C is a component of the inner kinetochore plate and is an essential component of active centromeres, although it is not sufficient to form a centromere (Sullivan & Schwartz, 1995; Fukagawa *et al.*, 1999). Disruption of the CENP-C gene results in the chromosomes failing to congress properly on the metaphase plate, and the cell arresting at the metaphase–anaphase transition (Kalitsis *et al.*, 1998; Fukagawa *et al.*, 1999).

CENP-D is a facultative or passenger protein, and not a permanent component of the centromere. It appears to be the same as RCC1 (regulator of chromosome condensation), which is a regulator of mitosis, but it is not known if it has any function at the centromere. On the other hand, CENP-E is an essential kinetochore protein that is needed for the metaphase–anaphase transition. It is located in the outer kinetochore plate and the corona, and is restricted to active centromeres (Sullivan & Schwartz, 1995; Cooke *et al.*, 1997). It does not appear at the kinetochore until prometaphase, remains associated with the kinetochore throughout most of mitosis and meiosis and is transferred to the mid-body at telophase. It is a kinesin-like motor protein that is required for the congression of chromosomes on to the metaphase plate (Wood *et al.*, 1997; Schaar *et al.*, 1997). MCAK (mitotic centromere-associated kinesin) and dynein are other kinetochore motor proteins, the latter, like CENP-E, being located in the outer plate and corona. CENP-F is needed for the assembly of hBUBR1 on to kinetochores, which in turn is required for the binding of CENP-E (Chan *et al.*, 1998). CENP-E requires the presence of hBUB1 before it can assemble on to the kinetochore (Jablonski *et al.*, 1998). CENP-E, hBUB1 and hBUBR1 give stronger

signals on unaligned chromosomes than on chromosomes that have aligned themselves on the metaphase plate, but are lost from the chromosomes by telophase. CLIP-170 (not one of the chromatid linking proteins) is found in unattached but not attached kinetochores. Proteins such as hBUB1, hBUBR1, hMAD, zw10 and others are involved in the metaphase–anaphase checkpoint (Section 2.3.2).

CENP-G is a DNA-binding protein that, like CENP-B, binds to alpha-satellite, more specifically to the α-1 subfraction that is rich in CENP-B boxes (He, D. *et al.*, 1998). Nevertheless, its binding sites are distinct from those of CENP-B, as is its ultrastructural location, in the inner kinetochore plate. Moreover, CENP-G is found at the centromere of the human Y chromosome, which does not bind CENP-B. CENP-H is another protein of the inner kinetochore plate, and is required for the localization of CENP-C to the kinetochore (Fukagawa *et al.*, 2001).

Another group of centromeric proteins are those located, at least in part, between the sister centromeres – the INCENPs (inner centromere proteins) and CLiPs (chromatid linking proteins), although these proteins also occur between the chromosome arms. The INCENPs appear to be required for the formation of the cleavage furrow, and have been regarded merely as passenger proteins, with no actual chromosomal function. Disruption of INCENP protein results in defective chromosome segregation (Cutts *et al.*, 1999), but there is no clear evidence that it has a function in holding sister centromeres together. The CLiPs have only been implicated in sister-centromere cohesion by their location (Rattner *et al.*, 1988). The cohesins, a subset of the SMC (structural maintenance of chromosomes) proteins, are known to be important for sister-chromatid cohesion, but a specific role for cohesins at the centromere has not yet been established (Biggins & Murray, 1999).

As described in Section 2.3, the separation of sister chromatids into daughter chromosomes at the beginning of anaphase requires two distinct functions: separation of DNA, and destruction of proteins that hold the chromatids together.

Although much has been learnt about the biochemistry of the anaphase-promoting complex (APC; Page & Hieter, 1999), it has not yet been localized on chromosomes, nor has the substrate for proteolysis been identified. The situation with separation of DNA is much clearer, although a lot of detail still needs to be worked out. Replicated DNA molecules remain intertwined (catenated) until acted upon by topoisomerase II (Topo II), which can cut one DNA molecule, pass another DNA molecule through the gap and then reseal the gap. Inhibition of Topo II in yeasts, *Drosophila* and mammals prevents or slows down the metaphase–anaphase transition (Section 2.3.1), and Topo II is found throughout the centromere at metaphase, but is lost at anaphase (Sumner, 1996) (Fig. 12.6). It is therefore present at the same site as the centromeric DNA until the sister chromatids have separated, after which it is lost (or inactivated).

12.4.2 Centromeric proteins in non-mammals

Unlike most other eukaryotes, the centromere of the budding yeast *S. cerevisiae* consists of specific, non-repeated DNA sequences (Section 12.2.3), yet has a number of proteins similar to certain mammalian centromeric proteins (Table 12.4; Fig. 12.7). Thus Cse4p is a histone variant similar to CENP-A, and Mif2p is similar to CENP-C (Dobie *et al.*, 1999), although other yeast centromeric structural proteins do not have obvious mammalian homologues. Yeast centromere proteins involved in sister-chromatid cohesion, the metaphase–anaphase checkpoint and chromosome segregation have also been identified; checkpoint proteins such as the BUBs and MADs were in fact first identified in yeasts, and only later were they identified in mammals. However, many yeast centromeric proteins have no clear structural homologues in mammals, although the same range of functions has been identified.

Centromeric proteins in fission yeast, *S. pombe* (Partridge *et al.*, 2000; Pidoux & Allshire, 2000), include two proteins known as Mis6 and Mis12 that are bound to the central region of the *S. pombe* centromere. The binding to the outer flanking domains of two chromodomain proteins, Swi6 (≡ HP1) and Chp1, is dependent on the proteins Rik1 and Clr4 (≡ Suvar39) (Section 7.3.2; Fig. 12.8). Two CENP-like proteins, Cnp1 (≡ CENP-A) and Cnp3 (≡ CENP-C), are also present but their precise location is unknown. Hypoacetylation of centromeric histones is required for correct centromeric functioning (Section 12.2.3). The protein Nuf2, which is conserved from yeast to humans, is involved in connecting the centromeres to the spindle microtubules (Nabetani *et al.*, 2001).

 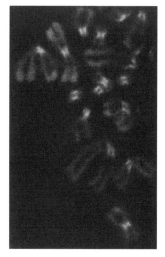

Figure 12.6 Immunofluorescence of topoisomerase II at the centromeres of CHO cells. (Left) Ethidium fluorescence to show total chromosomal DNA. (Right) Topoisomerase II immunofluorescence of the same chromosomes, localized as a line along the centre of each chromatid with a concentration at every centromere.

Table **12.4** *Saccharomyces cerevisiae* centromeric proteins.

Function	Name	Mammalian equivalent
Structural	Bir1	
	Cbf1, 3	
	Cep3 (p64)	
	Cse4	CENP-A
	Ctf13, 19	
	Mcm21	
	Mif2	CENP-C
	Mtw1	
	Ndc10 (p110)	
	Okp1	
	Skp1 (p23)	
	Slk19	
Sister chromatid cohesion	Scc1 (Mcd1 p)	
	SMCs	SMCs
Microtubule capture	Cbf5	
Metaphase–anaphase checkpoint	BUB1–3	hBUB1, hBUBR1
	MAD1–3	hMAD1–2
	Cdc20, 27	hCDC20
Segregation	Pds1	
	Esp1	
	Ase1	
	Clb2	
	Ndc80 p complex	HEC

Data from Clarke (1998), Pidoux & Allshire (2000) and Wigge & Kilmartin (2001).

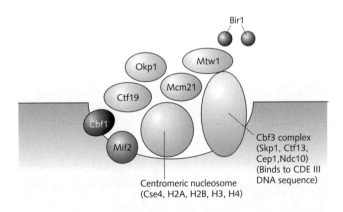

Figure 12.7 Structure and composition of the centromeric region of chromosomes from budding yeast, *Saccharomyces cerevisiae.*

The checkpoint proteins BUB and MAD and dynein and dynactin have been found in *Drosophila* as well as in mammals. *Drosophila* has a CENP-A homologue, called CID, which is required for normal kinetochore formation and function and for cell-cycle progression (Blower & Karpen, 2001). Protein CENP-meta, the *Drosophila* equivalent of CENP-E, remains attached to the kinetochore throughout the cell cycle (Yucel *et al.*, 2000). Loss of CENP-meta activity is lethal. In maize, CENP-C homologues have been found on standard kinetochores but

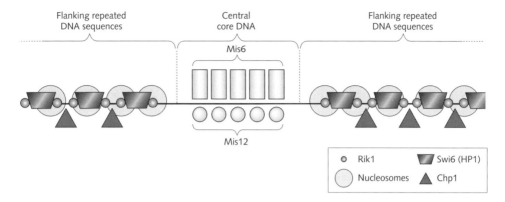

Flanking repeated DNA sequences   Central core DNA   Flanking repeated DNA sequences

Mis6

Mis12

| ⊙ Rik1 | ▨ Swi6 (HP1) |
| Nucleosomes | ▲ Chp1 |

Figure 12.8 Structure and composition of the centromeric region of chromosomes from fission yeast, *Schizosaccharomyces pombe*.

not at neocentromeres (Dawe *et al.*, 1999). Proteins similar to mammalian CENP-C, -E and -F, and to yeast centromeric proteins SKP1, CBF1 and CBF5, have been localized to barley (*Hordeum vulgare*) and bean (*Vicia faba*) chromosomes (ten Hoopen *et al.*, 2000).

So far, the functions and precise locations of many centromeric proteins are not established, nevertheless it is clear that sets of proteins responsible for both structural and functional aspects of centromeres have been identified. There may well be other such proteins. For example, certain heterochromatin proteins such as HP1 in *Drosophila* and its homologues in mammals (Section 7.3.2) may function in establishing centromeric structure; it is not always easy to distinguish between functions in heterochromatin and at centromeres, as centromeres are so often embedded in blocks of heterochromatin and are themselves heterochromatic.

12.5 Holocentric chromosomes

Most eukaryotes – or at least those that are familiar objects of cytogenetic study – have discrete kinetochores forming a distinct constriction on the chromosome, as described so far in this chapter. The localization of the kinetochores means that an individual kinetochore will only face one pole of the cell, and therefore only become attached to microtubules emanating

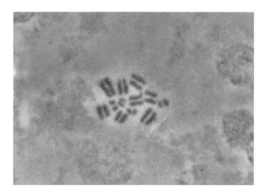

Figure 12.9 Holocentric chromosomes from a late metaphase cell of the aphid *Myzus persicae*. Note that there is no centromeric (or other) constriction, and that the two chromatids simply lie parallel to each other. Micrograph kindly provided by R.L. Blackman.

from one pole. Dicentric chromosomes (with two active centromeres) can, if the centromeres are sufficiently far apart, twist between the centromeres, so that each of the centromeres on the same chromatid can become attached to opposite poles of the cell, resulting in chromosome breakage or failure to segregate at anaphase (see Box 3.2). It is therefore surprising that a substantial number of plants and animals have holocentric (or holokinetic) chromosomes, in which spindle microtubules are attached throughout all or most of the length of the chromosome. Such chromosomes have no constriction, nor any localized region where the chromatids appear to be joined to each other; instead, they appear simply as a pair of rods lying side-by-side (Fig. 12.9), much like an acentric fragment (see Fig. 3.8c) from an organism with a localized centromere. It is not clear how holocentric chromosomes avoid the problems that occur with dicentric chromosomes, but it is clear that they do so, as they are found in many groups of organisms. These include various monocotyledonous plants, some protozoa, nematodes, some insects (Hemiptera and Homoptera) and at least one spider (Table 12.5). The term holocentric actually covers a variety of structures. Some organisms actually have polycentric chromosomes, with multiple discrete kinetochores along the length of the chromosome. In other species there is a single elongated kinetochore occupying all or most of the chromosome. Kinetochores may be of the ball-and-cup type, or trilaminar.

Although the kinetochores of the nematode *Caenorhabditis elegans* are holocentric, they have a trilaminar structure (Albertson & Thomson, 1982) and contain kinetochore proteins homologous with those of other organisms (Pidoux & Allshire, 2000). Both HCP-1 and -2 are homologues of CENP-F, and the former is located in a parallel line on the outer face of each chromatid; HCP-3 is a homologue of the histone H3 variant, CENP-A, and has a distribution similar to that of HCP-1; HCP-4 is equivalent to CENP-C and, like HCP-3, is needed to localize HCP-1 to the kinetochores (Moore & Roth, 2001). The HIM-10 protein (Howe *et al.*, 2001) is related to the Nuf2 kinetochore proteins that are conserved from yeasts to man (Wigge & Kilmartin, 2001). The PUMA1 protein of another nematode, *Parascaris univalens*, is also associated with the continuous kinetochore at mitosis (Pidoux & Allshire, 2000). Thus holocentric chromosomes appear to use much the same types of proteins to construct their kinetochores.

Because of crossing-over, use of elongated or multiple kinetochores would be disastrous at meiosis. Segments of the same (original) chromatid on either side of a chiasma, which should move to opposite poles, would both be pulled towards the same pole. To avoid this problem, holocentric chromosomes usually have only localized kinetochore activity at meiosis. Consistent with this, the *P. univalens* kinetochore protein PUMA1 is localized to the discrete spindle attachment sites in meiosis, rather than along the whole length of the chromosomes as in mitosis (Pidoux & Allshire, 2000). In several cases there is no evidence for a kinetochore plate at meiosis, even if one is present on mitotic chromosomes; instead, microtubules appear to be inserted directly into the body of the chromosome. In nematodes of the genus *Parascaris*, there is also a change in the extent of the kinetochores in somatic chromosomes that are subject to chromosome diminution, so that segments of chromatin that are to be eliminated lack any kinetochore (Pimpinelli & Goday, 1989).

Holocentric chromosomes are not only an interesting system in their own right, but could also furnish valuable information about kinetochore structure and function, and chromosome segregation. Changes in the extent of the kinetochores between mitosis and meiosis, or in chromatin diminution, should provide clues about the spatial regulation of kinetochore formation. Lack of a specific centromeric constriction should throw light on what holds sister chromatids together, and how they are separated at anaphase.

12.6 Kinetochores are essential for the functioning of chromosomes

Chromosomes are condensed into discrete, clearly visible bodies at mitosis and meiosis so that they can be distributed properly to daughter cells. Kinetochores are the chromosomal structures that ensure this distribution. Although they vary morphologically, they all function as sites of attachment of chromosomes to the spindle microtubules. It is not surprising, therefore, that kinetochore proteins are largely conserved from yeasts to mammals, but it was unexpected that there is no universal conserved centromeric DNA sequence. In those organisms with holocentric chromosomes, there is presumably no specific DNA sequence associated with

Table 12.5 Organisms with holocentric/holokinetic chromosomes.

Group	Species	Mitotic centromere structure	Meiotic centromere structure
Plants			
Monocotyledons	Cyperus	Multiple ball-and-cup	
	Luzula spp.	Multiple ball-and-cup	Ball-and-cup
Animals			
Protozoa	Nyctotherus ovalis (ciliate)	Trilaminar	
Nematoda	Ascaris lumbricoides	Kinetochore plate	No kinetochore structure
	Caenorhabditis elegans	Trilaminar	
	Meloidogyne hapla	Electron-dense plate	
	Parascaris equorum	Ladder-like kinetochore plate	
	Parascaris univalens	Ladder-like kinetochore plate	No kinetochore plate; microtubules attached to restricted region
Insecta			
Hemiptera	Euchistus spp.		Kinetochore activity restricted to limited region of chromosomes
	Oncopeltus fasciatus	Trilaminar, covering 75% of chromosome	No kinetochore structure, direct insertion of microtubules
	Philaenus		No kinetochore structure, direct insertion of microtubules
	Rhodnius prolixus	Trilaminar, covering 75% of chromosome	No kinetochore structure, direct insertion of microtubules
	Dysdercus	Multiple, trilaminar	No kinetochore structure, direct insertion of microtubules
Heteroptera	Graphosoma italicum	Trilaminar	No kinetochore plates
	Myrmus miriformis		Kinetochore activity restricted to telomeres
Arachnida	Tityus babiensis	Dense kinetochore plate	Kinetochore plates present

Data from Bostock & Sumner (1978) and Pimpinelli & Goday (1989).

the kinetochores at all, because the latter extend throughout the length of the chromosome. In organisms with localized kinetochores, centromeres appear to be able to form almost anywhere, although specific DNA sequences do generally seem to occur in centromeres. Kineto-chore localization may therefore be to some extent an epigenetic phenomenon. Although this might seem too uncertain a mechanism to ensure regular segregation of chromosomes, the survival and success of animals, plants and other eukaryotes show that it must be an effective strategy.

Telomeres

13.1 What is a telomere?

Whereas the chromosomes of prokaryotes are circular, those of eukaryotes are linear, and their ends, the telomeres, have special properties. Many years ago it was recognized that established ends of chromosomes behave differently from newly formed ends produced by chromosome breakage after treatment with radiation or clastogenic chemicals (Section 3.6). Newly formed ends are 'sticky', and readily join to any other such ends in the cell; established chromosome ends do not stick to each other or to newly broken ends. There must, therefore, be some special protective structure at the ends of chromosomes. When the mechanism of DNA replication was worked out, it became clear that with the standard mechanism (Section 3.4) it would be impossible to replicate right to the end of the molecule on both strands. Because DNA molecules, and therefore chromosomes, could not be allowed to shorten indefinitely, there must be some special mechanism for replicating telomeres. Such a mechanism was found, and it turned out to have important implications for senescence and immortalization of cells, leading to possible mechanisms for controlling cancers. In at least some organisms telomeres are involved in the interphase arrangement of chromosomes (Section 5.2) and in the pairing of meiotic chromosomes (Section 2.5.1) through their interactions with the nuclear envelope. Finally, in yeast and several other organisms, telomeres, or at least the telomeric regions, behave as a form of heterochromatin (Section 7.4.5), inducing position effect variegation and gene silencing in adjacent regions of the chromosome, even when there is no cytologically visible heterochromatin.

13.2 Telomeric DNA

Telomeres contain specific DNA sequences that are conserved throughout a vast range of organisms, although there are some exceptions. In many eukaryotes (e.g. vertebrates, some slime moulds, some protozoa) the telomeres consist of numerous copies of the hexanucleotide TTAGGG, but a variety of other short repeated sequences have been found in other species (Table 13.1). The sequences given refer to one strand, the G-strand, and are given in the 5′→3′ direction. This strand normally forms a 3′ single-stranded tail, which varies in length between and within different organisms and chromosomes. These telomeric sequences are not necessarily confined to telomeres, but can also be found in non-telomeric blocks of heterochromatin (Meyne et al., 1990): much of the heterochromatin of Chinese hamster chromosomes consists of the TTAGGG sequence (Bertoni et al., 1996). Sometimes blocks of TTAGGG repeats are present at sites of chromosome fusions, as in the human chromosome 2, which has been formed by the fusion of two chromosomes that are still separate in other apes.

Although most eukaryotes have short tandemly repeated telomeric sequences of the

Table 13.1 Telomeric DNA sequences.

Species	Sequence	Length	Overhang	Refs
Protozoa				
Euplotes	TTTTGGGG	28 bp	14 bases	
Oxytricha	TTTTGGGG	36 bp	16 bases	
Paramecium	TTGGG(G/T)			
Stylonychia	TTTTGGGG			
Tetrahymena	TTGGGG			
Crithidia	TTAGGG			
Plasmodium	TT(C/T)AGGG			
Trypanosoma	TTAGGG			
Yeasts				
S. cerevisiae	TG_{1-3}	~300 bp	>30 bp	
S. pombe	GGTTACA			
Slime moulds				
Dictyostelium	$G_{1-8}A$			
Didymium	TTAGGG			
Physarum	TTAGGG			
Algae				
Chlamydomonas	TTTTAGGG	4–9 kb		
Plants				
Aloe	? rDNA			1
Arabidopsis	TTTAGGG			
Alliaceae	375 bp satellite			2
	18S + 25S rDNA			
Nematodes				
Caenorhabditis	TTAGGC	4–9 kb		
Insects (many species)	TTAGG			3, 4
Diptera				
Anopheles	820 bp satellite			5
Chironomus	Long complex repeats,	200 kb		6
	176, 340 and 350 bp			
Drosophila	Retrotransposons HeT-A and TART			7, 8
D. virilis	370 bp satellite			9
Crustacea				
Gammarus	TTAGG			4
Vertebrates	TTAGGG			10
Xenopus	TTAGGG	10–50 kb		11
Mouse	TTAGGG	10–60 kb		
Human	TTAGGG	5–15 kb	45–275 bases	

Data from Blackburn (1991b), except where shown. See also text, and
resolution.colorado.edu/~nakamut/telomere/telomere.html for further information.

References: 1, Adams, S.P. *et al.* (2000); 2, Pich *et al.* (1996); 3, Okazaki *et al.* (1993); 4, Sahara *et al.* (1999); 5, Biessmann *et al.* (1996); 6, Kamnert *et al.* (1997); 7, Mason & Biessmann (1995); 8, Pardue *et al.* (1996); 9, Biessmann *et al.* (2000); 10, Meyne *et al.* (1989); 11, Bassham *et al.* (1998).

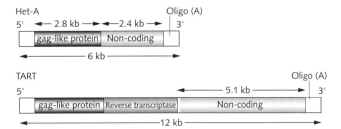

Figure 13.1 The structure of the *Drosophila melanogaster* telomeric retrotransposons HeT-A and TART.

type just described, there are at least two other distinct classes of telomeric sequence. In Chironomid flies (Diptera) telomeres consist of complex tandemly repeated sequences, 176–350 bp long according to species (Kamnert *et al.*, 1997). In *Chironomus pallidivittatus*, the repeats mainly belong to four subfamilies, of which only one forms the actual end of the DNA molecule. Individual repeat subfamilies are very effectively homogenized, probably by gene conversion. Like short tandem telomeric repeats, the telomeric repeat sequences in *Chironomus* spp. have a G-rich and a G-poor strand. Flies of the *Drosophila virilis* group, unlike *D. melanogaster* (below), use a satellite with a 370 bp repeat as their telomeres (Biessmann *et al.*, 2000). Members of the Alliaceae (onions and related plants) also lack short repeated telomeric sequences, and instead appear to use either a satellite DNA with a 375 bp repeating unit, or 18 + 25 S ribosomal DNA (Pich *et al.*, 1996).

The other exceptional type of telomeric DNA is found in *D. melanogaster*, in which the telomeres are formed by two retrotransposons, HeT-A and TART (Mason & Biessmann, 1995; Pardue *et al.*, 1996) (Fig. 13.1). Both retrotransposons have a 5′ segment containing an open reading frame (ORF) that codes for a gag-like protein, and a 3′ segment that is non-coding. There is no homology between the non-coding regions of HeT-A and TART. Retrotransposon TART also contains an ORF for a reverse transcriptase, which is lacking in HeT-A. Both have oligo (A) tails through which the retrotransposons attach themselves to pre-existing chromosome ends in a non-sequence-specific way.

As well as the specific telomeric sequences, there are usually characteristic sub-telomeric sequences, which are commonly repetitive (Pryde *et al.*, 1997). In *D. melanogaster*, for example, these comprise minisatellites with repeat length varying from 0.5 to 1.8 kb, and a proximal region with low copy-number sequences (Mason & Biessmann, 1995). Yeast (*Saccharomyces cerevisiae*) subterminal repeats are of two types: X, which varies from 0.3 to 3.75 kb in length, and Y′, which is either 5.2 kb or 6.7 kb long (Biessmann & Mason, 1992). Subterminal satellite DNAs have been found in a wide variety of species; they are often highly polymorphic in length, but no clear functions have been ascribed to them.

13.3 How do telomeres maintain chromosome length?

As already mentioned, 'conventional' DNA replication processes cannot replicate the very end of the lagging strand of a DNA molecule, and therefore other methods are required to ensure that the ends of chromosomes do not shorten indefinitely. Three methods have been identified: DNA synthesis using telomerase; recombination; and retrotransposition.

All organisms in which the telomeres consist of short highly repeated sequences appear to replicate them using telomerase (Fig. 13.2). Telomerases consist of a reverse transcriptase and an RNA template complementary to the sequence of the G-rich telomeric strand, and use this template to synthesize new telomeric DNA on the end of the existing molecule (Lingner & Cech, 1998; Collins, 1999, 2000; Pardue & DeBaryshe, 1999). Telomerases do not seem to require a specific DNA sequence from which to

Normal DNA replication

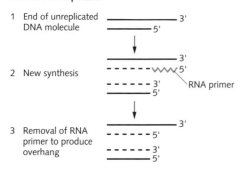

1 End of unreplicated DNA molecule

2 New synthesis

RNA primer

3 Removal of RNA primer to produce overhang

Synthesis of telomeric sequence

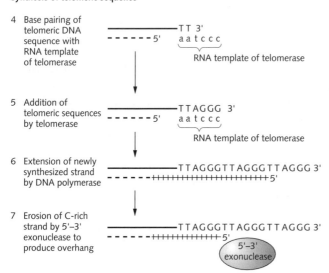

4 Base pairing of telomeric DNA sequence with RNA template of telomerase

RNA template of telomerase

5 Addition of telomeric sequences by telomerase

RNA template of telomerase

6 Extension of newly synthesized strand by DNA polymerase

7 Erosion of C-rich strand by 5'–3' exonuclease to produce overhang

5'–3' exonuclease

Figure 13.2 The replication of telomeres. Stages (1)–(3): replication of DNA using normal mechanisms results in new DNA strands that are shorter than the template. Stages (4)–(5): telomere replication using telomerases. Stages (6)–(7): the shorter, newly synthesized C-rich strand is extended and then shortened to leave an overhang.

start synthesizing new telomeric repeats, but the sequence must nevertheless be G-rich. Telomerase cannot bind to a blunt-ended double-stranded DNA molecule, but needs a single-stranded overhang of at least 4–6 nucleotides. A 3' overhang may be generated after synthesis of the leading strand by a 5'–3' nuclease (Lingner & Cech, 1998). As telomerase extends the G-rich strand, the complementary C-rich strand is synthesized using DNA primase and DNA polymerases α and δ (Diede & Gottschling, 1999). Surprisingly, because telomeres are not recognized by the cell as double-strand DNA breaks, several proteins involved in non-homologous end-joining also appear to be required for telomere maintenance (Gasser, 2000).

Telomerase can synthesize new telomeric sequences on broken chromosome ends as well as on existing telomeres; the former process has been demonstrated in yeasts, various protozoa and in humans, but is not necessarily as efficient as synthesis on pre-existing telomeres. There are also some developmental situations in which new telomeres are added to non-telomeric ends of DNA molecules. In the development of the micronucleus in ciliated protozoa, chromosome-sized DNA from the micronucleus is chopped into much smaller pieces, and new telomeres are synthesized on the ends of these DNA fragments (Blackburn, 1991a; Pardue & DeBaryshe, 1999) (Section 15.3). Similarly, in nematodes, the developmentally programmed process of chromatin diminution involves breaking up the chromosomes into a larger number of smaller chromosomes, each of which has new telomeric sequences added to its ends (Zetka & Müller, 1996).

Telomere length is characteristic of a species, even if somewhat variable (Table 13.1), so synthesis must be well regulated (Greider, 1996; Zakian, 1996). A number of proteins have been described that affect telomere length (Table 13.2; McEachern *et al.*, 2000; Shore, 2001), although the mechanisms by which they work are not yet clear. Some proteins, such as vertebrate TRF1 (Pardue & DeBaryshe, 1999), may regulate telomerase activity directly, but others, such as *S. cerevisiae* Tel2p, appear to act by binding to telomeric DNA sequences (Kota & Runge, 1999). Some cause the G-rich single-stranded overhang to fold back on itself, pairing by G–G bonds and thus presumably rendering it inaccessible to further telomerase action (Price, 1999b). Regulation of telomere length is a complex process, usually involving several proteins (McEachern *et al.*, 2000); for example, the human TIN2 protein interacts with TRF1 to affect telomere length. Tankyrase promotes elongation of human telomeres by ADP-ribosylating TRF1 and thereby inhibiting its negative regulation of telomere length (Smith & de Lange, 2000). The human orthologue of yeast Rap1, hRap1, appears to regulate telomere length through binding to TRF2 (Li *et al.*, 2000). Telomere length is very precisely controlled in the macronuclei of ciliates (Table 13.1), possibly connected with the peculiar state of these nuclei, which undergo neither mitosis or meiosis. In yeasts, telomere length is maintained in each cell generation, but in mammals, telomere extension is largely restricted to the germ cells, and there is little or no telomerase activity in somatic cells. Telomere length in somatic cells therefore decreases throughout the life of mammals, with implications for senescence (Section 13.6.1) and the development of cancer (Section 13.6.2).

As already mentioned, the length of *Drosophila* chromosomes is maintained by the addition of specific retrotransposons, HeT-A and TART. Unlike other retrotransposons, these only become incorporated into the chromosomes at the telomeres or, with greatly reduced efficiency, at broken chromosome ends (Pardue, 1995). The mechanism of attachment has not been established with certainty, but the first stage is believed to be attachment of the HeT-A RNA to the 5′ chromosome end through the oligo (A) tail of the RNA, a process mediated by the gag protein (Fig. 13.3). The RNA is then copied *in situ* by reverse transcriptase, and finally a second, complementary, DNA strand is synthesized and the new DNA ligated to the existing chromosome (Mason & Biessmann, 1995).

The presence of HeT-A and TART retrotransposons does not prevent *Drosophila* chromosomes from shortening, but it appears that new retrotransposons are added to the chromosomes at a sufficient rate to maintain average telomere length. Because the retrotransposons attach to telomeres that have been eroded to different extents, it seems that no specific sequence is required for their attachment. In each *Drosophila* generation, about 1% of the chromosomes get a new retrotransposon attached; this just balances the average loss of 75 bp of DNA from chromosome ends per generation (Mason & Biessmann, 1995). It should be noted that *Drosophila* chromosomes do not necessarily need telomeres to survive. Chromosomes with a terminal break have been produced, and have survived for many years (Pardue, 1995). Maintenance of telomeres by retrotransposition has not been reported in organisms other than *Drosophila*, but it is possible that something similar occurs in *Chironomus*. The telomeric sequences in *Chironomus* (Table 13.1) are not retrotransposons, and are in fact very much shorter. They are, however, transcribed (Kamnert *et al.*, 1997), so that an RNA sequence is available from which new telomeric sequences could be produced by a reverse transcriptase.

The mechanism of telomere maintenance in *Drosophila* may appear very different from that in most other organisms. However, in both cases new telomeric DNA is synthesized by a reverse transcriptase, and thus the mechanisms may be much more similar than it seems at first sight.

A third mechanism of telomere maintenance is recombination. It is an inefficient mechanism that can be used when telomerase is inactivated

Table 13.2 Telomeric proteins.

Protein	Species	Comments
Telomerase		A ribonucleoprotein reverse transcriptase
	Euplotes	230 kDa
	Oxytricha	
	Tetrahymena	
	S. cerevisiae	EST 2 (Ever-Shorter Telomeres)
	Mouse	
	Human	hTERT
TP1 (=TLP1)	Mammalian	290 kDa. Interacts with telomerase. ? Homologous with *Tetrahymena* p80
EST3	*S. cerevisiae*	Required for telomere function *in vivo*
rTP (replication telomere protein)	*Euplotes*	Telomere-bound replication factor (ssDNA-binding protein). Binds to both single- and double-stranded T_4G_4 repeats
Cdc 13p	*S. cerevisiae*	Binds to single-stranded 3′ telomere ends. Maintenance or synthesis of C-rich strand. ? Recruits telomerase to telomeres
Est1p	*S. cerevisiae*	Interacts with single-stranded telomere end and with telomerase
Ku	*S. cerevisiae*	Non-homologous end-joining protein. Protection of C-rich strand and recruitment of SIR 2, 3 and 4. ? Telomere length regulator. Attachment to nuclear envelope
Mlp1 and 2	*S. cerevisiae*	Attachment to nuclear envelope
Taz1	*S. pombe*	Telomere-binding protein. Homologous pairing at meiosis. Binds double-stranded region of telomeres. Orthologue of TRF (Li *et al.*, 2000)
Ndj1p	*S. cerevisiae*	Required for proper meiotic recombination and chromosome segregation
DNA pol α	*S. cerevisiae*	Extension of C-rich strand
DNA pol δ	*S. cerevisiae*	Extension of C-rich strand
hnRNP K	Human	Binds to C-rich strand (Lacroix *et al.*, 2000)
ASF/ASF2	Human	Binds to C-rich strand (Lacroix *et al.*, 2000)
TRF1	Vertebrates	Binds to double-stranded telomeric sequences. Negative regulation of telomeric length. ? Inhibits telomerase
hnRNPA1	Vertebrates	hnRNA-binding protein. ? Binds to G-rich strand overhang and recruits telomerase
5′–3′ exonuclease	*S. cerevisiae*	Production of G-rich strand overhangs
TRF2	Vertebrates	Binds to double-stranded telomeric sequence. ? Maintenance of G-rich strand overhang and prevents telomere fusion
Tankyrase	Vertebrates	142 kDa. Ankyrin-related poly(ADP-ribose) polymerase. ADP-ribosylates TRF1. ? Regulation of telomere length/interacts with TRF1
TIN2	Human	Binds to TRF1. Regulator of telomere length
Tel2p	*S. cerevisiae*	Telomere length regulator. Binds to single-stranded TG_{1-3} (3′ overhang)
Rap1p	*S. cerevisiae*	Major double-stranded telomere-binding protein. Regulator of telomere length. Also binds single-stranded TG_{1-3}. Mediates formation of structure held together by G–G interactions
hRAP1	Human	Binds to telomeres through TRF2 (Li *et al.*, 2000)
Telomere end-binding protein (TEBP)	*Oxytricha*	Alpha and beta subunits. Beta subunit *in vitro* folds ss T_4G_4 into four-stranded G-quartet. Caps chromosomes
Telomere-binding protein	*Euplotes*	Binds to single-stranded overhang

References: Kim *et al.* (1999); Kota & Runge (1999); Lingner & Cech (1998); McEachern *et al.* (2000); Pardue & DeBaryshe (1999); Price (1999b). See also text.

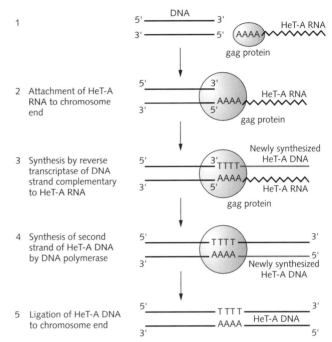

1

2 Attachment of HeT-A
   RNA to chromosome
   end

3 Synthesis by reverse
   transcriptase of DNA
   strand complementary
   to HeT-A RNA

4 Synthesis of second
   strand of HeT-A DNA
   by DNA polymerase

Figure 13.3 The transposition of HeT-A or TART retrotransposons to *Drosophila* chromosome ends. Redrawn with permission from Mason & Biessmann (1995) *Trends in Genetics* **11**, 58–62. © Elsevier Science.

5 Ligation of HeT-A DNA
   to chromosome end

(Pardue & DeBaryshe, 1999), and has been described in two species of yeasts in which telomerase components had been deleted. It is also a mechanism by which telomeres are maintained in immortalized human cell lines that do not express telomerase (ALT – alternative lengthening of telomeres; Section 13.6.2) (Dunham *et al.*, 2000). Recombination has been proposed as the mechanism of telomere elongation in the malarial mosquito *Anopheles*, and among plants in the Alliaceae (Biessmann & Mason, 1997), but it has yet to be confirmed that this is the usual mechanism. The telomeres of *Chironomus* could also be maintained by recombination, but evidence for this is lacking so far.

13.4 How do telomeres protect chromosome ends?

We have just seen (Section 13.3) that there are mechanisms to ensure that chromosome ends do not become progressively shorter as a result of failure to replicate to the very ends of DNA molecules. It is also clear that telomeres differ

from broken ends in not being 'sticky'; unlike freshly broken ends they are not recognized by the cell as double-strand DNA breaks, do not trigger cell-cycle checkpoints (Section 2.2.3) and are not subject to DNA repair mechanisms or degradation by nucleases. Telomeric chromatin must therefore have some special structure that differentiates it from ordinary chromatin.

Telomeric DNA is organized into a non-nucleosomal chromatin structure called the telosome. This structure is formed by the binding of some of the numerous telomeric proteins to each other and to the telomeric DNA, both the single-stranded overhang and the more proximal double-stranded DNA. Although several proteins have been identified that bind to different telomeric components (Table 13.2), and no doubt protect the telomeric DNA from degradation, so far the details of only one telomeric capping protein have been elucidated. This is the telomeric end-binding protein (TEBP) from *Oxytricha*. This protein consists of alpha and beta subunits that together form a groove in which the 3′ overhang of the telomere is buried, producing a very stable DNA–protein complex in which the

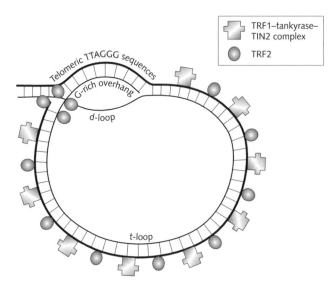

Figure 13.4 The structure of the ends of mammalian chromosomes. Redrawn from Shay (1999) *Nature Genetics* **23**, 382–383. © Nature America.

DNA is not accessible to nucleases (Price, 1999a). In spite of suggestions that the G-rich overhang might form unusual DNA structures with G–G pairing, the overhang in this structure is largely single-stranded, but with the last five nucleotides forming a loop.

Capping of mammalian chromosomes, which have a much longer single-stranded overhang (Table 13.1), is achieved in what seems to be a rather different way. The double-stranded telomeric DNA forms a large loop (the telomeric or t-loop) of as much as 23 kb, and the G-rich single-stranded overhang invades the double-stranded telomeric repeats, causing the formation of a single-stranded displacement loop (d-loop) and masking the end of the DNA molecule (Griffith *et al.*, 1999; Shay, 1999) (Fig. 13.4). A complex of the telomeric proteins TRF1, TRF2 and TIN2 associates with the telomere and may make the G-rich overhang inaccessible to nucleases. Access of telomerase is probably also prevented, so that these proteins, together with tankyrase and possibly others, also regulate the length of the telomeres. Loss of certain telomeric proteins, such as *Schizosaccharomyces pombe* taz, Rap1p of the budding yeast *Kluyveromyces lactis* or mammalian TRF2, allows 'uncapping' of the telomeres and fusion of telomeres of different chromosomes (Shore, 2001).

The non-nucleosomal region of telomeric chromatin is 80–130 bp long in *Euplotes*, and 250–400 bp long in *S. cerevisiae*. Proximal to these regions, the chromosomes are organized into nucleosomes, but nevertheless form a region in which the DNA is less accessible, and the histones are hypoacetylated (Gilson *et al.*, 1993; Zakian, 1995). This characteristic chromatin structure is probably responsible for the telomeric position effect (TPE) (Section 7.4.5), in which genes inserted near telomeres are generally silenced, and are late-replicating. Such position effects have been reported in the yeasts *S. cerevisiae* and *S. pombe*, and also in *Drosophila* (Zakian, 1995). In humans, a telomeric position effect on replication timing has been reported (Ofir *et al.*, 1999).

13.5 Telomeres and the spatial organization of nuclei

13.5.1 Interphase nuclei

There is abundant evidence that the arrangement of chromosomes in interphase nuclei is, in general, not random (Section 5.2). One aspect of this is that telomeres are frequently attached to the nuclear envelope. This has been reported in organisms as diverse as yeasts (*S. cerevisiae* and *S. pombe*), various plants, *Drosophila* (polytene chromosomes), salamanders, mice and humans (Vourc'h *et al.*, 1993; Strouboulis & Wolffe, 1996). A particularly striking example is in those organisms that have the Rabl organization in interphase nuclei, in which the telophase arrangement of chromosomes is maintained with the centromeres at one pole of the cell and the telomeres at the other. It is reasonable to suppose that these attachments are the result of a special affinity between telomeres and the nuclear envelope, although the evidence is rather circumstantial. In interphase nuclei of mouse lymphocytes, the position of the telomeres varies with the stage of the cell cycle (Vourc'h *et al.*, 1993). It is unlikely that some telomeres will ever be attached to the nuclear envelope. In many species, the nucleolus organizer regions (NORs) are subterminal (Section 11.2), and in interphase must be attached to the nucleolus, which is usually fairly centrally placed in the nucleus; the telomeres close to the NORs must therefore also be near the nucleoli. Attachment to the nuclear envelope may well be associated with the formation of condensed chromatin and gene silencing (Cockell & Gasser, 1999a).

In the budding yeast *S. cerevisiae*, proteins have been identified that attach telomeres to the nuclear envelope. The proteins SIR3 and SIR4 (silent information regulators), which are responsible for the gene silencing known as the telomeric position effect (TPE, Sections 7.4.5 and 13.4), are also required for the clustering of telomeres at or near the nuclear periphery (Cockell & Gasser, 1999a). Mutations in yeast Ku protein can prevent this clustering of telomeres; Ku is attached to the nuclear envelope through protein Mlp2 (Galy *et al.*, 2000).

In mammals, telomeres are found throughout the nucleus, and are associated with the nuclear matrix rather than with the envelope. One candidate for mediating interactions between telomeres and the nuclear matrix is the *ATM* gene product, which is defective in ataxia telangiectasia. There is evidence for differences in the binding of telomeres to the matrix in ataxia telangiectasia cells (Smilenov *et al.*, 1999),

although information on the exact nature of the defect is not yet available.

13.5.2 Mitosis and meiosis

Telomeres are not required to position chromosomes at mitosis. Nevertheless, defects in telomere function can prevent the separation of telomeres at anaphase (Hawley, 1997). The chromosomes pull apart, but sister telomeres remain attached to each other, so that the chromosomes are abnormally stretched and the division is abortive.

At meiosis, telomeres appear to play an essential role in bringing homologues together so that they can initiate synapsis (Section 2.5.1). Attachment of the telomeres to the nuclear envelope is part of this process; the telomeres can move over the envelope until they come into close proximity (Bass *et al.*, 1997). In the fission yeast *S. pombe* loss or mutation of the Taz1 protein impairs the clustering of telomeres and consequently reduces alignment of homologues and meiotic recombination, with increased missegregation of chromosomes (Price, 1999b). In *S. cerevisiae*, a meiosis-specific telomere-binding protein, Ndj1p, is necessary for recombination and segregation at meiosis. In mammals, mutation of the *Atm* gene (which is defective in ataxia telangiectasia) results in abnormal maintenance of telomere clustering with consequent defects in synapsis, and meiotic arrest (Pandita *et al.*, 1999). Although details have yet to be elucidated of how these various telomere proteins act in meiotic pairing and synapsis, proper telomere function is clearly necessary for normal meiosis.

13.6 Telomeres, ageing and cancer

Telomeres are believed to be closely involved in ageing and cancer. The hypothesis is that normal somatic cells can only divide for a limited number of divisions because their telomerase is not active, and so their telomeres shorten to a critical length at which no further growth is possible. In cancer cells, on the other hand, telomerase is active and cell growth is not inhib-

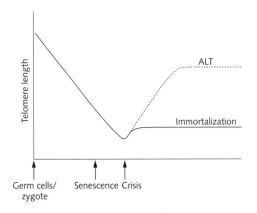

Figure 13.5 The changes in telomere length with age and senescence, transformation and immortalization.

ited because the telomeres are too short, which is why cancers can grow uncontrollably. As with any hypothesis, the details have turned out to be more complex, but it is nevertheless substantially correct that cell proliferation can be inhibited by preventing telomere replication, and cellular lifespan can be increased if telomerase is activated or if cells are transfected with telomerase (Bodnar *et al.*, 1998).

The basic observations behind the hypothesis are these (Greider, 1998; Lustig, 1999). In mammalian germ cells, telomerase is fully active, and telomeres are longer than in somatic cells. In somatic cells, however, telomerase is usually inactive, and telomere length decreases with the age of the individual. The same phenomenon can be seen in primary cell cultures: cells will grow for 40–50 generations, and then stop dividing, a stage known as senescence (Fig. 13.5). If senescent cells are activated by viral oncogenes, they can be induced to grow and divide again, with their telomeres still shortening, until they reach a stage known as crisis. At crisis, there are numerous chromosome abnormalities, and most cells die. About 1 cell in 10^{-7} survives crisis to become immortalized, and these immortal cells, like many cancer cells, have active telomerase and maintain their telomeres (although the telomeres often remain short, much shorter in fact than in senescent cells). If it were generally true that cancer cells express telomerase, and non-cancerous

somatic cells do not, telomerase should be an excellent target for cancer chemotherapy, with few significant side-effects on somatic cells. In fact, inhibition of telomerase does cause immortalized cells to die (Hahn *et al.*, 1999; Herbert *et al.*, 1999).

13.6.1   Telomeres and ageing

Introduction of telomerase into cultured cells that lack the enzyme can enable them to grow well past the stage at which normal cells senesce. For example, addition of telomerase to fibroblasts and retinal epithelial cells extended their life for some 200 population doublings beyond the stage at which telomerase-negative cells senesce (Bryan & Cech, 1999). In other cell types, however, telomerase does not prevent the onset of senescence at the usual time, and some cell types will senesce even though their telomerase is active and telomere length is maintained. Thus although telomere length is one factor in determining cellular senescence, it is probably not the only one.

Observations on telomerase knockout mice initially suggested that telomere loss was not important in this species. However, inbred mice strains have exceptionally long telomeres (Table 13.1), and in fact it was not until the sixth (mouse) generation that serious developmental problems and sterility occurred (Herrera *et al.*, 1999). Chromosomes lacked telomeric sequences, showed end-to-end fusions and cells were often aneuploid (Blasco *et al.*, 1997). Telomeres are much shorter in mouse species and subspecies recently derived from the wild, but lifespan is not correlated with telomere length (Hemann & Greider, 2000).

In experiments to clone mammals by nuclear transfer, it was found that the cloned sheep had significantly shorter telomeres than normal sheep (Shiels *et al.*, 1999). The telomere length depended on the age of the tissue from which the animal was cloned. This indicates that there was no restoration of telomere length in the cloned embryo, and it was suggested that this might lead to premature ageing in cloned animals. However, cattle cloned from adult or fetal cells showed no reduction in telomere length (Tian *et al.*, 2000), and serial cloning of mice for six generations did not result in any telomere shortening (Wakayama *et al.*, 2000). It is therefore not yet clear whether loss of telomeres is likely to be a serious problem in cloned animals.

If cells use the length of telomeres to decide when to stop growing, how do they do it? As shown in Fig. 13.5, the *average* telomere length at senescence (approximately 2–4 kbp, Henderson *et al.*, 1996) is still much longer than at crisis or in many immortal cells, and so cells can still grow with shorter telomeres. However, it is the length of the shortest telomere in the cell that results in loss of telomere function and cell viability (Hemann *et al.*, 2001). A chromosome without a telomere would presumably be recognized as having a double-strand DNA break, and activate the appropriate cell-cycle checkpoint. Different telomeres have different lengths (Henderson *et al.*, 1996) and shorten at different rates (between 50 and 150 bp per cell division; Blasco *et al.*, 1999).

13.6.2   Telomeres and cancer

With the exception of a few cell types, human somatic cells and benign tumours do not express telomerase activity, while most, but not all, human cancer cells do (Kim *et al.*, 1994; Harley & Villeponteau, 1995). Immortalized cells and tumour cells generally have stable, but short, telomeres. It therefore seemed that activation of telomerase could be an essential feature of malignant transformation, and that inhibition of telomerase could be a valuable therapy against cancers. In support of this, it has been shown experimentally that induction of differentiation of leukaemic cells inhibited telomerase activity (Sharma *et al.*, 1995), and that inhibition of telomerase inhibits the growth of human cancer cell lines, accompanied by shortening of telomeres and apoptotic cell death (Hahn *et al.*, 1999; Herbert *et al.*, 1999; Zhang *et al.*, 1999). Conversely, telomerase facilitates tumorigenesis by certain oncogenes (Zumstein & Lundblad, 1999), and some oncogenes can directly upregulate telomerase activity (Greider, 1999).

Nevertheless, although the basic hypothesis that tumour cells require stable telomeres seems to be established, there are a good number of exceptions that indicate that our knowledge of the situation is far from complete. Thus although induction of telomerase in normal human fibroblasts can immortalize these cells, it does not produce other changes associated with malignant transformation (Morales *et al.*, 1999). This should perhaps not be surprising, because there are so many other factors that are clearly involved in inducing cancer. Secondly, although telomerase may be expressed, telomere length is not necessarily stable in tumour cells, but may oscillate substantially (Jones *et al.*, 1998). This does not, of course, affect the idea that the presence of telomeres is necessary for tumour cells, and may indeed provide valuable information on the mechanisms of regulation of telomere length. Thirdly, evidence from telomerase knockout mice seems to indicate that not only is telomerase activity not necessary for cancer progression, but that lack of telomerase can lead to an increased susceptibility to cancers (de Lange & Jacks, 1999; Blasco *et al.*, 1999). Such results might seem to invalidate the correlation between telomerase and cancer, at least in mice, but it must be remembered that mice have much longer telomeres than humans, and that even after several generations of breeding, telomerase-negative mice might still have adequate telomeres on their chromosomes. The increased susceptibility to cancers in telomerase-negative mice might be explained by the greater ease of chromosome rearrangement between chromosomes that have lost their telomeres, as chromosome rearrangements are very common in cancers (Section 17.9.1).

Even if telomeres are required for immortalized tumour cells to survive, telomerase is not always necessary. As stated above, most human cancers do express telomerase; however, a proportion (11–83% of tumours, depending on the type) do not (Bryan *et al.*, 1997). Those cancers that do not express telomerase have unusually long telomeres, often >20 kb compared with lengths of ~2 kb found in telomerase-positive tumour cells. In such cells, telomeres are maintained by what is referred to as 'alternative lengthening of telomeres' (ALT), which, as in yeast, works through recombination (Dunham *et al.*, 2000), although there are still other possibilities (Blasco *et al.*, 1999). One is that the chromosome uses an existing telomeric t-loop (Section 13.4; Fig. 13.4) as a template to copy itself. Intriguingly, some 5–10% of nuclei from cells that show ALT have nuclear PML (promyelocytic leukaemia) bodies (Section 5.5.3) that contain telomeric DNA, certain telomeric proteins and some proteins involved in recombination, and it is tempting to speculate that in such cells the PML bodies could be involved in maintaining telomeres by a recombinational process.

Only time will tell if cancer therapy through telomerase inhibition is feasible. Compounds have now been synthesized that inhibit telomerase *in vitro* and *in vivo*, and produce telomere shortening without any acute toxicity (Damm *et al.*, 2001). Such compounds reduce the tumorigenic potential of tumour cells in mice. Evidently such therapy could not be applied to cancers that do not express telomerase, but the problem with those cancers that do express telomerase is that it might be many cell generations before the telomeres reached a short enough length to trigger cell death. Continuing work in this field should soon answer such questions and show whether this is a practicable approach to cancer therapy. Meanwhile, in addition to the importance of telomeres to ensure the replication of the ends of DNA molecules, and protecting them from degradation, their involvement in controlling the lifespan of cells, in both normal and cancerous growth, will ensure that they continue to be studied intensively for the foreseeable future.

Websites

TelDB contains links to a wide variety of sources of information on matters to do with telomeres, including numerous literature citations, and a telomere protein database:

www.genlink.wustl.edu/teldb/

The Telomere Club is a source of telomeric repeat sequences and telomere-related genes, as well as containing links to other topics concerned with telomeres:

resolution.colorado.edu/~nakamut/telomere/
telomere.html

Lampbrush chromosomes

14

14.1 What are lampbrush chromosomes?

Lampbrush chromosomes are chromosomes that have a particular morphological appearance and occur mainly at the diplotene stage of meiosis in female animals. The name lampbrush comes from a resemblance to the brushes used to clean the glass chimneys of oil lamps. These were familiar objects in the late nineteenth century when lampbrush chromosomes were discovered, but a more familiar analogy nowadays would be a bottle brush or test-tube brush. In fact, the structure of lampbrush chromosomes differs in two important ways from that of lampbrushes (or bottle brushes). Whereas lampbrushes have bristles, and these stick out in all directions from the central axis, lampbrush chromosomes have loops instead of bristles, and these loops occur in pairs, one on either side of the axis.

The loops of lampbrush chromosomes are engaged in vigorous RNA synthesis, and this distinguishes them from other chromosomes that may show lateral loops (Callan, 1986; Morgan, 2002). The RNA synthesis is necessary for the production of the materials required for early growth and development of the embryo following fertilization. Thus lampbrush chromosomes are found at the diplotene stage of meiosis in the vast majority of female animals. Exceptions are found in certain insects, some reptiles and mammals. In those insects that lack lampbrush chromosomes in their oocytes, the RNA and protein that will be required for growth and

development are provided by surrounding nurse cells: such ovaries are described as meroistic. A somewhat similar situation is found in certain reptiles, in which the cytoplasm of surrounding follicle cells is confluent with that of the oocyte, and the follicle cells provide the oocyte's requirements for RNA and protein. In mammalian embryos, RNA synthesis starts very early, at the two-cell stage, and this, combined with the slow growth rate, means that it is not necessary to furnish the mammalian oocyte with large stores of material to support early development (Callan, 1986).

In general, lampbrush chromosomes do not occur in male meiosis, which is consistent with a role in producing stores of RNA and protein for the developing embryo; apart from the paternal genome, spermatozoa normally contribute little or nothing to the zygote. The one exception is the Y chromosome of *Drosophila* and other higher Diptera, which do develop lampbrush loops and actively synthesize RNA and protein (Section 14.4). Lampbrush chromosomes are not found in plant meiosis either, no doubt at least partly because of the different mechanisms of early development in plants. The only plant known to have lampbrush chromosomes is the unicellular alga *Acetabularia* (Spring et al., 1975), in which they occur in the so-called 'primary nucleus', which may represent the diplotene stage in this organism (Callan, 1986).

Lampbrush chromosomes are of great intrinsic interest, because of their distinctive structure and, unusually for chromosomes in dividing cells, their intense synthetic activity. In addition,

Box 14.1 Preparation of lampbrush chromosomes

Lampbrush chromosomes are prepared most easily from oocytes of urodele amphibia (chapter 2 in Callan, 1986; Macgregor & Varley, 1988). In general urodeles have high nuclear DNA contents (C-values, Table 3.1) so that their chromosomes are large, and since lampbrush chromosomes are highly extended, it is not unusual to find that the largest ones are in the region of 1 mm (sic) long. This means that they are easy to prepare and visualize, and to manipulate experimentally.

Whereas most chromosome preparations are made in bulk from actively dividing tissue or cell cultures, yielding numerous metaphases from a single preparation, lampbrush chromosomes are dissected out from individual unfixed oocytes. The nucleus is extracted from the oocyte, its membrane is removed and the free lampbrush chromosomes allowed to settle on to a coverslip. Correct composition of the lampbrush chromosome isolation medium is vital to the success of the technique. Because of the skill involved, and no doubt because mammals, and humans in particular, do not have lampbrush chromosomes, the study of lampbrush chromosomes has always remained a specialized field that has been studied by only a few scientists.

Comprehensive protocols for preparing lampbrush chromosomes from urodeles, *Xenopus* and birds are available at: www.le.ac.uk/biology/lampbrush/protocols.htm

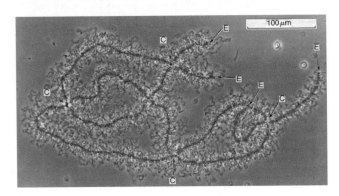

Figure 14.1 A lampbrush chromosome from the North American newt *Notophthalmus viridescens*, showing the chromosome axes, lateral loops, chiasmata (C) and the chromosome ends (E). Reproduced with permission from Macgregor (1993) *An Introduction to Animal Cytogenetics.* © Kluwer Academic Publishers.

however, study of lampbrush chromosomes has helped to elucidate many general points of chromosome organization, particularly the uninemy of chromosomes, and their organization into an axis from which loops radiate. These aspects of chromosome structure were in fact established in lampbrush chromosomes long before it became clear that they also applied to mitotic chromosomes (Sections 6.2 and 6.3).

14.2 Lampbrush chromosome structure

Lampbrush chromosome structure has been studied mainly in urodele amphibia (Box 14.1),

from which the following description is largely derived, although the essential features are the same in all organisms. The subject has been reviewed by Macgregor (1980, 1993) and Callan (1982, 1986) and most recently by Morgan (2002).

Lampbrush chromosomes are diplotene bivalents, and therefore consist of two axes, connected at the chiasmata, and the loops extend on both sides of the axes (Fig. 14.1). The axes of a lampbrush chromosome consist of a series of dense granules, 0.25–2.0 µm in diameter and 1–2 µm apart; there may be in the region of 5000 of these granules in the haploid genome. These granules consist of deoxyribonucleoprotein and are often, confusingly, referred to as chromo-

Table 14.1 Lampbrush loop lengths and DNA C-values.

Species	C-value (pg)	Loop length*
Gallus gallus domesticus	2.5	2–3
Xenopus laevis	3.0	5–10
Ascaphus truei	8.2	4
Plethodon cinereus	20.0	8.35
Triturus cristatus	29.0	11
Tarichia granulosa	29.0	14.9
Ambystoma mexicanum	35.0	12
Plethodon dunni	38.8	17
Bolitoglossa subpalmata	87	22

*Measured from the chromosome axis to the furthest point of the loop. Values are in μm. In some cases the loops may have been stretched somewhat during preparation.
Data from Macgregor (1980), León & Kezer (1990) and Solovei *et al.* (1993).

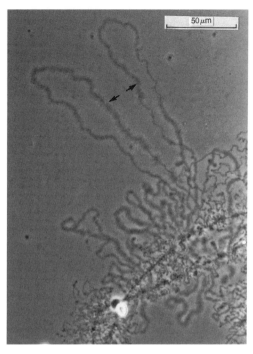

Figure 14.2 Part of a lampbrush chromosome from *Notophthalmus viridescens*, showing a pair of large fluffy loops (arrows) that are much longer than most of the other loops. Reproduced with permission from Macgregor (1993) *An Introduction to Animal Cytogenetics.* © Kluwer Academic Publishers.

meres. These granules are clearly not the same as the chromomeres of pachytene chromosomes (Section 6.4.2), which are much larger and many fewer in number. On the other hand, the number of these granules is similar to that of the 'chromomeres' of polytene chromosomes (Section 15.2); however, the numerical similarity is almost certainly misleading, as the chromomeres of polytene chromosomes appear to contain the genes, whereas, as we shall see below, the chromomeres of lampbrush chromosomes do not.

A few of the chromomeres have no loops attached to them, and others bear multiple pairs of loops; perhaps most bear only a single pair of loops. The loops extend, at their maximum, for between 5 and 50 μm from the axis, and occasionally even more (Table 14.1). There is a rough correlation between the size of the loops and the C-value of the species, which perhaps reflects the underlying organization of the genome. However, lampbrush chromosomes induced to form from *Xenopus* sperm chromatin when it was injected into germinal vesicles of a newt formed loops whose size was characteristic of those of the newt rather than of *Xenopus*. The *Xenopus* lampbrush chromosomes in the newt oocytes also bore newt rather than *Xenopus* proteins, sug-

gesting that it is the host proteins rather than the donor DNA sequence that determines loop morphology (Morgan, 2002).

Most of the loops ('normal loops') are of the same general type, but a small proportion have a very distinctive morphology and provide fixed 'landmarks' that help to identify the chromosomes (Fig. 14.2). These latter include giant fusing loops, giant granular loops and other loops with much more material attached to them than normal loops (Fig. 14.3). Other distinctive structures are the spheres – spherical bodies 7–10 μm in diameter that are homologous to coiled bodies (Gall, 2000; Morgan, 2002; Section 5.5.2). Unlike the loops, the spheres are not paired, but instead there is only a single sphere per chromosome, or occasionally the spheres on the two homologues that make up the bivalent fuse. They have been shown to contain RNA polymerase II, as well as

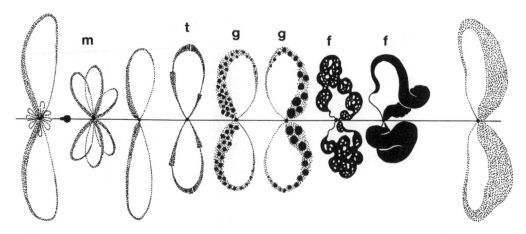

Figure 14.3 Different types of loops on lampbrush chromosomes. Note the increase in thickness of the matrix along the loops; multiple transcription units (t); multiple loops emanating from the same chromomere (m); granular loops (g); and fusing loops (f). Reproduced with permission from Macgregor (1993) *An Introduction to Animal Cytogenetics.* © Kluwer Academic Publishers.

a number of other proteins required for polymerase II transcription (Morgan, 2002).

The centromeres of lampbrush chromosomes are chromomeres that lack lateral loops, and in some species are flanked by condensed regions known as axial bars, which are the pericentromeric heterochromatin, contain highly repetitive satellite DNA and lack loops. In birds, characteristic protein bodies are associated with centromeres, but so far little is known about their function or composition (Morgan, 2002).

How are lampbrush chromosomes put together? The basic unit of the lampbrush chromosome is a pair of deoxyribonucleoprotein (DNP) fibres that run together along the axis of the chromosome, parting company at the chromomeres where each one runs round one of the pair of loops, and the two fibres re-unite at the other end of the loop. Evidence for this comes from two sources. If lampbrush chromosomes are stretched, some of the chromomeres break transversely, so that the chromosomes are held together only by the fibres that form the loops. The DNP in the chromosome axis and the loops is thus continuous, but it is not continuous across the chromomeres. The second bit of evidence comes from experiments in which lampbrush chromosomes were digested with DNase. To produce a break in a double-stranded DNA

(dsDNA) molecule requires two adjacent cuts, and therefore the rate of production of breaks is proportional to the square of the length of time of digestion; for two dsDNA molecules, it would be proportional to the fourth power of the digestion time. These experiments showed that the loops contained a single dsDNA fibre, whereas the axis contained two dsDNA molecules as expected, because the chromosomes have replicated and therefore have two chromatids. In fact, the two chromatids can occasionally be distinguished, but more often the axis is single. These experiments, therefore, do not merely support the view that the DNA fibres are continuous through the loops, but also show that the chromosomes are unineme, that is, there is only a single DNA fibre running throughout their length. This conclusion gained further support from measurements of the diameter of the axis between the chromomeres. This is only 3–6 nm, sufficient only to accommodate two DNA molecules, one for each chromatid.

The normal loops of lampbrush chromosomes are clearly visible by light microscopy, and therefore must be much thicker than a single DNA molecule or even a 30 nm chromatin fibre. In fact, their thickness is due to a large amount of ribonucleoprotein (RNP), which is the direct result of transcription from DNA in the loops.

This RNP matrix is asymmetrically distributed on the loop; it is thin at one end of the loop, and gets progressively thicker (up to 3 μm) as one proceeds round the loop. This is a visible manifestation of RNA synthesis, with the length of the transcript (complexed with protein) increasing as synthesis proceeds along the loop, although some material may be lost from the ends of the longer transcripts so that there is no longer a uniform increase in their length. It is probable that the matrix consists of a series of RNA polymerase molecules attached to the DNA of the loop, and that thousands of individual transcripts, emanating from the polymerase molecules, and complexed with proteins, form the matrix. The DNA of the loop axis is also complexed with proteins, but does not form nucleosomes. In the description just given, there is continuous transcription round the loop, which thus forms a single transcription unit. However, examples have been described of loops that contain two or more transcription units, and the different transcription units within the same loop do not necessarily have the same orientation. So just as there are more loops than chromomeres, there are more transcription units than there are loops.

Just what sort of RNA is transcribed from lampbrush chromosomes? The loops are the most prominent features of lampbrush chromosomes, but in fact they only make up a few per cent of the total chromosomal DNA, perhaps less than 2.5% (Macgregor, 1980; Morgan, 2002). It was once supposed that the loops were spun out at one end and drawn back into the axial granule at the other, so that a large part of the total DNA might appear in the loops at some time during the lampbrush stage, but this is now known not to be so (although the loops as a whole are extended from the chromosomes at the beginning of the lampbrush stage, and retracted again at the end of it). Nevertheless, in most eukaryotes the genes form only a small proportion of the total DNA, so perhaps the transcribed regions on the loops could correspond to the genes. There is no doubt that a lot of the DNA that is transcribed is single-copy DNA, and although very few genes have been identified so far, it is reasonable to suppose that much of

the transcribed material is messenger RNA (mRNA). However, a substantial amount of moderately repetitive and highly repeated satellite DNA is also transcribed. Many of these sequences are transcribed as a result of 'read-through'; that is, transcription can be initiated in the normal way on lampbrush loops, but apparently does not stop until it reaches some physical obstacle such as the granule where the loop rejoins the chromosomal axis. Transcription of satellite DNAs has been described in newts (Varley et al., 1980) and in birds (Solovei et al., 1996), and in birds the telomeric repeats are also transcribed (Solovei et al., 1994).

14.2.1 Proteins of lampbrush chromosomes

The immense amount of synthetic activity in the loops, the structural differentiation between the loops and the differences between the loops and the axis of lampbrush chromosomes mean that information on the proteins of lampbrush chromosomes would be very valuable, providing as they do a means of studying structural and functional differentiation at high resolution. Although many immunofluorescence studies have been done on lampbrush chromosomes, in most cases the antigens involved were not adequately characterized, and therefore such studies merely serve to emphasize the visible morphological differentiation.

Among RNP proteins, some are bound to all transcripts, whereas others are restricted to a very few (Sommerville et al., 1978); one heterogeneous nuclear RNA protein has been identified that is specific for giant landmark loops (Piñol-Roma et al., 1989). A protein involved in pre-mRNA splicing was also found in the lateral loops (Roth et al., 1991). Nucleoplasmin is associated with the sites of transcription on all lampbrush loops, but is never present on the chromosome axis (Moreau et al., 1986). Surprisingly, acetylated histone H4 was present only at very low levels in the RNP matrix of the normal loops, in the spheres and in the matrix of the marker loops, but the chromomeres on the axis were heavily labelled by the antibody

(Sommerville *et al.*, 1993). Induced overexpression of histone deacetylase causes retraction of the loops (Morgan, 2002). Injection of histone H1 into oocytes also causes loop retraction; lampbrush chromosomes are free of histone H1 and its variants (Morgan, 2002). Proteins HMGN1/2 are also absent from lampbrush chromosomes, although they are ubiquitous in somatic nuclei and are believed to enhance transcription (Morgan, 2002). The DNA in the transcriptionally inactive chromatin of the chromosome axis was rich in 5-methylcytosine, but 5-methylcytosine was absent from the vast majority of loops, except for the untranscribed spacers seen in some loops (Angelier *et al.*, 1986).

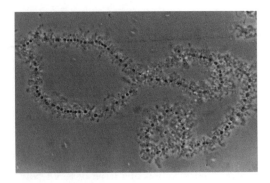

Figure 14.4 A lampbrush bivalent from *Bipes*, showing prominent axial granules. Reproduced with permission from Macgregor & Klosterman (1979) *Chromosoma* **72**, 67–87. © Springer-Verlag.

14.2.2 Landmarks and longitudinal differentiation of lampbrush chromosomes

As mentioned above, certain loops have distinctive structures that allow them to be used as landmarks in the mapping of lampbrush chromosomes (Callan, 1986, pp. 66–85; Morgan, 2002), and there are other distinctive structures such as spheres, axial bars, axial granules and so on. What is the significance of such structures? It might be tempting to suppose that at least some of the landmark loops could be the sites of specific types of genes, such as those for the highly repeated ribosomal and 5S RNAs, but it has to be admitted that the functions of such loops are not yet known. In some species of *Triturus*, there is in any case no transcription from the ribosomal DNA loci, all the activity taking place in the thousands of free nucleoli that are present in the nucleoplasm at this stage (Section 11.5.3). In those species where the chromosomal copies of the ribosomal genes are active, such as Plethodontid salamanders, they form a nucleolus as on any other (non-lampbrush) chromosome (Callan, 1982). In *Xenopus*, the nucleolus organizer region (NOR) appears as an axial granule (Callan *et al.*, 1988). The 5S RNA genes appear to be on thin loops with only a small amount of matrix (Morgan, 2002). The histone genes are on normal loops but are adjacent to the spheres, which are homologous to coiled bodies and are involved in processing histone mRNA (Gall, 2000; Section 5.5.2).

The main characteristic of complex loops is that their axes follow a contorted path within the matrix (Morgan, 2002). They also contain specific proteins that are absent from the majority of loops, or alternatively they lack proteins that are found in normal loops. However, information on these points is still rudimentary so it is not yet possible to explain why some loops form complex morphologies or to what extent this might be due to specific DNA sequences or proteins.

Some of the chromomeres along the axis of lampbrush chromosomes of urodeles are particularly large, and are known as axial granules (Fig. 14.4). These appear to have a core of DNA, which is surrounded by protein. One of the constituents of the axial granules is topoisomerase II (Topo II), which is apparently absent from the majority of the chromomeres along the axis of the chromosomes (Hock *et al.*, 1996). Axial granules have a tendency to fuse with one another, either with homologous or non-homologous granules, and it has been speculated that they might be involved in pairing and recombination between homologues. *Xenopus* lampbrush chromosomes, which lack axial granules, do not have any immunocytochemically detectable Topo II either. It should be noted that these observations support morphological ones, which suggests that there is no continuous scaffold in lampbrush chromosomes.

The telomeres of the lampbrush chromosomes from birds show an unusual structure (Fig. 14.5). In amphibia the telomeres appear simply as a granule, but in the birds studied (except quail) these chromomeres have loops attached (Solovei et al., 1994). The morphology of these loops varies from one chromosome end to another, but the most unexpected finding is that in some cases the 'loops' are only attached to the telomeric granule at one end. The free end of the 'loops', not the telomeric granule, appears to carry the telomeric DNA sequences, and these are transcribed only from the C-rich strand.

14.2.3 Heterozygosity, heteromorphism and sex chromosomes

Structures on lampbrush chromosomes do not always show identical size or structure on the two chromosomes that make up the lampbrush bivalent. Homologous loops may differ in size or structure, or more extensive regions may differ between the two homologues: the latter situation is observed in species with differentiated sex chromosomes, but also occurs in other situations. In its simplest form, heterozygosity can be seen simply as a difference in loop size, and is most easily seen where specific loops are stained with silver (Varley & Morgan, 1978) or with fluorescently labelled antibodies. In the newt *Triturus cristatus carnifex*, individuals may be homozygous or heterozygous for certain giant fusing loops. In hybrids between *T. c. carnifex* and *T. c. cristatus*, there is heterozygosity of giant granular loops. Such observations show that loop morphology is an intrinsic property of a specific loop (Callan, 1982).

Heterozygosity for other lampbrush chromosomes has also been reported. In *Triturus cristatus*, most subspecies do not have axial bars flanking their centromeres; however, the subspecies *T. c. karelinii* does. Crosses between different races therefore show heterozygosity for the presence or absence of axial bars (Callan, 1982), which contain highly repeated satellite DNAs (Baldwin & Macgregor, 1985) and correspond to C-banded centromeric heterochromatin (cf. Section 7.2).

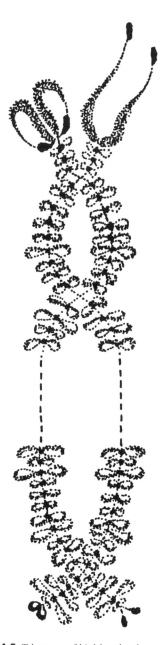

Figure 14.5 Telomeres of bird lampbrush chromosomes. The chicken chromosome 1, showing at the top the two different appearances of the telomeric loops. (Left) bow-like telomeric loops; (right) open-ended loops. Reproduced with permission from Solovei et al. (1993) *Chromosome Research* **2**, 460–470. © Kluwer Academic Publishers.

A developmentally significant heteromorphism of the long arms of chromosome 1 is found in *T. cristatus* and *T. marmoratus* (Sims *et al.*, 1984). This heteromorphism is not associated with sex determination, but occurs in all animals of both sexes. In fact, homozygotes for either form of chromosome 1 invariably die at the tailbud embryo stage.

Heteromorphic lampbrush sex chromosomes have been identified in the salamander *Pleurodeles poireti*, where the differential segment is interstitial (Callan, 1982), and in birds (Solovei *et al.*, 1993). The latter have been described in some detail, and present a number of points of interest. The W chromosome is largely heterochromatic, and consists mainly of satellite DNA; as a lampbrush chromosome it is disproportionately condensed, with a thicker axis and very small loops (Fig. 14.6). Although in general the lengths of lampbrush chromosomes are similar to the relative lengths of the mitotic chromosomes in the same species, the lampbrush W is much shorter in proportion than the mitotic W. There is also differential condensation of the arms of the lampbrush Z chromosome, no doubt because of the presence of a large block of heterochromatin in the long arm. Like the heterochromatic W chromosome, this heterochromatic region on the Z carries much smaller loops than the remainder. A final feature is a pair of distinctive giant lumpy loops at or very near the point where the chiasma between the Z and W chromosomes is, although it is not yet known if the presence of these loops at this point is coincidental, or whether they have some special significance for chiasma maintenance (for example).

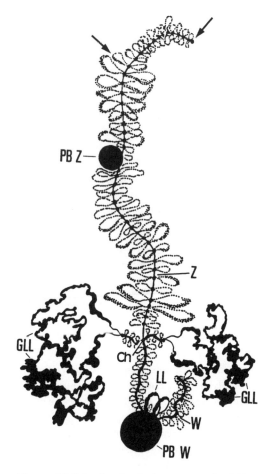

Figure 14.6 Sex lampbrush chromosomes: the ZW bivalent from a pigeon. Ch, chiasma between the Z and W chromosomes; PB, protein body; LL, lumpy loops; GLL, giant lumpy loops. Note the smaller loops on the heterochromatic W chromosome (W) and on the heterochromatic part (arrows) of the Z chromosome (Z). Reproduced with permission from Solovei *et al.* (1993) *Chromosome Research* **1**, 153–166. © Kluwer Academic Publishers.

14.3 What have we learnt from oocyte lampbrush chromosomes?

Lampbrush chromosomes have provided important evidence for two general aspects of chromosome organization: the uninemy of chromosomes, and their organization in loops extending from an axis. It is nevertheless important to note that it is also clearly adapted to the specific requirements of lampbrush chromosomes, which evidently need to be highly extended to permit the intense RNA synthesis that they perform. Apart from their immense extension, they seem to differ from other forms of chromosomes in lacking any sort of scaffold or continuous matrix (Callan, 1986, p. 181; cf. Sections 2.5.2 and 6.3) that forms a basis on which to arrange the chromosomal DNA. It may indeed be that because the chromosome is so extended, there is no need for a scaffold, and that the DNA is all that is needed to determine the form of the lampbrush chromosome.

The most distinctive feature of oocyte lampbrush chromosomes is their intense and apparently indiscriminate RNA synthesis, however. Although it is generally accepted that large quantities of RNA have to be produced to provide a store for early development, this surely cannot be the role of the repeated, non-coding RNAs that are also produced. One suggestion is that all this material is just bulk, produced to swell the germinal vesicle (oocyte nucleus) to an appropriate size (Cavalier-Smith, 1978), although there are a number of difficulties with this hypothesis (Macgregor, 1980). In addition, some of the RNA synthesized on lampbrush chromosomes is degraded, so that after a certain point the amount of RNA reaches a constant level, and it may in fact be the protein translated from the RNA that is important for early development (Callan, 1982).

It has been tempting to equate loops, or the chromomeres on the axis of lampbrush chromosomes, with genes. We have seen that this must be at least an oversimplification, as there are sometimes several loops attached to one granule, and several transcription units in the same loop. Further evidence against a correlation between genes and loops comes from a study of related salamanders with very different nuclear DNA contents (C-values). Such species would be expected to have very much the same number of genes, but the number of loops, far from being similar in the species studied, is actually roughly proportional to the amount of chromosomal DNA, and varies by a factor of >1.5 (Vlad & Macgregor, 1975).

14.4   Lampbrush Y chromosomes in *Drosophila* spermatocytes

The lampbrush Y chromosomes of *Drosophila* spermatocytes are in many ways similar to oocyte lampbrush chromosomes, but nevertheless have some distinctive features of their own, of which the most obvious is that only one chromosome is involved; the autosomes and the X chromosome do not form lampbrush loops. The Y chromosome, of course, has no homologue with which to pair, and is therefore not part of a bivalent. The number of loops on the *Drosophila* Y (six in *D. melanogaster* and five in *D. hydei*) is also very much smaller than has been reported on any oocyte lampbrush chromosome, but this must be at least in part due to the nature of the *Drosophila* Y. In somatic cells the Y chromosome is wholly heterochromatic, and moreover has no role in sex determination: *Drosophila* males without a Y are fully viable and perfectly normal except that they are completely sterile. Mapping studies have shown that the sites of lampbrush loops on the Y correspond to the sites of fertility factors, and a deficiency in any of the loops causes arrest of spermiogenesis.

The structure and behaviour of the lampbrush Y chromosomes of *Drosophila* have been reviewed by Hennig (1985) and Hackstein & Hochstenbach (1995). The species that has been studied most intensively is *D. hydei*, largely because of the ease of studying its lampbrush chromosomes; as with oocyte lampbrush chromosomes, the choice of species to study has been determined to a very great extent by the ease with which the chromosomes can be obtained. Nevertheless, structures that appear to be lampbrush chromosomes have been identified in the spermatocytes of well over 50 species of *Drosophila*. In *D. hydei*, there are five pairs of lampbrush loops, with names that are descriptive of their morphology (Fig. 14.7). The short arms of the Y chromosome bear two pairs of loops known as nooses, and these are the ones that most closely resemble the normal loops of oocyte lampbrush chromosomes. The long arms bear two pairs of 'tubular' ribbons, so called because of their fine structure, of which the more proximal ones bear 'clubs' – bodies resembling the material on the giant granular loops of amphibian lampbrush chromosomes. The most distal pair of loops consists of condensed and diffuse regions and carries a rounded body, the pseudonucleo-

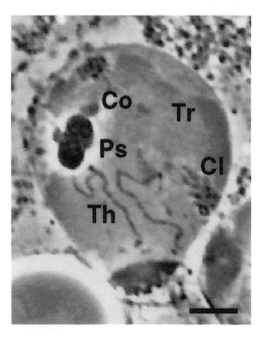

Figure 14.7 A primary spermatocyte nucleus from *Drosophila hydei*, showing the Y chromosome lampbrush loops. Cl, clubs; Co, cones; Ps, pseudonucleolus; Th, threads; Tr, tubular ribbons. Scale bar = 10 μm.
Reproduced with permission from Reugels *et al.* (2000) *Genetics* **154**, 759–769. © Genetics Society of America.

lus. As with oocyte lampbrush chromosomes, the form of the Y chromosome lampbrush loops is autonomous. When two species are crossed that have different loop morphology (e.g. *D. hydei* and *D. neohydei*), the loop morphology of the male parent is always maintained in the spermatocytes of the hybrid.

The *Drosophila* lampbrush loops are composed almost entirely of simple-sequence, highly repetitive DNA, retrotransposons and other middle repetitive DNA sequences, and all these sequences are transcribed (Hackstein & Hochstenbach, 1995). However, at least one loop, *Threads*, contains a gene for dynein, which forms the outer arms of the microtubules in the sperm tails. The large size of the loops is explained by the enormous size of the introns in this gene (Reugels *et al.*, 2000); in the absence of crossing-over in male *Drosophila*, there is no mechanism that prevents the rapid growth of clusters of satellite DNA to produce such large introns (Kurek

et al., 2000). Whereas the dynein mRNA is transported into the cytoplasm, the RNA transcribed from the repeated DNA sequences remains in the nucleus. It had been suggested that, rather than being a source of mRNA, the loops of *Drosophila* Y lampbrush chromosomes might function by segregating and storing proteins, but the discovery of a gene in one of the loops indicates that this is not their primary function. Nevertheless, a specific protein that is essential for spermatogenesis has been shown to associate with a specific loop (Heatwole & Haynes, 1996). As for oocyte lampbrush chromosomes, there is still much to be learnt about *Drosophila* spermatocyte lampbrush chromosomes. Although such studies could potentially benefit from the vast amount of information on *Drosophila* genetics, the study of these lampbrush chromosomes remains highly specialized work undertaken by only a few scientists, and progress is therefore inevitably going to be slow.

Like oocyte lampbrush chromosomes, those of *Drosophila* Y chromosomes are not only of intrinsic interest, but can help to illuminate more general questions of chromosome organisation. In *Drosophila*, it is clear that the individual loops correspond to specific genes, a situation that is probably not true in amphibia (Section 14.3). *Drosophila* lampbrush chromosomes are possibly better material than those of amphibia to investigate the factors affecting loop morphology, because the *Drosophila* loops are so few and their functions known (although as yet imperfectly). On a more general level, *Drosophila* lampbrush chromosomes have helped us to understand better the nature of heterochromatin (Section 7.4.2), questions of transcription and the evolution of genes and repetitive DNA sequences when isolated on a chromosome such as the *Drosophila* Y (Hackstein & Hochstenbach, 1995; Kurek *et al.*, 2000).

Websites

A great variety of information on lampbrush chromosomes is available on the University of Leicester website, as follows:

Introduction
www.le.ac.uk/biology/lampbrush/intro.htm

Preparation protocols
www.le.ac.uk/biology/lampbrush/protocols.htm

Publications (this claims to include every publication on lampbrush chromosomes since their discovery)
www.le.ac.uk/biology/lampbrush/pubs.htm

People (bibliographies and photos of researchers who have studied or are at present studying lampbrush chromosomes)
www.le.ac.uk/biology/lampbrush/people.htm

Polytene chromosomes

15.1 What are polytene chromosomes?

Many, perhaps most, organisms have a proportion of cells whose nuclei are polyploid. In a few tissues in certain animals, the nuclei may contain up to half-a-million times the DNA of normal (haploid) cells, although values are usually much lower (see Nagl, 1978, for compilations of polyploid DNA values in plants and animals). There are several ways in which polyploidy can be produced. It may be caused by nuclear restitution, in which the chromosomes enter mitosis but division is not completed, so that the two daughter sets of chromosomes remain in the same nucleus. Thus the number of chromosomes doubles as the amount of DNA doubles. In endocycles, there is no attempt at chromosome segregation, no spindle is formed and G and S phases alternate. In endomitosis, chromosome condensation can be seen during the endocycle, whereas in endoreduplication the increase in DNA occurs without any visible chromosomes being formed; in either case there is a progressive increase in the amount of nuclear DNA. The amount of DNA usually increases in geometrical progression, being proportional to 2^n, where n is the number of cycles of replication. However, deviations from this simple pattern occur, as a result of either under-replication or extra rounds of replication of specific DNA sequences.

In many polyploid nuclei, it is not known how the chromosomes are organized, because they never condense and the individual chromosomes are not visible. In a few cases, however, the products of successive rounds of DNA replication remain together to form a giant polytene ('multi-threaded') chromosome that is easily visible with a low-powered microscope. The best known polytene chromosomes are those of Dipteran flies (Section 15.2), especially those of *Drosophila*, *Chironomus* and *Rhynchosciara*; the clarity of the banded structure of these chromosomes, combined with the immense knowledge of the genetics of these flies, made such chromosomes valuable objects of study and allowed correlations to be drawn between chromosome structure and genetics. Because polytene chromosomes are interphase chromosomes, and are therefore transcribed, Dipteran polytene chromosomes provide an opportunity to study transcription by direct observation, and transcriptional responses to specific stimuli can be observed.

Polytene chromosomes have been described in at least four other groups. Among insects, the Collembola, a group not closely related to the Diptera, have polytene chromosomes that are very similar to those of the Diptera, although they have not been studied intensively (Cassagnau, 1974). Three other groups have polytene chromosomes that show distinctive structural and functional features. In certain ciliate protozoa (Section 15.3), the formation of polytene chromosomes is part of the process of DNA amplification that leads to the formation of macronuclei. The polytene chromosomes of mammalian trophoblast (Section 15.4) and plant

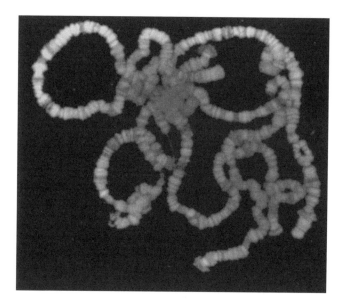

Figure 15.1 Polytene chromosomes from *Drosophila melanogaster*, with the banding pattern revealed by propidium iodide fluorescence for DNA. Micrograph kindly provided by C.E. Sunkel and P. Coelho.

antipodal and suspensor cells (Section 15.5) are less clearly defined structurally.

15.2 Polytene chromosomes in Diptera

The polytene chromosomes of Diptera are too well known to need detailed description: long, fat chromosomes consisting of alternating dense bands and diffuse interbands, arranged in characteristic patterns (Fig. 15.1). The patterns of bands (which are in no way related to those produced by banding techniques on mitotic chromosomes; Section 10.2) are reproducible, so individual chromosomes and parts of chromosomes can be identified by their patterns, and individual bands can be identified by their size and structure. To confuse matters, the bands are sometimes referred to as chromomeres, although they are probably not homologous with the 'chromomeres' of lampbrush chromosomes (Section 14.2), and are certainly not the same thing at all as the chromomeres of pachytene chromosomes at meiotic prophase (Section 6.4.2). The amount of DNA in the interbands is much lower than in the bands: published values range from 0.8% to 25% (Laird *et al.*, 1981; Sorsa, 1982), with the usual

values probably being somewhere between these extremes. No doubt the ratio of DNA concentration between bands and interbands varies a good deal anyway. In favourable preparations examined by electron microscopy, polytene chromosomes are seen to consist of numerous parallel chromatin fibres. Polytene chromosomes may be as long as 0.5 mm, and up to 20 μm in diameter (Nagl, 1978, p. 52).

In this section, the tissue distribution of polytene chromosomes will be listed, the relationship between genes and bands discussed and the differential replication of DNA and the transcription of RNA from polytene chromosomes will be described. Finally, an outline will be given of what is known about mechanisms of formation and stabilization of polytene chromosomes. Much more detailed information on the polytene chromosomes of *Drosophila* is given by Zhimulev (1996, 1998, 1999).

15.2.1 What tissues are polytene chromosomes found in?

In Diptera, polytene chromosomes are a phenomenon of terminally differentiated cells. Although their occurrence in salivary glands is perhaps best known, they can be found in at least eight different tissues. Many of these are larval tissues, such as Malpighian tubules, fat bodies and various parts of the gut (e.g. Rasch, 1970; Smith & Orr-Weaver, 1991), but others, such as the trichogen cells of *Calliphora erythrocephala* (e.g. Ribbert, 1972) and the footpads of *Sarcophaga bullata* (e.g. Samols & Swift, 1979), are active in the pupa, forming adult tissues and degenerating shortly after emergence of the adult fly. Ovarian nurse cells retain polytene chromosomes in adult flies (e.g. Ribbert, 1979; Hartman & Southern, 1995).

The patterns of bands on polytene chromosomes in a particular species are essentially the same, regardless of the tissue in which they are found. Apart from purely technical factors, and variations in the quality of banding from one tissue to another, the main cause of apparent differences between tissues is the development of puffs, which are regions of chromatin decondensation associated with RNA synthesis (Section 15.2.4). Patterns of puffing do vary between tissues, and would thus contribute to tissue-specific differences in banding patterns of polytene chromosomes. One apparent exception to the rule of constancy of banding patterns has been found in the Mediterranean fruit fly *Ceratitis capitata*. In this species the polytene chromosomes from most tissues (salivary glands, fat bodies, hind gut) show similar patterns of bands if allowance is made for differential puffing. However, the orbital bristle trichogen cells appear to have a different pattern (Zacharopoulou *et al.*, 1991).

15.2.2   Genes and bands

Even in the 1930s, when the systematic study of polytene chromosomes was just beginning, and the true nature of genes was still unknown, it was postulated that a polytene chromosome band might be equivalent to a gene. Even now, however, when the nature of genes is much better understood, and complete genomic sequences of various organisms are beginning to become available, this question has still not been answered definitively. To do so, information is required on a number of points. How many bands are there? How many genes? Are they in the same place? Is there enough DNA in a band to form a gene?

The first comprehensive band count in *Drosophila melanogaster* was made in the 1930s by C.B. Bridges, who counted 5059 bands (Lefevre, 1976; www2.hawaii.edu/bio/Chromosomes/poly/poly.html). Although this seems a very precise figure, there are many uncertainties about it. Although the principal bands are easily recognizable, subsequent workers have not always been able to identify many of the minor bands. Certain 'doublet' bands may be artefacts of preparation. And perhaps most significant, detailed counts of bands by electron microscopy (EM; www.helsinki.fi/~saura/EM/) usually reveal many more bands than can be seen by light microscopy. Depending on which segment of the polytene chromosome was studied, 25% or more extra bands can be seen by EM (Sorsa *et al.*, 1984); in *D. hydei* 40–50% more bands can be seen by EM (Kalish *et al.*, 1985). This increase is not merely a result of the greater resolution of the electron microscope; the method of preparing the chromosomes for EM stretches the chromosomes more, thereby revealing extra bands. Thus there could be as many as 6000–7000 bands in *D. melanogaster*. Counts of the band numbers in the genomes of other flies are also available (Table 15.1), and are generally lower

Table 15.1 Numbers of bands in genomes of different species of Diptera.

Species	Number of bands	Ref.
Acricotopus lucidus	2216	Staiber & Behnke (1985)
Chironomus	~2000	Pelling (1972)
Drosophila hydei	~2000	Berendes (1965)
Drosophila melanogaster	5059	Lefevre (1976)
Drosophila virilis	~1560	Kress (1993)

than those for *D. melanogaster*. It seems inconceivable that different species of flies, even in the same genus, should have grossly different numbers of genes when their level of morphological and functional complexity must be very similar. Taken at face value, then, the very different numbers of bands in different species would be strong evidence against a one-to-one relationship between bands and genes, but considering the technical uncertainties in obtaining accurate band counts, described above, these figures should clearly not be taken too literally.

Now that it is possible to sequence whole genomes, and information has become available for *D. melanogaster*, a figure of about 13 600 genes has been estimated (Adams, M.D. *et al.*, 2000). This would indicate about two genes per band in this species but, until functions have been assigned to all the sequences that have been identified by computer programs as possible genes, there is no certainty that the actual number of genes is as high as 13 000.

Indirect approaches to clarifying the relationship between genes and bands have used mutation induced by X-rays and chemicals to define the number of complementation groups (i.e. genes) in specific chromosomal regions (Beermann, 1972). As early as 1937, Alikhanian had estimated 968 genes in the X chromosome of *D. melanogaster*, remarkably close to the number of bands on this chromosome, which is 1012. More recently, Judd *et al.* (1972) did a much more detailed study on a restricted region of the X chromosome and found 12 'functional units' in a region containing 12 bands, although these methods would fail to detect genes that are not lethal when mutated. Nevertheless, such experiments do point to a one-to-one relationship between genes and bands (Beermann, 1972).

15.2.3 Differential DNA replication in polytene chromosomes

Although the amount of DNA roughly doubles with each round of replication during the formation of polytene chromosomes, there are regions that do not replicate at all, some that replicate less than the main body of the chro-

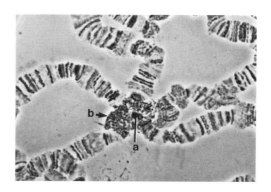

Figure 15.2 The chromocentre of *Drosophila melanogaster* polytene chromosomes, showing the dense α-heterochromatin (a) and the surrounding, more diffuse β-heterochromatin (b). Reproduced with permission from Gall *et al.* (1971) *Chromosoma* **33**, 319–344. © Springer-Verlag.

mosome and some segments that have extra rounds of replication (Spradling & Orr-Weaver, 1987).

Constitutive heterochromatin is often significantly under-replicated. In *D. melanogaster* it is not replicated at all (Gall *et al.*, 1971) so neither the centromeric heterochromatin nor the wholly heterochromatic Y chromosome can be detected in polytene nuclei (Lefevre, 1976). As in most Drosophilids, the polytene chromosomes are all joined together at their centromeres to form a chromocentre. In theory the centromeric heterochromatin must form a small region in the middle of the chromocentre, known as α-heterochromatin, which is surrounded by the diffuse, fuzzy β-heterochromatin (Fig. 15.2), which does not show any bands. β-Heterochromatin was defined as material lying between the α-heterochromatin and the euchromatic parts of the polytene chromosomes, but the situation is not as simple as that. The gene density in β-heterochromatin is claimed to be similar to that in euchromatin (Miklos & Cotsell, 1990), and thus it should perhaps not be regarded as heterochromatin at all. More intriguing is the finding that β-heterochromatin contains a high level of middle repetitive sequences, in particular the transposable *P*-elements. The *P*-elements are interspersed with the satellite DNA sequences

Table 15.2 Amplification of heterochromatin in polytene chromosomes of Diptera.

Species	Amplification	Ref.
Chironomus	C-Bands present in polytene chromosomes	1
Chrysomya bezziana	Variable degree of under-replication	2
Drosophila melanogaster	Satellite DNA: none	See text
	P-Elements: as euchromatin	
Drosophila nasutoides	Under-replication	3
Lucilia cuprina	Under-replication of sex chromosomes	4
Prodiamesa olivacea	Under-replication	5
Sarcophaga bullata	~10-fold amplification	6
Sciara coprophila	1% of polytene chromosomes compared with 20% in mitotic chromosomes	7
Simulium spp.	C-Bands present in polytene chromosomes	8

References: 1, Hägele (1977); 2, Bedo (1994); 3, Zacharias (1993); 4, Bedo (1982); 5, Zacharias (1979); 6, Samols & Swift (1979); 7, Eastman et al. (1980); 8, Bedo (1975).

that make up the centromeric heterochromatin of the mitotic chromosomes; however, when poly-tenization occurs, these moderately repetitive sequences are replicated to the same level as euchromatic regions of the chromosomes (Elgin, 1996). This interspersion of non-amplified satellite DNAs and amplified middle repetitive sequences no doubt accounts for the unusual cytological appearance of the chromocentre and of β-heterochromatin; a strict linear order of the DNA sequences is not maintained in this situation.

Less information is available about the degree of replication of the centromeric heterochromatin in other species of flies, but it is clear that it is very variable (Table 15.2), from perhaps one-hundredth of the amount of replication of the euchromatin in *Sarcophaga bullata*, about one-twentieth in *Sciara coprophila*, up to some situations where there is only a very small degree of under-replication, although definitive evidence for replication of centromeric heterochromatin to the same level as that of the euchromatin is lacking.

Regions of polytene chromosomes subject to position effect variegation (PEV, Section 7.4.5) are also commonly under-replicated, but the degree of under-replication varies widely, from about 3% to about 75% of the level found in the rest of the polytene chromosomes (Umbetova

et al., 1991; Wallrath *et al.*, 1996). Nucleolus organizer regions (NORs) are also under-replicated, at least in *D. hydei* (Hennig & Meer, 1971), where the ploidy of the ribosomal genes is about 128X, compared with 1024X for the rest of the chromosome. In *D. melanogaster* the *Ubx* sequence is under-replicated in salivary gland polytene chromosomes and forms a constriction, but in fat body polytene chromosomes this sequence is not under-replicated and there is no constriction (Lamb & Laird, 1987).

Although it is usual to refer to sequences such as those just described as 'under-replicated', there is really no convincing evidence for this. It merely seems to be the most likely explanation. Glaser *et al.* (1992) have proposed, on the basis that stalled replication forks are not found at the boundaries of under-replicated segments, that at least in the case of heterochromatin, the reduction in chromatin might be due to DNA elimination instead.

As well as under-replication of DNA, extra rounds of DNA replication occur at certain loci on polytene chromosomes of flies in the family Sciaridae during the late larval stages, and are manifested as the so-called DNA puffs (Fig. 15.3). Such loci have been reported from *Sciara* (Crouse & Keyl, 1968), *Rhynchosciara* (Pavan & da Cunha, 1969) and *Bradysia* (Coelho *et al.*, 1993), and represent a two-, four- or eightfold amplifi-

Figure 15.3 Deoxyribonucleic acid puffs (arrows) on polytene chromosomes of *Rhynchosciara baschanti*. Larvae were injected with ³H-thymidine, incorporated especially strongly in the DNA puffs, which have undergone extra rounds of replication. Scale bar = 10 µm. Micrograph kindly provided by A.J. Stocker.

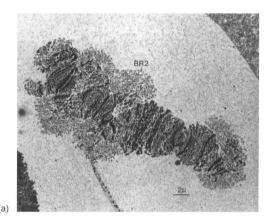

(a)

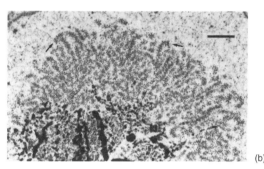

(b)

Figure 15.4 (a) Balbiani rings (giant RNA puffs) on polytene chromosome IV of *Chironomus tentans*. The chromosome banding pattern is also visible in this electron micrograph. Scale bar = 2 µm. Reproduced with permission from Daneholt (1975) *Cell* **4**, 1–9. © Cell Press. (b) Higher power electron micrograph of a Balbiani ring, showing the loop structure (arrows). Scale bar = 1 µm. Reproduced with permission from Daneholt (1992) *Cell Biology International Reports* **16**, 710. © Academic Press.

cation of the DNA at these loci. As well as active DNA synthesis, there is RNA synthesis at these loci, and it has been shown that the DNA puffs do contain amplified gene sequences (Glover *et al.*, 1982; Coelho *et al.*, 1993).

15.2.4 Puffs and transcription

Although DNA puffs are apparently confined to members of the Sciaridae, RNA puffs (Fig. 15.4) are found in all Dipteran polytene chromosomes; the largest ones are known as Balbiani rings. They are a manifestation of intense RNA synthesis, and patterns of puffing are specific both to the tissue from which the chromosomes are

derived and to the stage of development. As well as developmental changes in puffing patterns, puffs can be induced or caused to regress by a variety of experimental stimuli: ecdysone, juvenile hormone, heat shock and various other agents (e.g. Ashburner, 1972; Berendes, 1972). Such experiments provide clear evidence of a physiological response to the various stimuli, leading to increased or decreased RNA synthesis, followed by a corresponding change in synthesis of specific proteins. In salivary glands, the formation of puffs and Balbiani rings has been correlated with the requirement for the synthesis of large amounts of salivary gland proteins. For a detailed review of genes found in Balbiani

Table 15.3 Characteristics of DNA and endoreduplicated chromosomes in ciliates (after Ammermann, 1987).

Species	Giant chromosomes	No. of giant chromosomes	No. of bands	DNA size in macronucleus
Subclass Gymnostomata				
Loxophyllum	Polytene			
Subclass Vestibulifera				
Bursaria	Oligotene			High MW, >20 kb
Subclass Hypostomata				
Chilodonella cucullulus	Polytene			High MW
Chilodonella steini	Polytene			High MW
Chilodonella uncinata	Oligotene			
Subclass Suctoria				
Ephelota	Oligotene			
Subclass Spirotricha				
Order Heterotricha				
Nyctotherus cordiformis	Polytene	1		
Order Hypotrichida				
Stylonychia lemnae	Polytene	70	~10 000	Gene-sized
Oxytricha	Polytene	~120	~10 000	Gene-sized
Euplotes	Polytene			Gene-sized

MW, molecular weight.

rings in *Chironomus tentans*, see Wieslander (1994).

Puff formation involves loosening of the polytene chromosome structure over a region of the chromosome that may involve several bands and interbands. Chromatin fibres are looped out from the body of the chromosome to make a more diffuse structure in which individual chromatin fibres can be distinguished. The loops are highly decondensed, forming a fibre only 5 nm in diameter that is free of nucleosomes at the upstream end. In the region where transcription occurs, RNA polymerase particles are arranged along the fibre, with 20 nm ribonucleoprotein (RNP) fibres growing from them, which as they elongate reorganize themselves into RNP globules up to 50 nm in diameter (Andersson *et al.*, 1980, 1982; Ericsson *et al.*, 1989). As transcription declines, the RNP particles are lost, and the 5 nm fibre condenses to one of about 25 nm. These structures therefore appear to be very similar to

RNA-synthesizing loops on lampbrush chromosomes (Section 14.2).

15.3 Polytene chromosomes and macronucleus formation in ciliates

The other polytene chromosome system that has been studied in detail is found in certain ciliates (*Euplotes*, *Oxytricha*, *Stylonychia*, etc.; see Table 15.3), in which the formation of polytene chromosomes is one stage in the production of a new macronucleus after conjugation. Old macronuclei degenerate, and one of the new micronuclei undergoes a complex series of changes to produce a new macronucleus. The micronucleus is transcriptionally inactive, contains normal chromosomes and can undergo mitosis and meiosis. All the transcription that occurs in the cell takes place in the macronucleus, which instead of normal chromosomes contains pieces

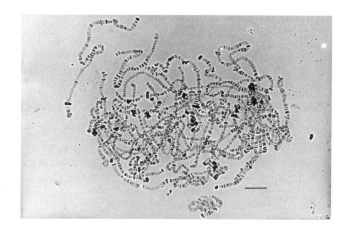

Figure 15.5 Polytene chromosomes from a macronucleus of the ciliate *Stylonychia lemnae*. Scale bar = 20 μm. Reproduced with permission from Kraut *et al.* (1986) *International Review of Cytology* **99**, 1–28. © Academic Press.

of chromatin that each contain only a single gene. The macronucleus cannot divide by mitosis but instead divides amitotically, when the multiple copies of each gene are randomly distributed to the daughter macronuclei. Various aspects of these processes have been reviewed by Ammermann (1987), Klobutcher & Jahn (1991), Prescott (1992, 2000), Coyne *et al.* (1996) and Lipps & Eder (1996).

15.3.1 Formation and degradation of polytene chromosomes

Immediately after conjugation and mitosis to form two diploid nuclei, the nucleus that is selected to become the macronucleus (the 'macronucleus anlagen') undergoes many rounds of replication. In certain species (Table 15.3) giant polytene chromosomes result; in others, either the number of rounds of replication is fewer, or the newly replicated chromatids do not remain together, and 'oligotene' chromosomes result, with only a few parallel chromatin strands. Yet other species (e.g. *Paramecium*, *Tetrahymena*) produce neither polytene nor oligotene chromosomes although they do amplify their DNA. The reason why some species form polytene chromosomes and others do not is unknown. Species that form polytene chromosomes are widely scattered among different groups of ciliates, yet within the same genus (*Chilodonella*; Table 15.3) some species have polytene chromosomes but another has oligotene chromosomes.

Ciliate polytene chromosomes look remarkably similar to Dipteran polytene chromosomes (Fig. 15.5), at least by light microscopy. In some species, over-replication of certain bands occurs, and DNA puffs have been observed in *Chilodonella cucullulus*. However, there is no transcription from ciliate polytene chromosomes, and so there are no RNA puffs. At the ultrastructural level, ciliate polytene chromosomes look rather different from those of Diptera: the interbands consist of parallel 10 nm chromatin fibres, but the bands ('chromomeres') consist of aggregates of loops of 30 nm chromatin fibres. Polytene chromosomes in ciliates are a transitory phase, as they are chopped up into small segments by the formation of proteinaceous septa (Fig. 15.6). The chromosomes are degraded, the DNA is reduced to 'gene-sized' pieces and these are then amplified again to produce the mature macronucleus, which of course shows no sign of polytene chromosome structure.

15.3.2 Excision of unwanted DNA

These changes in chromosome morphology and nuclear DNA content are accompanied by a series of changes in the composition of the macronuclear DNA. Multiple rounds of replication are needed to produce the polytene chromosomes, but even at this stage some DNA – the internal eliminated sequences (IESs) – is removed (Prescott, 2000). Once the polytene chromosomes have been formed, transposon-like

190   Chapter 15

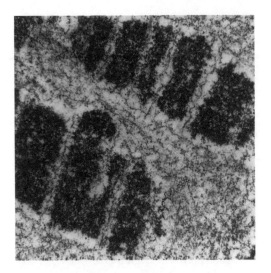

Figure 15.6 Macronucleus from the ciliate *Oxytricha nova*, showing the polytene chromosomes in the process of being divided up by the growth of septa between the bands. Reproduced with permission from Prescott (1992) *Bioessays* **14**, 317–324. © John Wiley & Sons.

sequences as well as IESs and non-coding parts of genes are excised and destroyed. In *Oxytricha nova* all the repetitive sequences (40% of the total DNA) and 95% of the unique sequences are removed (Prescott, 1992). In *Stylonychia lemnae*, only 20% of the chromosomes form polytene chromosomes, the other 80% being destroyed without ever being amplified; of the DNA that does form polytene chromosomes, more than 90% is destroyed.

In *Euplotes*, the excised sequences, Tec1 and IESs, are bounded by short direct repeats that appear to be markers for the sites of cutting. Just one copy of the repeat sequence is retained at the ends of the 'gene-sized' sequences that are ultimately produced (Tausta *et al.*, 1991). Conserved sequences (E-Cbs) near the fragmentation sites have a core sequence, 5′-TTGAA-3′, that is recognized by telomerase and can form a substrate for the synthesis of new telomeres at the ends of the gene-sized fragments (Klobutcher *et al.*, 1998). Other species do not appear to have the same strict sequence requirements for telomere addition, however (Coyne *et al.*, 1996). The new telomeres are synthesized considerably longer than they will eventually become,

and are subsequently pared down to the short, tightly regulated lengths found in the mature macronucleus (Roth & Prescott, 1985; see also Table 13.1).

15.3.3 Reshuffling of genes

Many of the genes in hypotrichous ciliates cannot function in the form in which they are found in the micronucleus because different segments of the genes are in the wrong order, and are interrupted by other sequences. As a result of the processes of amplification, cutting, elimination and splicing, the different segments of a gene can be spliced together to produce functional genes in the macronucleus (Prescott, 2000).

The result of all these processes is macronuclei that contain many millions of gene-sized DNA molecules, each with telomeric sequences at both ends. In *Euplotes*, these molecules average ~1830 bp in length, although the molecules that contain ribosomal RNA genes are ~7400 bp long. Each gene is present on average in about 1000 copies, although different genes have different copy numbers: the ribosomal RNA genes of *Euplotes* are present in 10^5 copies, for example (Prescott, 1992).

15.4 Mammalian polytene chromosomes

Among mammals, polytene chromosomes with a ploidy between 32C and 2048C are found mainly in the giant trophoblast cells of the placenta (Fig. 15.7; Zybina & Zybina, 1996). These polytene chromosomes only have a transitory existence, after which they break down into smaller fragments that appear to contain an exact multiple of the diploid amount of DNA. Morphologically, mammalian polytene chromosomes do not have the highly organized banded structure found in Diptera and ciliate polytene chromosomes, and, except in heterochromatic regions and around the NORs, the chromatids are often separated instead of being closely bound together. The failure of the chromatids to cohere closely may be the reason why bands are not generally developed. Unlike Dipteran polytene chromosomes, there is no indication that heterochromatin is under-replicated in giant trophoblast cells. Occasionally puff-like structures can be seen, but it is not clear if these are RNA

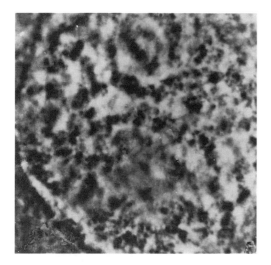

Figure 15.7 Polytene chromosomes of rabbit giant trophoblast cells. Reproduced with permission from Zybina & Zybina (1996) *International Review of Cytology* **165**, 53–119. © Academic Press.

or DNA puffs. Blocks of heterochromatin are often surrounded by loops of DNA, but these are hardly likely to correspond to puffs.

Trophoblast giant chromosomes appear in two different forms: one with the chromatin decondensed and dispersed throughout the cytoplasm, and the other in which they appear as condensed bundles of chromatin fibres. The former represent the S phase, and the latter the G phase. Such a morphologically conspicuous alternation of phases does not seem to have been reported in Diptera or ciliates.

15.5 Polytene chromosomes in plants

Polytene chromosomes have been found in a variety of flowering plants, both monocots and dicots (Table 15.4). In many ways they resemble mammalian polytene chromosomes: they are generally in tissues associated with embryonic development, they generally lack clear banding patterns, the chromatids are often only loosely held together (Fig. 15.8) and they are held

Table 15.4 Polytene chromosomes in plants.

Species	Cell type	Ploidy
Monocots		
Allium spp.	Antipodal cells	
	Endosperm haustoria	
	Synergids	
Alisma plantago-aquatica	Suspensor	512C
Clivia miniata	Antipodal cells	32C
Scilla bifolia	Antipodal cells	1024C
Triticum spp.	Antipodal cells	196C
Zea mays	Endosperm	24C
Dicots		
Aconitum spp.	Antipodal cells	128C
Bryonia dioica	Anther hairs	256C
Dicentra spectabilis	Antipodal cells	
Eruca sativa	Suspensor	75C
Papaver rhoeas	Antipodal cells	128C
Phaseolus spp.	Suspensor	2048–8192C
	Endosperm	96–192C
Rhinanthus spp.	Endosperm haustoria	384C
Thesium alpinum	Endosperm haustoria	384C
Tropaeolum majus	Suspensor	2048C

Data from Nagl (1978).

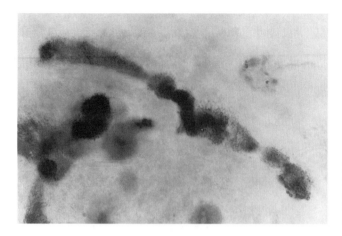

Figure 15.8 Isolated polytene chromosome from the plant *Phaseolus coccineus*, showing the lack of clear bands. Micrograph kindly provided by A. Pedrosa.

together mainly by heterochromatin (Nagl, 1978). During replication the chromosome structure becomes more diffuse (Brady & Clutter, 1974). Both α- and β-heterochromatin have been reported (Brady & Clutter, 1974), although the nature of the latter is not clear: it replicates after the euchromatin but before the α-heterochromatin. Heterochromatin is well replicated in at least some plant polytene chromosomes, as it still appears as large blocks using staining methods for heterochromatin (Schweizer, 1976).

Puffs have been described in plant polytene chromosomes, just as in polytene chromosomes from organisms other than ciliates. An extreme case is the NORs, where the individual chromatids have become very widely spaced throughout the nucleolus (Schweizer, 1976; Nagl, 1978; Schweizer & Ambros, 1979).

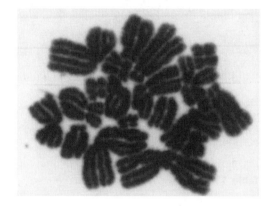

Figure 15.9 Diplochromosomes from a Chinese hamster ovary (CHO) cell. These are the result of two rounds of chromosomal replication without an intervening separation of the chromatids. Repetition of this process would produce polytene chromosomes.

15.6 Mechanisms of polytenization

How do chromosomes become polytene? We have already seen that there are mechanisms to prevent re-replication of DNA until the cell has divided (Section 2.2.2.2); however, it is not in fact too difficult to override such mechanisms, and a variety of treatments can induce the formation of diplochromosomes, which are the result of two rounds of DNA replication without the chromosomes separating (Fig. 15.9). In diplochromosomes the sister chromatids are

clearly separate, so a second question is: what holds the chromatids together? Research in these fields has only recently begun, so it is not yet possible to give comprehensive answers to these questions.

As *Drosophila* cells enter endocycles, cyclins A and B are lost, and pulses of cyclin E drive the S phases of the endocycles (Follette *et al.*, 1998). Cyclin A normally blocks replication, so in its absence the endocycle can proceed. Cyclin E induces DNA synthesis, but is inhibited by a feedback mechanism: high cyclin E represses cyclin E activity so that there is an oscillation in its level, and replication proceeds in cycles (Sauer et al., 1995). Minichromosome maintenance (MCM) proteins are also required for DNA replication, and in *Drosophila* polytene chromosomes they occur either associated with or dissociated from the chromosomes (Su & O'Farrell, 1998). Cyclin E causes the association of MCM proteins with DNA, and DNA replication causes the dissociation of MCM proteins. In maize endosperm, endoreduplication requires the inhibition of MPF (M-phase promoting factor) and the induction of S-phase-related protein kinases (Grafi & Larkins, 1995). Thus the normal mitotic cell cycle controls are modified so that DNA replication is not inevitably followed by mitosis (cf. Chapter 2).

15.7 What is the point of polytene chromosomes?

Why do polytene chromosomes develop in certain tissues of certain organisms? The diverse situations in which they are found suggests that there is unlikely to be a single answer, except insofar as all the cells that have polytene chromosomes are terminally differentiated and do not divide mitotically. This applies even in ciliates, where the polytene chromosomes are simply one stage in macronucleus formation; the macronuclei cannot go on dividing indefinitely, but need to be regenerated from the micronuclei from time to time. In any case, it would probably be difficult for a cell with chromosomes as large as polytene chromosomes to segregate them properly at a mitotic division.

It has been suggested that the complex series of amplifications and eliminations of non-genic DNA during the formation of the ciliate macronucleus are a means of getting rid of unnecessary material and producing a more efficient nucleus designed solely for transcription. It would be ironic if this were so, given the complexity of the processes involved in macronucleus formation. In any case, ciliates probably need to develop a large macronucleus, with multiple copies of genes, to provide for the needs of a very large cell.

In multicellular organisms with polytene chromosomes, reasons for polyteny are less obvious, although there could be some important nucleotypic effect that requires a large nucleus (Cavalier-Smith, 1978). Another possibility is that differential polytenization allows a disproportionate increase in the dosage of genes required for cell-specific functions, and a disproportionate decrease in the dose of genes only needed in small quantities in the differentiated cell (Nagl, 1978). Whatever the reasons, polytene chromosomes are certainly not disadvantageous, as Dipteran flies are one of the more successful groups of organisms in the world. Meanwhile, they have provided scientists with a wonderful material with which to study many intriguing aspects of chromosome organization and function.

Websites

Drosophila polytene chromosome maps
www2.hawaii.edu/bio/Chromosomes/poly/poly.html
www.helsinki.fi/~saura/EM/

Plant polytene chromosomes, particularly those of Phaseolus
Beanref: www.ba.cnr.it/Beanref/polytene.htm

Chromosomes, the karyotype and evolution

16

16.1 Chromosomes and evolution

Different organisms have very different sets of chromosomes (or karyotypes), and in general the karyotypes of closely related species are more similar to each other than they are to the karyotypes of distantly related species. Changes in chromosome size and morphology are therefore evidently characteristic of the evolutionary process, and it is possible to describe the numerous ways in which chromosomes and whole genomes change during evolution (Section 16.3). But is chromosome change just a chance accompaniment of evolution, or is it a directed process? The variety of chromosomal changes during evolution, and the enormous differences in the amount of change that has occurred between closely related species, certainly seem to indicate that no specific changes are required for speciation, although particular types of changes may occur quite commonly in particular groups. On the other hand, certain kinds of chromosomal change may tend to promote speciation, because of reduced fertility in hybrids (King, 1993; Section 16.4). Before going on to a consideration of these points it is, however, worth considering what constraints, if any, there are on the karyotype of any organism (Section 16.2). Are there maximum and minimum limits on the numbers of chromosomes that an organism could have, beyond which mitotic and meiotic division would cease to function effectively? Are there constraints on the size or morphology of individual chromosomes?

16.2 Constraints on chromosome size, shape and number

Chromosome numbers per cell range from 1 to over 600 pairs. A single chromosome pair is found in an ant (Crosland & Crozier, 1986) and in the nematode *Parascaris univalens*. The latter is somewhat heterodox as the single chromosome pair is only found in the germ line and very early embryo; the chromosomes fragment and heterochromatin is lost during differentiation of somatic cells (Section 7.4.1). At the other extreme, a fern has been described with over 630 pairs (Otto & Whitton, 2000). Thus there appear to be no fundamental problems that affect the mechanisms of cell division caused by either low or high chromosome numbers. Nevertheless, if a chromosome were too long, it might not be possible to pull the daughter chromosomes far enough apart at anaphase before the cell and the chromosome were cut across by the cleavage furrow (Schubert & Oud, 1997). *Parascaris univalens* does not have this problem as its chromosomes are holocentric and are attached to the spindle throughout their length (Section 12.5); thus there are no lagging chromosome arms in this species. Nevertheless, it seems probable that most chromosomes are well below the maximum size that can be accommodated by the cell.

A different problem might arise with a very large number of chromosomes. If each one is to be segregated properly at a normal mitosis, it must be attached to at least one spindle microtubule. Each microtubule is about 25 nm in diameter, so the microtubules necessary to attach to 500 chromosomes would occupy a cross-sectional area about 0.625 μm in diameter. This may seem quite a modest bulk, but room must also be found for other cellular components, and for cytoplasm between the microtubules. Nevertheless, this is obviously an arrangement that works. However, it seems highly unlikely that the gene-sized chromosome fragments in ciliate macronuclei (Section 15.3) could be segregated mitotically, because the number of gene-sized pieces runs into millions. A million microtubules would occupy a block nearly 30 μm in diameter, which is clearly too many for even a very large cell to accommodate comfortably. So again there must be an upper mechanical limit to the number of chromosomes that can be segregated efficiently, but it has probably not been reached even in the organisms with the highest known chromosome numbers.

It is possible to imagine other problems with high chromosome numbers, such as a greater tendency to loss at anaphase. It could be that in cells with large chromosome numbers the checkpoint that prevents progression to anaphase until all the chromosomes are attached to the spindle (Section 2.3.2) might be less efficient, because the signal from only one unattached chromosome out of a very large number might not be adequate to delay mitotic progression. Such a hypothesis would predict greater chromosome loss in cell division in species with high chromosome numbers but, because we have few or no reliable data on chromosome loss in any species, we are not yet in a position to test it.

Chromosome size is, of course, inversely related to the number of chromosomes in an organism, but is also directly related to the amount of nuclear DNA. Organisms with very large genomes and therefore very large chromosomes seem to have no difficulty in segregating their chromosomes. On the other hand, there are always potential problems with segregating very small chromosomes, particularly at meiosis. In chiasmate meiosis, the formation of a chiasma is usually necessary for proper chromosome segregation. Small chromosomes normally form only one chiasma, and if this should fail to form, non-disjunction and aneuploidy will result. In the achiasmate

meiosis of male *Drosophila*, there is a small but significant rate of loss of the smallest (4th) chromosome (Section 7.4.4). Small size itself may be a significant factor in this loss. Minichromosomes are lost at mitosis and particularly at meiosis in *Drosophila*, mammals and the bean *Vicia faba* (Schubert, 2001). Very small yeast artificial chromosomes (YACs) are often lost at cell division, but larger ones are more stable (Section 18.3.1). There may well be, therefore, a lower limit on the size for a chromosome to be transmitted efficiently from one cell generation to the next.

There appears to be no evidence that chromosome shape is of any great significance. The centromere can and does occur in any position from the middle to the end of the chromosome; some organisms have all acro- or telocentric chromosomes, others have all meta- or submetacentric chromosomes, and others have a mixture of types (Box 16.1). In general it is difficult to see that any particular type of chromosomal morphology might have any advantage over any other. An exception to this rule might be very large acro- or telocentric chromosomes, which might suffer the problem mentioned above of being cut across by the cleavage furrow before they have been properly separated at anaphase; a metacentric of the same total length would have arms of only about half the length of the long arm of an acro- or telocentric, and would thus be much more likely to survive intact.

16.3 Types of chromosome change during evolution

The discussion in the previous section seems to indicate that there are no particularly strong factors that might determine the form that the karyotype might take. In this section we shall discuss the actual changes that have been observed, or rather the differences that are found between related species. For much more detailed reviews of these matters than can be given here, see White (1973) and King (1993). In general, it is not easy to determine the direction of chromosomal change, although comparison with an outgroup can be helpful. Thus an apparent addi-

Box 16.1 Chromosome shape and centromere position

All the chromosomes in the same species have much the same width, and therefore chromosome shape is determined by length, and by the positions of the centromere (primary constriction) and the nudeolus organizer region (secondary constriction). Chromosomes can be classified by the position of the centromere (Fig. 1). *Metacentric* chromosomes have the centromere at or close to the middle of the chromosome, and *acrocentric* chromosomes have their centromeres close to one end. Intermediate conditions can be referred to as *sub-metacentric* when the centromere is some way from the middle of the chromosome but closer to the middle than to the end, and as *sub-acrocentric* when the centromere is closer to the end than to the middle. *Telocentric* chromosomes were originally defined as chromosomes with a strictly terminal centromere

produced by centromere misdivision or breakage within the centromere. In the light of current knowledge of telomere organization and behaviour, such chromosomes would be unstable. However, mouse chromosomes in which the centromeric DNA is linked directly to the telomeric sequences can reasonably be regarded as telocentric (Kipling *et al.*, 1991).

Chromosome shape can also be defined in terms of the centromeric index or the arm ratio. The centromeric index is the length of the shorter arm divided by the total chromosome length, and thus varies from 0.5 for a truly metacentric chromosome to zero for a telocentric one. The arm ratio is the length of the long arm divided by the length of the short arm, and thus ranges from unity for a truly metacentric chromosome to infinity for a truly telocentric chromosome.

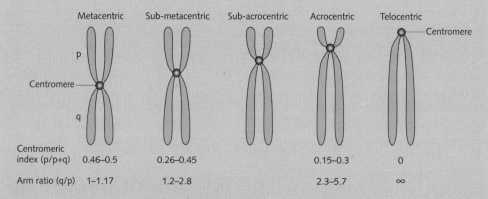

Figure 1 Chromosome shape and centromere position. See text for further explanation.

tion of heterochromatin in some species might really be a loss in other species, or what is described as chromosome fusion might really be chromosome fission.

16.3.1 Methods of studying chromosome homology between species

Evolutionary studies of chromosomes require methods for identifying the same (that is, homol-

ogous or homoeologous) chromosomes or chromosome segments in different organisms. A surprisingly large amount of work was done before the invention of chromosome banding and chromosome painting, but even then much of it involved *Drosophila* and other species with polytene chromosomes (Chapter 15) whose detailed banding patterns are obviously ideal for studying chromosome change and chromosome relatedness. Banding techniques (see Boxes 10.1–10.4)

Box 16.2 Conserved synteny and conserved chromosome segments

Synteny is the occurrence of two or more genes on the same chromosome. Synteny differs from linkage because two genes on a large chromosome may be separated by sufficient crossover events that they do not appear to be linked; they are nevertheless syntenic. *Conserved synteny* is when two or more genes that are found on a single chromosome in one species are also found on a single chromosome in another species. Syntenic genes lie in a *conserved segment* when the linear order of the genes has been maintained between species without rearrangements or the insertion of non-syntenic segments.

are obviously limited to comparisons where sufficient of the pattern is conserved for it to be recognizable in different species; banding allows quite precise location of breakpoints, and is also valuable for studying changes in heterochromatin. Chromosome painting (see Box 5.2) identifies homologous chromosome segments in different species by labelling the non-satellite DNA sequences that they have in common. The method can detect small chromosome segments, right down to the limit of resolution of the microscope and the limit of detectability of the chromosome paint. Identification of breakpoints must depend on length measurements, and can only be correlated with banding patterns by double staining or by assuming that there is no differential condensation of the chromosome. In species in which gene mapping is sufficiently advanced, homologous segments in different species can be identified by conserved synteny (Box 16.2).

16.3.2 Speciation with little or no chromosome change

Gross chromosomal changes do not necessarily accompany speciation, and many examples can be given of groups in which there is remarkable similarity of chromosomes between species (Sumner, 1990, p. 313; King, 1993). Among mammals this is true of cats and seals, certain primates, some marsupials and various other groups. Birds are generally very conservative karyotypically, and four species of gulls were reported to have indistinguishable karyotypes. Among Hawaiian *Drosophila* species, 67 fall into 18 homosequential groups; that is, groups that have identical banding patterns on their polytene chromosomes (Carson, 1981).

Even when whole karyotypes have not been maintained unaltered, it may be possible to recognize individual chromosomes that appear to have remained unchanged during the divergence of species, even between different orders of mammals (Chowdhary *et al.*, 1998). Several human chromosomes have been identified unaltered in cats (O'Brien *et al.*, 1997). Some chromosomes have apparently been maintained over extraordinarily long periods of time; among turtles, certain chromosomes have been conserved for something like 200 Myr (Bickham, 1981). A degree of conservation can even be found between certain human and chicken chromosomes, which must have diverged between 300 and 350 Myr ago (Nanda *et al.*, 1999; Chowdhary & Raudsepp, 2000). It remains to be shown that there are no differences at all between apparently conserved chromosomes in distantly related groups, and in the case of the human–chicken homology the chicken chromosome is known to be smaller (see also Section 16.3.6).

16.3.3 Chromosome rearrangements

In spite of the examples given in the previous section, karyotypes usually differ between organisms, even closely related ones, and many of these differences are due to chromosomal rearrangements. These include translocations and inversions (both pericentric and paracentric), duplications and tandem fusions. Robertsonian fusions and fissions are special types of rearrangements that occur commonly and are considered in the next section (16.3.4). All these types of rearrangements have occurred commonly in evolution (White, 1973; King, 1993). As a result of translocations, material that comprises a single chromosome in one organism can become distributed among two or more chromosomes in another (Fig. 16.1). Some spectacular rearrangements have occurred among rodents. Material that forms a single chromosome in rats or mice may be distributed over several different human chromosomes, and vice versa (O'Brien *et al.*, 1999; http://www.sciencemag.org/feature/data/ 1044631.shl; IHGSC, 2001). Each rodent chromosome may contain segments homologous with parts of as many as 6, 7, 8 or even 9 human chromosomes, and human chromosomes may contain segments homologous with parts of up to 6 or 7 mouse chromosomes. Human and mouse gene mapping shows that there are about 183 chromosome segments that are conserved between these species (IHGSC, 2001).

Pericentric and paracentric inversions are not detectable using whole chromosome paints, but have been demonstrated by banding in a wide variety of groups including reptiles, birds and mammals (Sumner, 1990, p. 319; King, 1993, pp. 80–84).

Duplications, varying in size from 1 to 400 kb, have turned out to be important in the evolution of primates, including humans (Eichler, 2001; Bailey *et al.*, 2002; van Geel *et al.*, 2002). About 5% of the human genome and nearly 11% of chromosome 22 consist of duplications. Duplications can be derived from non-homologous chromosomes, or occur within a specific chromosome. The chromosomal distribution of duplicated segments is non-random, with concentrations in the pericentric and subtelomeric regions (van Geel *et al.*, 2002).

Tandem fusions have been reported in the chromosomes of various species (Sumner, 1990, p. 320), but the most famous example is that of the Indian muntjac. The Chinese muntjac (*Muntiacus reevesi*) has $2n = 46$ chromosomes, which is a typical mammalian number, and all its chromosomes are acrocentric. The Indian muntjac (*M. muntjak vaginalis*) has only six (female) or seven (male) chromosomes (having an XY_1Y_2 sex chromosome system) yet can form viable hybrids with *M. reevesi*. The reduction in chromosome number has occurred largely by tandem fusions, although other types of rearrangement have occurred as well (Yang *et al.*, 1997). Small segments of centromeric satellite DNA (Frönicke & Scherthan, 1997) (Fig. 16.2) and telomeric DNA (Lee *et al.*, 1993) remain at the sites of fusion.

Human chromosome 2 is also the result of end-to-end fusion. Separate chromosomes homologous to the long and short arms of human chromosome 2 are found in gorilla, chimpanzee, pygmy chimpanzee and orang-utan. Telomeric sequences are still present at the point of fusion in human chromosome 2 (Azzalin *et al.*, 2001). These consist of two arrays of TTAGGG orientated in opposite directions and flanked by low-copy number repeats derived from the subtelomeric regions (Fig. 16.3). Intrachromosomal telomeric repeats on human chromosome 1 may also have originated from chromosome fusion (Azzalin *et al.*, 2001).

16.3.4 Robertsonian fusion and fission

Robertsonian fusion, in which the centromeric regions of two acro- or telocentric chromosomes fuse to form a single meta- or submetacentric chromosome, is a very common evolutionary change and has been reported in most groups of organisms. It can occur sporadically in humans (at a frequency of about 1 in 1000 births; Choo, 1990) and in other organisms. Robertsonian fission – the splitting of a metacentric chromosome at the centromere to form two telocentrics – has been reported much less frequently, although as already pointed out, it is often not possible to be certain of the direction of evolutionary changes in chromosomes. The terms centric fusion and fission have also been used to describe such rearrangements.

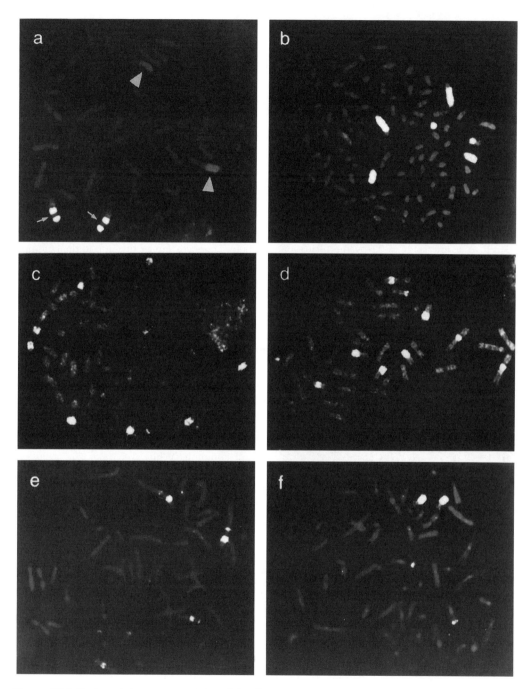

Figure 16.1 Reciprocal chromosome painting of dog, fox and human chromosomes: *in situ* hybridization of paints for (a) dog chromosomes 1 (arrowheads) and 18 (arrows) to fox chromosomes; (b) fox chromosome 1 to dog chromosomes; (c) human chromosome 3 to dog chromosomes; (d) human chromosome 3 to fox chromosomes; (e) dog chromosome 10 to human chromosomes; (f) dog chromosome 9 to human chromosomes. In all cases material that forms a single chromosome in one species is distributed among more than one chromosome or chromosome segment in another species. Reproduced with permission from Yang *et al.* (1999) *Genomics* **62**, 189–202. © Academic Press.

Among mammals, sheep differ from goats largely in having three pairs of metacentrics whose banding patterns correspond to those of six acrocentrics in the goat (Evans *et al.*, 1973), and many other examples of centric fusion have been reported in the Bovidae (Buckland & Evans, 1978; Bunch & Nadler, 1980). Centric fusion is a process that is continually occurring, and different populations of the house mouse provide extraordinary examples of this process. Most populations have a karyotype consisting of 40 acrocentrics, but numerous populations exist, often in isolated places such as alpine valleys, in which the chromosome number has been reduced to as low as $2n = 22$ by the formation of metacentrics from acrocentrics (Nachman &

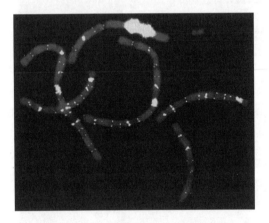

Figure 16.2 Indian muntjac chromosomes: *in situ* hybridization with centromeric satellite DNA shows small segments at intervals along the chromosomes that represent the remnants of centromeric satellite from the individual chromosomes that have fused to make the small number of very large chromosomes in this species. Reproduced with permission from Frönicke & Scherthan (1997) *Chromosome Research* **5**, 254–261. © Kluwer Academic Publishers.

Searle, 1995; Searle, 1998). (In $2n = 22$ mice, all the autosomes except one pair have fused to form metacentrics). Each race is characterized by its own combinations of acrocentrics to form metacentrics; one, the tobacco mouse ($2n = 26$) has been regarded as a separate species (*Mus poschiavinus*). The speed with which such centric fusions can accumulate is illustrated by populations of mice on the island of Madeira. It seems probable that the house mouse reached Madeira in the fifteenth century; there are now six separate chromosomal races that have sets of Robertsonian fusions that differ from each other and from races elsewhere (Britton-Davidian *et al.*, 2000). There is a similar situation in the common shrew (Searle & Wójcik, 1998). Numerous examples of centric fusions in other groups could also be described (White, 1973; King, 1993), but the house mouse has long been of importance for genetic studies, and among mammals banding techniques have allowed the easy identification of the chromosomes involved in the fusions.

16.3.5   Changes in heterochromatin

Differences in the quantity, position and properties of heterochromatin among related species are very common (White, 1973; Sumner, 1990, pp. 314–318; King, 1993, pp. 84–86). Examples can be found in plants, insects and vertebrates (including mammals) and the reader is referred to the books just cited for more details. Here only the different types of changes in heterochromatin will be discussed. Among many rodents and some birds, heterochromatic short arms of chromosomes occur in some species and not others, with a corresponding difference in the total amount of nuclear DNA. Loss of hete-

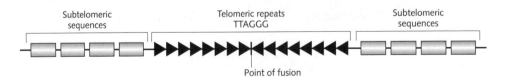

Subtelomeric sequences   Telomeric repeats TTAGGG   Subtelomeric sequences

Point of fusion

Figure 16.3 Arrangement of telomeric and subtelomeric sequences at the fusion point in human chromosome 2 (2q13). Telomeric sequences form two arrays with opposite orientations, indicated by the direction of the arrowheads, and are flanked by subtelomeric low-copy number repeats.

rochromatin often occurs when metacentrics are formed by the fusion of two acrocentrics, as in sheep (Section 16.3.4). Among grasshoppers, there can be substantial differences between different populations in the amount of heterochromatin (King, 1993, pp. 84–85).

In other cases, particularly in *Drosophila* and rodents, differences between species have been described in the properties of heterochromatin. Many of the reported differences have been differences in staining properties, which, although pointing to some underlying chemical differences, do not give any clear information on their nature. Fry & Salser (1977) proposed that differences in the nature of heterochromatin between species could be because these species share a library of DNA sequences, some of which may be amplified to form a block of heterochromatin in one species, while different sequences might be amplified in another species. Something of this sort has clearly occurred in mice. In the house mouse (*Mus musculus*) the large blocks of centromeric heterochromatin are composed of 'major' satellite, while the 'minor' satellite is restricted to the centromere proper. In other species of *Mus*, however, the proportions of the two satellites in the heterochromatin can be very different (Wong *et al.*, 1990; Garagna *et al.*, 1993).

Another proposed mechanism of change in heterochromatin is euchromatin transformation (King, 1980), in which a segment of euchromatin becomes converted to heterochromatin without any change in the total amount of DNA. The possibility of such a mechanism would give rise to many questions, and in fact certain cases of

euchromatin transformation have turned out to be addition of heterochromatin (Sumner, 1990). If the euchromatin did become heterochromatin, what would happen to the genes in the euchromatin? Loss of all the genes in a segment of euchromatin large enough to become visible as a block of heterochromatin would surely be lethal. Although a process something like euchromatin transformation probably occurs in the formation of some Y chromosomes (Section 8.2), convincing evidence that it may be a general process remains to be produced.

16.3.6 Widespread gain or loss of DNA

The genomes of closely related organisms may differ substantially in the amount of DNA they contain, often without substantial changes (other than size) in their karyotypes. We have already seen (Section 14.3) that salamanders of the genus *Plethodon* have very similar karyotypes, but differ by a factor of >1.5 in the amount of DNA they contain. Among the Bovidae, cattle have over 10% more DNA than sheep or goats, yet all three species have very similar chromosomal banding patterns (Sumner & Buckland, 1976). Parts of the genome of the pufferfish *Fugu rubripes* have the same gene order as the homologous segments of the human genome, although the *Fugu* genome is about 7.5 times smaller (Miles *et al.*, 1998). Only the spacing of the genes differs (Fig. 16.4). The genomes of cereals (rice, wheat, barley, maize, etc.) all have much the same gene order, yet the wheat genome is 40 times larger than that of rice (Moore *et al.*, 1995). Most of the extra

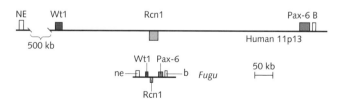

Figure 16.4 Scale comparison of the WAGR region of chromosomes of human (top) and *Fugu rubripes* (bottom). The same genes are present in the same order in each species, but they are much closer together in *Fugu*; *Rcn1* is transcribed in the opposite direction to the other genes in both species; *NE* and *B* are expressed sequence tags that had not been identified as genes. Reproduced with permission from Miles *et al.* (1998) *Proceedings of the National Academy of Sciences of the USA* **95**, 13068–13072. © 1998 National Academy of Sciences, USA.

DNA consists of moderately repetitive sequences, in particular retrotransposons (Moore, 1995; Petrov, 2001). In *Drosophila* species, which have small genomes, the quantity of retrotransposons is small and pseudogenes are virtually absent (Petrov *et al.*, 1996). In fact, the rate of DNA loss in *Drosophila* species is about 40 times greater than in a cricket, which has a genome that is 11 times larger, and is also greater than the rate of loss in mammals (Petrov *et al.*, 2000). Gain and loss of redundant DNA sequences is therefore an important mode of genome evolution in both plants and animals (Capy, 2000).

16.3.7   Polyploidization

Between 30% and 80% of angiosperm plants are polyploid (Moore, 1995; Otto & Whitton, 2000), and polyploidy is therefore a very important mode of genome evolution in this group, especially as it allows sympatric speciation (Otto & Whitton, 2000). Polyploidy is also frequent in ferns, but is largely absent from fungi and gymnosperms, for example (Otto & Whitton, 2000). Among animals, polyploidy is more sporadic (Otto & Whitton, 2000), probably in part because of the problems it can cause in sexually reproducing organisms (Section 16.4). Nevertheless, examples of polyploidy are known in many groups, both parthenogenetically and sexually reproducing (Otto & Whitton, 2000), although polyploidy is conspicuously absent from mammals (with one exception; Gallardo *et al.*, 1999) and birds. It is believed that the vertebrate genome arose by polyploidization, vertebrates perhaps being octoploid compared with their ancestral deuterostome (Meyer & Schartl, 1999). Evidence for this comes from the number of copies of various genes and gene clusters. There may have been another round of doubling in actinopterygian fish, which would therefore be 16-ploid. Among the fish, salmonids are often held to be polyploid (Hartley, 1987; Johnson *et al.*, 1987).

The result of polyploidization is not simply having twice as many genes for everything. Both in autopolyploids (consisting of two copies of a single species' genome) and in allopolyploids (produced by combining genomes from two or more species) a number of changes may occur very soon after polyploidization (Pikaard, 2001). Gene expression can be altered, changes occur in DNA methylation patterns and low-copy number sequences can be eliminated, sometimes as early as the first generation after the polyploidization event. In autopolyploids, each chromosome has three (in a tetraploid) or more (in higher polyploids) homologues to pair with at meiosis, with the result that some chromosomes pair with more than one homologue or remain unpaired. The same problems can also occur in allopolyploids, so that polyploidization can easily result in chromosome loss and unbalanced karyotypes. It might therefore be that only a small proportion of polyploidization events is successful and leads to the formation of new species.

16.3.8   Hybridization

If a species is regarded as an interbreeding population separated from other species by a reproductive barrier, it may seem surprising that new species can arise by hybridization. After all, interspecific hybrid individuals, if they occur at all, are often abnormal, and development may cease at an early stage; or if morphologically normal hybrids are produced, they are often (but not invariably) sterile. Nevertheless, there are some good species that have arisen by hybridization. Particularly among plants, hybrid species may arise through allopolyploidy; that is, the different parental genomes in the hybrid both double, so that at meiosis each can pair normally with its own homologue. Such a species is wheat, *Triticum aestivum*, which is actually an allohexaploid made up of three genomes: that of diploid wheat, and those of two species of *Aegilops*. These species are actually sufficiently closely related that their chromosomes can pair at meiosis with those of the other species that form the hybrid, although this is normally prevented by a genetic mechanism that restricts pairing to chromosomes derived from the same parental species (John, 1990, pp. 268–269).

Animal species produced by hybridization are fewer, but nevertheless over 3000 hybrid species combinations are known in fish (Otto &

Whitton, 2000). Examples of hybridogenesis, that is, species that are produced entirely by repeated hybridization in each generation, have been found among frogs and fish. The edible frog *Rana esculenta* is always a hybrid between *R. lessonae* and *R. ridibunda*, and has a set of chromosomes derived from each parent (Heppich, 1978). Before the start of meiosis the *R. lessonae* genome is eliminated, and the *R. ridibunda* genome is duplicated and goes through a normal meiosis to produce *R. ridibunda* gametes (Heppich *et al.*, 1982), although in other populations it may be the *R. ridibunda* genome that is eliminated (Vinogradov *et al.*, 1990). A similar system operates in hybridogenetic fish of the genus *Poeciliopsis*, in which one parental set of chromosomes fails to attach to the spindle at the pre-meiotic division (Heppich, 1978).

16.4 Chromosome changes and speciation

In the preceding section (16.3) the differences in chromosomes that can be found between species have been reviewed briefly. Are these changes just a chance accompaniment of speciation, or can chromosome change be an essential cause of speciation? It must be emphasized that even if chromosomal changes are required for speciation in some cases, there must be other situations in which they are not, because speciation can occur without any significant change (Section 16.3.2); purely genetic or behavioural factors can produce reproductive barriers between species.

Chromosome changes are most likely to produce reproductive barriers when they cause problems at meiosis in heterozygotes, leading to reduced fertility. Changes within a chromosome, such as inversions, may or may not cause problems at meiosis, and small insertions or inversions, although initially failing to pair properly and form loops, eventually resolve so that even the non-homologous regions are paired (King, 1993, pp. 80–84). Changes in heterochromatin are unlikely to cause problems either, especially as heterochromatin often does not pair at meiosis

(Section 7.4.3). The difficulties arise with translocations, tandem fusions and centric fusions or fissions, when a chromosome from one parental genome will be homologous to two (or more) chromosomes from the other parental genome, and a trivalent will form (Fig. 16.5). There are a number of ways in which a trivalent can be disjoined, and only in some cases will a balanced karyotype result in the gametes; the aneuploid gametes will, of course, give rise to aneuploid zygotes, which in most species have reduced viability (Sections 17.2 and 17.3). More complicated rearrangements produce quadrivalents, higher multivalents and chains of chromosomes, with even more chance of producing aneuploidy. Studies on hybrids between mice with and without centric fusions show that there is increased non-disjunction and reduced fertility, although heterozygotes with a single Robertsonian translocation have almost normal fertility (e.g. Redi & Capanna, 1988; Wallace *et al.*, 1992; Hauffe & Searle, 1998).

In other hybrids (e.g. mules) there is almost complete absence of meiosis, and the testes are almost devoid of meiotic cells. Extreme divergence of karyotypes might give rise to pairing difficulties, which in turn would tend to result in meiotic breakdown, most probably at pachytene, but other factors could be involved in such cases.

Polyploidy can also give rise to meiotic problems. A cross between a diploid and a tetraploid produces a triploid, which may well be viable (though not in mammals). However, orderly meiosis is not possible, as there must inevitably be many unpaired or partly paired chromosomes. Triploids are therefore generally sterile, unless they have managed to adopt a parthenogenetic mechanism of reproduction.

It is clear that in many cases the chromosomal differentiation that occurs between species can give rise to reproductive barriers, although this depends on the nature of the chromosomal changes (King, 1993). It is, however, rarely possible to know if the chromosomal changes are the factor (or one of the factors) causing speciation, or whether they might have arisen subsequently.

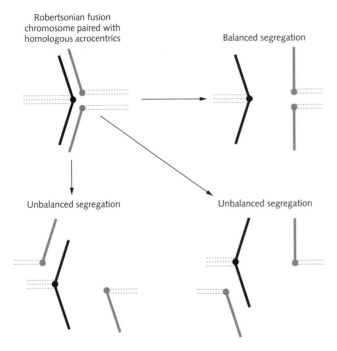

Robertsonian fusion
chromosome paired with
homologous acrocentrics

Balanced segregation

Unbalanced segregation

Unbalanced segregation

Figure 16.5 Segregation of a trivalent at meiosis to give balanced or unbalanced products.

16.5 Nucleotypic effects

It has already been suggested that the size and shape of individual chromosomes in the karyotype is rarely of any great significance (Section 16.2). On the other hand, there is good evidence that the total size of the genome may have various influences on the cell and the organism, independently of the effects of individual genes (Cavalier-Smith, 1978, 1982; Bennett, 1985). Such phenotypic influences are known as nucleotypic effects, the nucleotype being those aspects of DNA and chromatin quantity and bulk that affect the organism's phenotype independently of the action of any gene. Many phenotypic characteristics have been correlated with the amount of nuclear DNA in an organism, particularly in plants (Table 16.1); many of these correlations can be attributed ultimately to the correlation between nuclear DNA content, nuclear and cell size and cell-cycle time. For example, plants and animals with small genomes, because they have shorter cell cycles, can grow faster and thus occupy ephemeral habitats or live in polar or alpine habitats with a shorter growing

Table 16.1 Correlations between genome size and cellular and organismal phenotypes.

Character	Correlation with nuclear DNA amount
Nuclear size	Positive
Cell size	Positive
Cell-cycle length	Positive
Length of S phase	Positive
Length of mitosis	Positive
Length of meiosis	Positive
Rate of development/growth	Negative
Basal metabolic rate	Negative
Body size	Positive
Annual/perennial plants	Positive
Size of xylem cells (tracheids)	Positive
Number of chloroplasts per cell	Positive
Seed weight	Positive
Latitude of growth (crop plants)	Positive/negative
Life at high latitudes/altitudes	Positive
Pollen grain size	Positive
Leaf size	Positive/negative
Brain complexity (amphibia)	Negative

Data from: Cavalier-Smith (1978); Bennett (1985); Vinogradov (1995, 1997); Caceres *et al.* (1998); Chung *et al.* (1998); Gregory & Hebert (1999); Otto & Whitton (2000); Petrov (2001).

season. As a general rule, large genome size results in larger cell and body size and slower growth. Some characters are positively or negatively correlated with genome size, depending on the organism; refer to the references quoted for details of these apparent anomalies.

The most interesting point is that genome size can be influenced indirectly by selection. In maize, *Zea mays*, different varieties have been bred to grow further and further north, requiring faster growth in the shorter growing season. As a result, the size of the genome has become reduced, and this is at least partly a result of loss of heterochromatin from the chromosomes (Rayburn *et al.*, 1985). Thus we have an example of selection affecting chromosome morphology. Although the selection is artificial, and the effect is on the whole karyotype, it shows clearly that, contrary to what may have been suggested earlier in this chapter, chromosome morphology can be subject to selection.

16.6 Chromosomal change is a concomitant of evolution

Studies of the effects of chromosome alterations in individuals of particular species have given rise to the idea that chromosome change is normally deleterious (Chapter 17), and indeed in most species the karyotype remains remarkably stable. On the other hand, comparative studies show clearly that in the course of evolution an enormous amount of chromosomal change has taken place. Many types of chromosomal rearrangements, such as inversions and Robertsonian translocations, pose no particular problems in somatic cells, but can give rise to pairing difficulties at meiosis. The occurrence of such rearrangements as polymorphisms in natural populations (e.g. mice and shrews) indicates that they are perhaps not as deleterious as has been supposed. Some offspring may be inviable or sterile, but sufficient survive for the chromosomal rearrangement to be preserved. In other cases, the presence of the rearrangement may be much more deleterious, but the evidence of evolution is that it must have occurred. As long as a few such alterations survive, they can give rise to new species. Chromosomal events that give rise to a new species are undoubtedly rare within the lifetime of a species and the lifetime of the scientists who study them. Over the period of evolutionary time and the vast number of species that exist today and have existed in the past, chromosomal events involved in speciation could and indeed have occurred numerous times.

Chromosomes and disease

17

17.1 The significance of chromosomal disease

In previous chapters we have examined the structure of DNA and chromosomes, and the ways in which they are maintained and divided equally between daughter cells and transmitted to future generations. Many of the processes involved, such as DNA replication and repair, or mitosis and meiosis, are very complicated, and it is not surprising that from time to time they go wrong. In this chapter, therefore, chromosomal diseases will be described; that is, situations in which defects in some aspect of chromosome organization or behaviour leads to a disease state. Many of these diseases are, thankfully, very rare, but others are surprisingly common. If one considers a process such as mitosis, it is clear that any serious defect will be lethal, because if the cells cannot divide properly, the organism cannot grow. Conversely, problems in meiosis are fully compatible with normal life, although the individual in which such problems occur may be sterile, or may produce abnormal offspring. The chromosomal diseases described in this chapter have been studied mainly in humans, although some are also known to occur in domestic animals. Similar defects could potentially occur in all eukaryotes, but because they would inevitably have reduced fitness, they would be eliminated rapidly by natural selection. Reduced selection pressure and high standards of health care allow humans with certain chromosomal defects to survive into adulthood, but others have defects so severe that death is inevitable at an early stage, often before birth.

17.2 Numerical chromosome defects – errors in cell division

Numerical chromosome defects include trisomies, in which there is an extra chromosome of a particular type, and polyploidies, either triploidy or tetraploidy, with three or four sets of chromosomes. Monosomies – the presence of only a single chromosome of a particular type – are unknown in humans, except for the X chromosome, and as mosaics, in which there is a cell population with a normal chromosome complement as well as the monosomic population. As it is expected that monosomic cells would be produced at much the same rate as trisomic cells, it is believed that monosomy for autosomes is invariably lethal at such an early stage that pregnancies with monosomic embryos are never recognized. Of recognized pregnancies that are spontaneously aborted, about 50% have chromosomal abnormalities (Jacobs & Hassold, 1995). Mosaics can arise by non-disjunction at any stage, but complete trisomies and polyploids are the result of errors at meiosis; it is possible to work out in which parent the error occurred, and whether it happened at the first or the second meiotic division.

Table 17.1 Human autosomal trisomies.

Chromosome	Frequency		Liveborn	Comments
	Clinically recognized pregnancies	Spontaneous abortions		
1	Nil	Nil	Nil	
2	0.16%	1.1%	Nil	
3	0.04%	0.3%	Nil	
4	0.12%	0.8%	Nil	
5	0.02%	0.1%	Nil	
6	0.04%	0.3%	Nil	
7	0.14%	0.9%	Nil	
8	0.12%	0.8%	Nil	
9	0.10%	0.7%	Nil	
10	0.07%	0.5%	Nil	
11	0.01%	0.1%	Nil	
12	0.02%	0.2%	Nil	
13	0.18%	1.1%	0.005%	Patau's syndrome: severe abnormalities, die shortly after birth
14	0.14%	1.0%	Nil	
15	0.26%	1.7%	Nil	
16	1.13%	7.5%	Nil	
17	0.02%	0.1%	Nil	
18	0.18%	1.1%	0.01%	Edwards' syndrome: severe abnormalities, die shortly after birth
19	Nil	Nil	Nil	
20	0.09%	0.6%	Nil	
21	0.45%	2.3%	0.12%	Down's syndrome: 75% spontaneously aborted, but many survive to adulthood
22	0.40%	2.7%	Nil	

Data from Jacobs & Hassold (1995).

17.2.1 Autosomal trisomies

Trisomies occur at a very high rate in humans, but the vast majority are lethal during early pregnancy. Something like 4% of all human conceptions are trisomic, but the frequency and outcome of recognized trisomies for different chromosomes are very variable (Table 17.1). Only trisomies 13, 18 and 21 are compatible with live birth, but all show multiple abnormalities (Czepulkowski, 2001, pp. 111–113). Only individuals with trisomy 21 (Down's syndrome, mongolism) survive longer than a few days, and most of these survive into adulthood, although of those that are conceived over 75% are spontaneously aborted (Jacobs & Hassold, 1995). A small proportion (4–5%) of individuals with Down's syndrome are a result of Robertsonian translocation between chromosome 21 and another acrocentric chromosome (Fig. 17.1); if the other chromosome is also 21, there is a 100% risk of producing a child with Down's syndrome, but if the translocation is between 21 and a different acrocentric, the risk is only about 1% (Czepulkowski, 2001). Trisomy 16 is one of the commonest trisomies in spontaneous abortions, but is never found in liveborns.

What gives rise to this high rate of trisomy in humans? The incidence of all autosomal trisomies increases with the age of the mother (Hassold & Hunt, 2001), and in women over the age of 40 years it has been estimated that at least 20% of all oocytes have chromosomal abnormalities (Warburton, 1997). The vast majority of trisomies

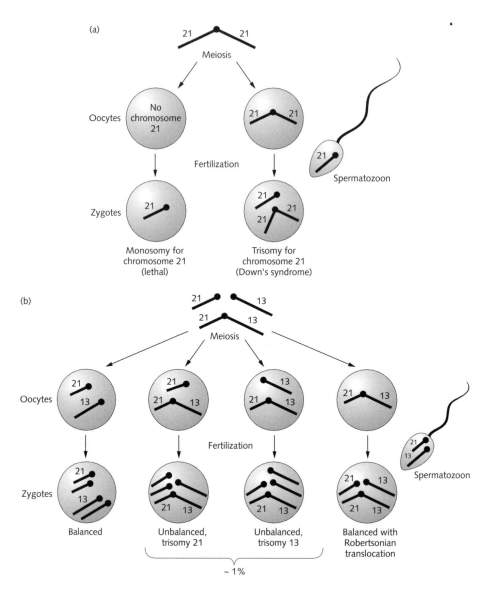

Figure 17.1 Trisomy 21 resulting from Robertsonian translocation in a parent. (a) The parent has a Robertsonian translocation involving the two no. 21 chromosomes [rob (21;21)]. At meiosis this must segregate to produce gametes (here shown as oocytes) containing either none or two no. 21 chromosomes. On fertilization by a normal sperm, either a monosomic zygote (lethal) or a trisomic zygote (Down's syndrome) is produced. (b) The parent has a Robertsonian translocation involving chromosome 21 and another acrocentric (here shown as chromosome 13, but it could also be 14, 15 or 22). Potentially four types of gamete could be produced at meiosis, but production of unbalanced gametes is substantially rarer than production of balanced gametes.

are the result of errors in maternal meiosis, with meiosis I errors being about three times commoner than meiosis II errors (Hassold & Hunt, 2001). These observations indicate that the errors must be connected somehow with the very long time (up to 50 years or more) for which human female meiosis can be arrested at diplotene. Two factors have been implicated (Hassold & Hunt, 2001): a lack or abnormal distribution of chiasmata (Fig. 17.2), and a lack of an efficient

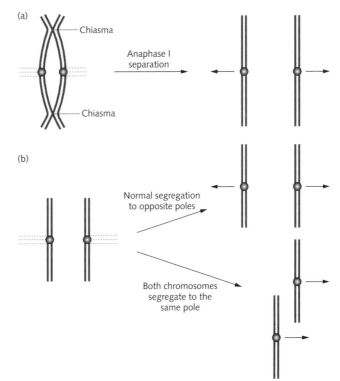

Figure 17.2 The role of chiasmata in ensuring proper chromosome segregation at meiotic anaphase I. (a) Holding the bivalent together with chiasmata ensures that the centromeres are attached to spindle microtubules leading to opposite poles of the cell, ensuring regular segregation. (b) If no chiasmata are formed, each chromosome can orientate independently; in some cases normal segregation will occur (upper), but in others both daughter chromosomes will segregate to the same pole of the cell (lower).

anaphase checkpoint in female meiosis. In trisomy 21, nearly half of all errors in the meiosis I leading to non-disjunction were due to complete lack of chiasmata (Lamb *et al.*, 1997; Warburton, 1997). In most of the rest of the cases, the chiasmata were clustered near the telomeres, but for meiosis II errors the chiasmata were centromerically clustered, suggesting that the location of the chiasmata may render the chromosomes prone to non-disjunction (Lamb *et al.*, 1997; Hassold & Hunt, 2001). It has been proposed that with distal chiasmata insufficient tension on the kinetochores may develop, leading to instability and reorientation of the bivalent on the metaphase plate (Wolstenholme & Angell, 2000; Fig. 17.3). Combined with reduced chromatid cohesion as a result of ageing, the chromatids might then segregate independently, leading to the possibility of non-disjunction at the first or second meiotic division. The length of meiosis II is not correlated with maternal age, and it was therefore surprising that trisomies

resulting from meiosis II errors should increase with maternal age. It turns out that, for technical reasons, non-disjunction may be classified as occurring at meiosis II when in fact the error really occurred at meiosis I (Warburton, 1997), consistent with an apparent absence of univalents at metaphase II (Angell, 1997).

In mitotic cells, there is a checkpoint that prevents the cell from moving into anaphase until all the chromosomes are properly attached to the spindle (Section 2.3.2). This mechanism does not seem to operate efficiently in female meiosis in mammals, so non-disjunction, leading to aneuploidy, can readily occur (LeMaire-Adkins *et al.*, 1997). This checkpoint is reported to operate effectively in male meiosis, but in spite of this, and although there is no arrest of male meiosis at diplotene, a significant though smaller proportion of trisomies arise from errors in the father, so other factors must be involved that are not yet clearly identified.

Humans seem to be unique in the high level

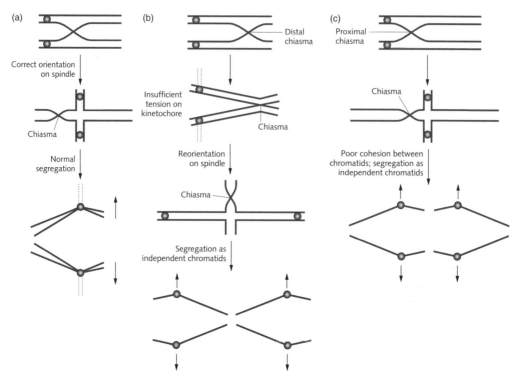

Figure 17.3 (a) Normal meiosis with correct orientation of chromosomes on the spindle and proper cohesion of chromatids, leading to segregation of bivalents as whole chromosomes. (b) Chromosomes with a distal chiasma: as a result of mis-orientation on the metaphase plate and failure of chromatid cohesion, the chromatids can segregate independently, leading to possible aneuploidy. (c) Chromosomes with a proximal chiasma: again failure of chromatid cohesion allows independent segregation of chromatids.

of meiotic non-disjunction and aneuploidy that they experience, perhaps an order of magnitude higher than that in other mammals (Warburton, 1997; Hassold & Hunt, 2001). Nevertheless, autosomal trisomy has been recorded and studied in laboratory mice and domestic animals, with various degrees of abnormality and fetal wastage.

17.2.2 Sex chromosome aneuploidies

Sex chromosome aneuploidies are relatively common (Table 17.2) and are fully compatible with life, although most suffer some intrauterine mortality (Jacobs & Hassold, 1995). They produce relatively minor abnormalities, but in some cases are sterile. They have, of course, been studied

intensively in humans and mice, but have also been reported from various domestic animals (Table 17.2); sex chromosome aneuploidies are an important cause of infertility in racehorses. Some sex chromosome aneuploidies are associated with reduced intelligence, but in general individuals with sex chromosome aneuploidies are much more normal than those with autosomal aneuploidies. This is no doubt because of the inactivation of all but one X chromosome (Section 8.4.3), and the low level of activity of the Y chromosome; sex chromosome aneuploidies do not, therefore, give rise to large-scale changes in the dosage of genes on the chromosomes involved. Nevertheless, a few genes of the X chromosome do escape inactivation (Section

Table 17.2 Sex chromosome aneuploidies in humans and other mammals.

Karyotype	Syndrome	Frequency	Fertility	Other features	Occurrence in other species
45,X	Turner's	1–2% of conceptions 1 in 5000 females at birth	Sterile	99% die before birth; remainder live normal lives (? mosaic). Short stature. Normal intelligence. Characteristic anatomical abnormalities	Mouse, cat, pig, horse, rhesus monkey, sheep, black rat
47,XXX	'Super-female'	1 in 1000 females	Usually normal (infertile in horse)	Sometimes retarded mental development	Horse
47,XXY	Klinefelter's	1 in 750 males	Sterile	Tall. Slightly reduced intelligence. Gynaecomastia in a minority	Mouse, Chinese hamster, cat, dog, sheep, ox, pig
47,XYY		1 in 1000 males	Often fertile	Tall. Generally normal; a small minority with criminal tendencies	Mouse

References: Chandley (1984); Zinn *et al.* (1993); see www.angis.su.oz.au for references to further examples in mammals and other vertebrates.

8.4.3), and it is presumably differences in the dosage of such genes that produce the observed phenotypes. This is most clearly seen in human XO females with Turner's syndrome, who clearly differ from normal XX females (Zinn et al., 1993). In mice, many fewer genes escape X inactivation, and XO mice are much more similar to normal female mice than Turner's women are to normal women. The XO mice are fertile, but with a reduced reproductive span and some growth retardation (Zinn et al., 1993; Disteche, 1995). In fact, Turner's syndrome has many unusual features: unlike the other chromosomal syndromes discussed so far, the 45X karyotype is the result of the loss of one X chromosome from a normal 46XX embryo at an early division, and as a result nearly 30% of women with Turner's syndrome are mosaics (Zinn et al., 1993).

17.2.3 Triploidy and tetraploidy

Mammalian triploids and tetraploids are highly abnormal, and virtually all die before birth; the few that do survive to birth die very shortly afterwards. This is in contrast to the situation in most groups of organisms, both animals and plants, in which triploids and tetraploids are perfectly normal and viable, although triploids are commonly sterile because of the difficulty of obtaining regular chromosome segregation at meiosis. Polyploidy is, in fact, a significant mode of chromosome change in evolution (Section 16.3.7).

In humans, triploids comprise about 1% of all recognized conceptions and about 10% of all spontaneous abortions (Zaragoza et al., 2000). Most are of maternal origin, resulting from the fertilization of a diploid oocyte produced by defective segregation at meiosis I or II; the remainder are of paternal origin as a result of dispermy (fertilization of an oocyte by two sperm) (Baumer et al., 2000). The reason why triploidy is lethal in mammals is undoubtedly because a triploid karyotype is unbalanced, both as a result of X chromosome inactivation (Section 8.4.3) and imprinting (Chapter 9). Triploid females have either one or two inactive X chromosomes; in either case the ratio of active X chromosomes to autosomes will be incorrect. Information on the imprinting status of triploids does not seem to be available, but similar imbalances would be expected for imprinted chromosomal regions. Because neither dosage compensation by X inactivation nor imprinting occurs in non-mammalian vertebrates, such considerations do not apply to them, and triploids are essentially normal but are usually sterile unless they can reproduce parthenogenetically. Tetraploidy in humans is rarer than triploidy, forming about 6% of all the chromosome abnormalities in spontaneous abortions.

17.3 Diseases produced by chromosome deletions and duplications

Monosomy – the absence of one chromosome of a pair – is lethal (Section 17.2) but there are several genetic diseases that are the result of deletion of a specific small part of a chromosome (Table 17.3). A few of special interest have been studied intensively. In retinoblastoma and the WAGR syndrome, deletion of the appropriate chromosome segment removes a 'good' allele of a tumour suppressor gene, leaving only a single copy that predisposes to tumour formation if mutated (Macleod, 2000). Wilms' tumour also involves anomalies of imprinting, as does Prader–Willi syndrome (Section 17.6). Jacobsen's disease (deletion of 11q) is sometimes the result of breakage at a fragile site (Section 17.5).

A principal method of producing deletions is by unequal crossing-over between region-specific low copy-number repeat sequences that flank the deleted regions (Lupski et al., 1996; Chen et al., 1997; Lupski, 1998; Shaikh et al., 2000; Fig. 17.4; Section 3.3.1.3), and this implies that there should also be diseases caused by duplications. Malformations due to chromosomal duplications have long been known in Drosophila (Lupski et al., 1996) and have now been discovered in humans. Charcot–Marie–Tooth disease type 1A is the result of a duplication in 17p12, and hereditary neuropathy with liability to pressure palsies (HNPP) results from

Table 17.3 Some diseases resulting from chromosomal deletion.

Deletion	Syndrome	Phenotype
5p	Cri-du-chat	Mewing cry, multiple physical abnormalities
7q11.23	Williams	Short stature, mental handicap, hypercalcaemia
8q22–24	Langer–Giedion (tricho-rhinopharyngeal type I)	Mental retardation, bulbous nose, thin lips, sparse hair
11p13	WAGR	Wilms' tumour, aniridia, genitourinary malformation
11p15	WAGR	Wilms' tumour, aniridia, genitourinary malformation
11q23–tel	Jacobsen's	Growth and mental retardation, etc.
13q14	Retinoblastoma	Childhood eye tumours
15p11	Prader–Willi	'Happy puppet' syndrome
17p11.2	Smith–Magenis	Brachycephaly, hyperactivity, mental and growth retardation
17p12	HNPP	Hereditary neuropathy with liability to pressure palsies
17p13.3	Miller–Dieker lissencephaly	Lissencephaly, microcephaly
20p12	Multiple endocrine neoplasia	
22pter-q11	Cat-eye	
22q11.2	DiGeorge	Congenital heart defect, facial dysmorphism

References: de Grouchy & Turleau (1986); Tassabehji *et al.* (1999); Tunnacliffe *et al.* (1999); Czepulkowski (2001).

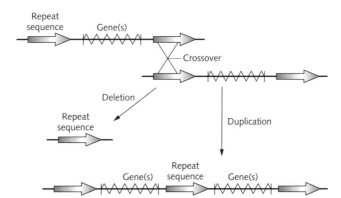

Figure 17.4 Production of deletions and duplications by unequal crossing-over between region-specific low copy-number repeats (shown as arrows).

the corresponding deletion. A 1.5 Mb segment of DNA is duplicated or deleted, respectively, as a result of crossing-over between flanking 24 kb repeats that have the same orientation (Lupski, 1998). A duplication in 17p11.2 has also been identified that is the reciprocal of the deletion that causes Smith–Magenis syndrome (Potocki *et al.*, 2000).

17.4 Chromosome breakage syndromes – failures in DNA repair

In a number of diseases there is a high incidence of chromosome breakage as a result of defects in DNA repair (Table 17.4). The study of such dis-

eases is important not only for clinical reasons, but also because they have provided valuable insights into mechanisms of DNA repair. The types of chromosome damage that occur in the different syndromes are characteristic, and are an important diagnostic aid.

Some of the main symptoms of the various human chromosome breakage syndromes are listed in Table 17.4, and more details are given in the Atlas of Genetics and Cytogenetics in Oncology and Haematology (www.infobiogen.fr/services/chromcancer). The chromosome instability in these syndromes arises from several different causes, and takes different forms. In many of the diseases there is apparently random breakage and rearrangement, although specific

Table 17.4 Chromosome breakage syndromes.

Disease	Sensitive to	Symptoms	Frequency	Type of chromosome damage	DNA repair deficiency	Gene
Xeroderma pigmentosa (XP)	UV	Skin and other cancers		No spontaneous instability	Excision repair	
Cockayne's syndrome (CS)	UV; oxidation-induced damage	Growth failure; poor neurological development		No spontaneous instability	Transcription-coupled repair	
Trichothiodystrophy (TTD)	UV	As XP			Excision repair	
SCID (severe combined immunodeficiency)	X-rays	Immunodeficiency			Double-strand break repair	
Ataxia telangiectasia (AT)	Ionizing radiation	Tumour susceptibility Immunodeficiency	1 in 4–10000	Breaks, multiradials; specific rearrangements	Radiation damage (checkpoint failure)	ATM
Bloom's syndrome (BS)		Immunodeficiency Cancer susceptibility High SCE frequency	1 in 50000	Increased frequency of SCEs; quadriradials	DNA helicase	BLM
Fanconi's anaemia (FA)	DNA cross-linking agents	Susceptibility to leukaemias No immunodeficiency	1 in 40000	Breaks, multiradials		
Nijmegen breakage syndrome (NBS)		Immunodeficiency Radiosensitivity		Breakage of 7p13, 7q35, 14q11, 14q32	Double-strand break repair (checkpoint failure)	Nibrin
Werner's syndrome		Cancer susceptibility Premature ageing		? Slightly increased level of chromosome breakage	DNA helicase/exonuclease	

SCE, sister-chromatid exchange; UV, ultraviolet light.
Data from: http://www.infobiogen.fr/services/chromcancer.

chromosomal regions are involved in certain diseases. In both Nijmegen breakage syndrome (NBS) and ataxia telangiectasia (AT), sites on chromosomes 7 (7p13 and 7q35) and 14 (14q11 and 14q32) are preferentially affected; these are the sites of immunoglobulin heavy-chain and T-cell receptor genes. Rearrangements at these sites must be related to a failure to produce fully functional immunoglobulins and T-cell receptors, resulting in the immunodeficiency characteristic of these closely similar diseases. However, immunodeficiency is not confined to these two diseases, and because production of mature functional immunoglobulins and T-cell receptors requires DNA breakage and recombination, any defect in these processes is liable to lead to immunodeficiency.

In Bloom's syndrome, sister-chromatid exchange (SCE) occurs at a greatly increased frequency (about 90 per cell, or about ten times more than in normal cells) (Fig. 17.5). A characteristic of Fanconi's anaemia (FA) is the formation of large numbers of triradials, quadriradials and more complex figures (Fig. 17.6), although these also occur at a lower frequency in other chromosome breakage syndromes. The FA cells are particularly sensitive to DNA crosslinking agents. Unlike other chromosome breakage syndromes, FA patients do not suffer from immunodeficiency (Joenje & Patel, 2001). Fanconi's anaemia protein, FANCA, interacts with the SWI/SNF chromatin remodelling complex, raising the possibility that deficiencies in FANCA may affect functions such as transcription and DNA repair (Otsuki *et al.*, 2001).

All the chromosome breakage syndromes are associated with cancer, which is hardly surprising in view of their DNA repair deficiencies, which would result in mutations being allowed to persist, including those that predispose to cancer. A high rate of chromosome rearrangement is characteristic of cancers (Section 17.9), as it is of the chromosome breakage syndromes, and at least some of the rearrangements in cancer give rise directly to mutated genes that cause cancer. In AT at least, chromosome rearrangement results in activation of an oncogene (Shiloh, 1997).

Figure 17.5 Metaphase from a patient with Bloom's syndrome, showing a greatly increased level of sister-chromatid exchanges (SCEs). Micrograph kindly provided by I.P. Kesterton.

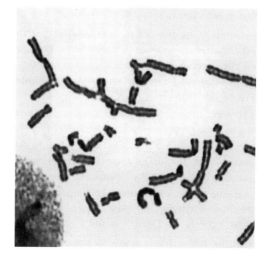

Figure 17.6 Chromosomes from a patient with Fanconi's anaemia, showing the characteristic spontaneous alterations found in this syndrome: triradials, quadriradials, etc. Micrograph kindly provided by I.P. Kesterton.

Deficiencies in most of the known types of repair mechanisms have been identified among the chromosome breakage syndromes. Xeroderma pigmentosum (XP) always involves deficiencies in nucleotide excision repair, but there are seven types of XP with mutations in different parts of the excision repair system. One of these, XP group D, involves mutations in a DNA helicase (XPD) that is part of the TFIIH transcription factor complex, which is essential for both transcription and repair (Winkler & Hoeijmakers, 1998). However, other mutations in the XPD helicase can produce Cockayne's syndrome (CS) or trichothiodystrophy (TTD), which share ultraviolet hypersensitivity with XP but have their own sets of symptoms that probably result from progressively increased effects on transcription. In fact, CS is the result of defects in transcription-coupled repair (Section 3.6.3), which deals with lesions that prevent proper transcription (Hanawalt, 2000). Deficiency of repair of double-strand breaks in DNA, as in SCID, AT and NBS, not only leads to sensitivity to ionizing radiation, which is energetic enough to produce double-stranded DNA breaks, but predictably results in immunodeficiency, because V(D)J recombination involves breakage and ligation of double-stranded DNA molecules. However, it appears not to be the repair system itself that is defective in AT and NBS, but the checkpoint controls that normally prevent the cells from proceeding through the cell cycle if any unrepaired DNA damage remains (Shiloh, 1997). The genes that may be mutated in AT – *ATM* and *ATR* – code for protein kinases and apparently regulate p53 (Brown & Baltimore, 2000). At meiosis both are localized on the synaptonemal complex, though at different sites (Section 2.5.2). Both the Bloom's syndrome (BS) protein, BLM (Neff *et al.*, 1999), and the Werner's syndrome protein, WRN, are DNA helicases, and mutations in them inhibit DNA repair by their failure to unwind the DNA molecule and separate the two strands. The considerable differences between BS and Werner's syndrome may well be because WRN also has a $3' \rightarrow 5'$ exonuclease activity (Huang *et al.*, 1998; Shen & Loeb, 2000). The BLM is localized on the synaptonemal complex in meiotic cells (Moens *et al.*, 2000), and Bloom's syndrome patients are generally infertile, consistent with the observation that both BLM and WRN are believed to be involved in recombination.

17.5 Fragile sites and triplet repeat diseases

Fragile sites are locations on chromosomes that have a tendency to break or appear as a gap or constriction when cells are grown under appropriate conditions (Fig. 17.7). They might have remained little more than an academic curiosity if it had not turned out that one fragile site was associated with the commonest form of X-linked mental retardation, which showed peculiar non-Mendelian inheritance and was caused by amplification of trinucleotide repeats. Several of these triplet-repeat expansion diseases have now been identified, and although most do not manifest fragile sites, it is nevertheless appropriate to describe them briefly here.

Fragile sites are classified as common (found in virtually all people) or rare (found in less than

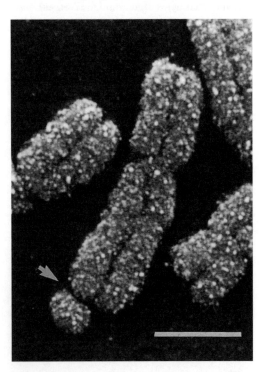

Figure 17.7 Scanning electron micrograph of a fragile site (FRAXA) on a human X chromosome (arrowed). Scale bar = 2 μm.

Table 17.5 Properties of human rare fragile sites.

	Folate-sensitive	BrdU-induced	Distamycin A-induced
Location	Most	FRA10B*	FRA16B*
Unit of amplification	CCG trinucleotide	~42 bp, A+T-rich	33 bp, A+T-rich
Size of amplified sequence	0.6–5.5 kb	5–100 kb	10–70 kb
Length instability	Yes	Yes	Not known

*FRA indicates a fragile site; the number is the number of the chromosome; the final letter differentiates fragile sites on the same chromosome.
After Sutherland *et al.* (1998).

one person in 20). Most common fragile sites are induced by culture in the presence of aphidicolin, a DNA polymerase inhibitor, and most rare fragile sites are induced by reduction in levels of folate, which is a co-factor for conversion of uridine monophosphate to thymidylate (Sutherland *et al.*, 1998); a few rare fragile sites are induced by bromodeoxyuridine (BrdU) or by distamycin (Sutherland *et al.*, 1998). Rather little is known about the common fragile sites, and they do not appear to be associated with any disease condition; most of this section will therefore be concerned with the rare fragile sites, which are all associated with expansions of trinucleotides, or of other micro- or minisatellites (Hewett *et al.*, 1998), and which in several cases are associated with disease. Complete lists of human fragile sites have been compiled (Hecht *et al.*, 1990). Although fragile sites have been studied predominantly in humans, they have also been described in a wide variety of mammals (e.g. Elder & Robinson, 1989; Smeets & van de Klundert, 1990). The DNA sequences that are amplified in rare fragile sites are of three types, which correlate with the agents used to produce them (Table 17.5). The induction of both common and rare fragile sites is probably a consequence of late replication, induced both by the amplified DNA repeats and by the agents used to demonstrate the fragile sites (Sutherland *et al.*, 1998; Le Beau *et al.*, 1998). The late replication would, in turn, delay or prevent chromosome condensation in these regions, and thus render them susceptible to breakage.

The presence of a fragile site, FRAXA, on the X chromosome was associated with X-linked mental retardation many years ago, and was the prototype both for diseases linked to fragile sites

and for triplet-repeat expansion diseases (Table 17.6). Fragile X is the commonest inherited form of mental retardation, with a frequency of about 1 in 1500 males and 1 in 2500 females (Oostra & Willems, 1995). The principal features of fragile X syndrome include moderate to severe mental retardation, a long face with prominent ears and macro-orchidism (de Vries *et al.*, 1998). Normal people have 6–54 copies of the CGG repeat in the 5′ untranslated region of the FMR-1 gene, and this number of copies is stable. Carriers of the pre-mutation, who are normal, have 43–200 copies, and this number is unstable and liable to increase in each generation. Affected individuals have more than 200 CGG repeats, and these repeats, and their flanking regions, become methylated. It is this methylation, and not the repeat expansion, that switches off the FMR-1 gene and causes the symptoms of fragile X syndrome.

The expression of the fragile site FRAXA is correlated with the number of CGG repeats, although even in affected individuals it is rare for as many as half of the chromosomes to show a fragile site (de Vries *et al.*, 1993). The chance of expansion of the repeats from the pre-mutation to the full mutation also increases with the number of repeats: with <55 triplet repeats, the risk is close to zero, but it is 20–30% with 56–65 repeats; if, however, there are more than 90 repeats, the risk of expansion to the full mutation is almost 100%. These expansions are much more likely to occur in females, and affected females receive the fragile X site from their mothers, not from their fathers. The proportion of affected sons of carrier mothers is 0.4, not 0.5 as expected for normal Mendelian segregation. Fragile X carrier daughters of normal transmitting males

Table 17.6 Triplet-repeat expansion, fragile sites and disease.

Disease	Fragile site	Mode of inheritance*	Gene product	Triplet repeat	Location†	Number of copies		
						Normal	Carrier (pre-mutation)	Affected
Loss of function, large expansion								
Fragile X (Martin–Bell)	FRAXA	XD	FMRP	CGG	5'-UTR	6–54	43–200	200 to >2000
Fragile X	FRAXE	XR	?	GCC	5'-UTR	6–25	43–200	>200
No disease	FRAXF			CGG				
Myotonic dystrophy		AD	Myotonin	CTG	3'-UTR	5–37	50–180	200 to >2000
Jacobsen	FRA11B		CBL2	CCG	5'-UTR	11‡	80–100	400–1000§
Friedreich's ataxia		AR	Frataxin	GAA	Intron	7–27	40–60	200 to >1000
No disease	FRA16A			CGG				
Gain of function, moderate expansion								
Huntington's		AD	Huntingtin	CAG	Exon	6–34		36–121
Dentatorubral-pallidoluysian atrophy		AD	Atrophin	CAG	Exon	7–25		49–88
Spinobulbar muscular atrophy (Kennedy's disease)		XR	Androgen receptor	CAG	Exon	11–34		40–62
Spinocerebellar ataxia (SCA)								
Type 1		AD	Ataxin-1	CAG	Exon	6–39		41–81
Type 2		AD	Ataxin-2	CAG	Exon	15–29		35–59
Type 3 (Machado–Joseph disease)		AD	Ataxin-3	CAG	Exon	13–36		68–79
Type 6		AD	Ataxin-6	CAG	Exon	4–16		21–27
Type 7		AD		CAG				
Type 8		AD		CTG				
Type 12		AD	Protein phosphatase regulatory subunit	CAG	5'-UTR	7–28		66–78
Synpolydactyly				GCG				
Pentamer repeat expansion								
Spinocerebellar ataxia type 10				ATTCT	Intron	10–22		Up to 4500
Dodecamer repeat expansion								
Myoclonus epilepsy (EPM-1)		AR	Cystatin B	G+C-rich	Promoter	2–3	12–17	50–75

*AD, autosomal dominant; AR, autosomal recessive; XD, X-linked dominant; XR, X-linked recessive.

†3'-UTR, 3' untranslated region; 5'-UTR, 5' untranslated region.

‡Non-expression of fragile site.

§Expression of fragile site.

References: Mandel (1997); Mitas (1997); Holmes et al. (1999); Koob et al. (1999); Jones et al. (2000); Matsuura et al. (2000).

are also normal, and have a low risk of producing fragile X children (Richards & Sutherland, 1992). These and related phenomena of inheritance of the fragile X syndrome are known as the Sherman paradox, and are explained by repeat expansion occurring when the trinucleotide repeats pass through female meiosis, but not when they pass through male meiosis. The phenomenon of anticipation, in which the disease appears earlier and with greater severity in successive generations, is explained by the increase in the number of trinucleotide repeats in affected individuals from one generation to the next (de Vries *et al.*, 1998).

The fragile X protein, FMRP, is important for proper synaptogenesis, and its absence or mutation leads to abnormal dendritic spines that have not matured properly. FMRP is an RNA-binding protein that associates with actively transcribing polyribosomes and also binds to messenger RNAs (mRNAs) that contain the G-quartet motif, in which four guanines occur in a planar conformation. Such mRNAs may require FMRP to transport them to ribosomes, thus regulating their expression (Kaytor & Orr, 2001).

None of the other triplet-repeat diseases listed in Table 17.6 show exactly the same details as fragile X syndrome. One group, including fragile X syndrome, myotonic dystrophy (MD) and Friedreich's ataxia, have large triplet-repeat expansions in affected individuals and asymptomatic carrier individuals with unstable pre-mutations. The repeats are always in non-translated regions of genes, and cause loss of gene function. The CTG expansion in myotonic dystrophy may cause loss of function of the protein kinase gene *DMPK*, but it may also affect the function of a downstream homeobox gene DMAHP; another possibility is that the transcript of the CTG repeats may interfere with RNA processing (Cummings & Zoghbi, 2000). In Friedreich's ataxia, the GAA repeats can form a triple helix (Gacy *et al.*, 1998) that interferes with transcription (Bidichandani *et al.*, 1998). The CCG repeats in Jacobsen's syndrome (Section 17.3) have been implicated in the breakages that lead to the deletions in this syndrome (Jones *et al.*, 2000).

In the second main group of triplet-repeat expansion diseases, which includes Huntington's disease and several forms of spinocerebellar ataxia (SCA), the repeated triplet is CAG, which is always in an exon and is translated as polyglutamine. This group of diseases shows a moderate number of repeats (21–121) in affected individuals, and the conditions are generally inherited as autosomal dominants (Table 17.6), and cause loss of specific groups of neurons in the brain (Reddy & Housman, 1997). The polyglutamine expansions cause a change in protein function, resulting in the formation of intracellular inclusions and causing cytotoxicity and neuronal degeneration (Rubinsztein *et al.*, 1999; Marsh *et al.*, 2000).

There are a few diseases that do not fit easily into this simple classification. Although the repeated triplet in SCA12 is CAG, it is found in the 5′ untranslated region, and not in an exon. In SCA8, there is a CTG expansion instead of a CAG, while in synpolydactyly the GCG, coding for alanine, is expanded. Finally, the expanded sequence does not even need to be a nucleotide triplet: in myoclonus epilepsy, a G+C-rich dodecamer in the promoter of the cystatin B gene is amplified (Lalioti *et al.*, 1997). Repeating units of 33 bp and 42 bp have been implicated in the formation of certain fragile sites (Hewett *et al.*, 1998), but without causing any disease. Trinucleotide repeats also occur in other mammals and lower vertebrates, but the number of trinucleotide repeats is always lower than in humans, they contain nucleotide substitutions more commonly and are less prone to expansion (Sutherland & Richards, 1995; Djian *et al.*, 1996).

How and why do nucleotide triplets expand? Simple sequence repeats can be expanded by replication slippage (Section 3.3.1.3), and this seems to be the most likely mechanism in triplet-repeat expansion diseases. Expansion is promoted by the tendency of those triplets that are involved in repeat expansion diseases (but not other triplets) to form stable hairpin structures, and the larger the hairpins (i.e. the greater the number of triplet repeats) the more stable they are and thus more likely to promote expansion (McMurray, 1995; Gacy & McMurray, 1998). The secondary structure of the trinucleotide repeats may also inhibit the processing of displaced Okazaki fragments during replication (Spiro et al., 1999). Finally, in the normal population, the CGG repeats at the FMR-1 locus are interrupted by single AGG triplets at intervals, which appears to lower the stability of the hairpins dramatically (McMurray, 1995); such AGG triplets are missing in patients with fragile X syndrome that have undergone triplet-repeat expansion. Similar interruptions have also been reported in myotonic dystrophy and SCA. Thus the molecular properties of triplet repeats are linked to the expansion phenomena seen in triplet-repeat expansion diseases, which in turn account for some of the symptoms seen in such diseases. The formation of fragile sites appears to be related to late replication, although it is not yet clear why they only appear in a small proportion of triplet-repeat diseases.

17.6   Diseases of imprinting

A small number of diseases result from problems with imprinting (Table 17.7). The main features of imprinting have been described in Chapter 9. Angelman syndrome results from the loss of one or more maternally expressed genes in human chromosome region 15q11–q13, and Prader–Willi syndrome (PWS) from the loss of paternally expressed genes in the same region. In both cases, the most common mechanism is deletion of a 4 Mb chromosomal region (see Fig. 9.2), although uniparental disomy, gene mutation or other changes are the cause in a minority of cases (Mann & Bartolomei, 1999). A microdeletion is also a major cause of Beckwith–Wiedemann syndrome (BWS) (see Fig. 9.3), but a few cases result from changes in methylation leading to an altered imprinting pattern (Reik et al., 1995) or from uniparental disomy (Robinson, 2000). Both of the genes that have been implicated in producing the BWS phenotype – $p57^{KIP2}$ and $IGF2$ – are involved in cell-cycle regulation (Caspary et al., 1999). These genes act antagonistically, and symptoms of BWS can be produced by loss-of-function mutation of $p57^{KIP2}$ or by gain-of-function mutations of $IGF2$. Wilms' tumour

involves two chromosomal regions: 11p13, at which loss of heterozygosity for the Wilms' tumour gene WT1 occurs; and 11p15, the same region that is involved in BWS. Wilms' tumour associated with 11p15 appears to involve loss of imprinting at this site (Feinberg, 1994). Thus imprinting diseases can result from deletions that leave only the imprinted, inactive allele, or from alterations in the imprinting itself, either as a result of uniparental disomy or by changes in methylation. In the latter cases, the result is that the two alleles no longer show differential activity according to their parental origin.

17.7   DNA methylation and disease

Apart from cancer (Section 17.9), in which DNA methylation levels often differ from those in normal cells, there are at least three diseases in which there are defects in DNA methylation, and the level of methylation may be significant in Down's syndrome.

In ICF syndrome (immunodeficiency, instability of centromeric heterochromatin, and facial anomalies), the constitutive heterochromatin of chromosomes 1 and 16, and more rarely that of chromosome 9, is decondensed in lymphocytes and is often fused to form multiradial figures (Smeets et al., 1994) (Fig. 17.8). It is the paracentric heterochromatin that is affected, and not the centromeres themselves (Sumner et al., 1998). These chromosome abnormalities are no doubt the cause of lagging chromosomes at anaphase, chromosome fragmentation and micronucleus formation that are seen in ICF patients (Stacey et al., 1995). The satellite DNAs that form the paracentromeric heterochromatin are undermethylated in ICF syndrome (Miniou et al., 1994), and the chromosomal abnormalities in ICF syndrome are very similar to those seen when chromosomes are demethylated experimentally using 5-azacytidine (Schmid et al., 1983a). In fact, the demethylation in ICF syndrome occurs throughout the genome, and also involves single-copy and *Alu* sequences (Schuffenhauer et al., 1995; Miniou et al., 1997). The failure of methylation has been traced to a

Table 17.7 Diseases of imprinting.

Syndrome	Chromosome region	Frequency	Causes	Symptoms	Genes affected	Ref.
Angelman (AS)	15q11–q13 (loss of maternal)	1 in 15000	~4Mb chromosome deletion (75%) / Paternal uniparental disomy (2%) / Mutation of UBE3A gene (5%)	Cognitive impairment, seizures, ataxia, inappropriate laughter, etc.	UBE3A	1
Prader–Willi (PWS)	15q11–q13 (loss of paternal)	1 in 15000	~4Mb chromosome deletion (75%) / Maternal uniparental disomy (25%)	Neonatal hypotonia, hyperphagia with obesity, mental retardation, etc.	SNURF-SNRPN	
Beckwith–Wiedemann (BWS)	11p15.5	1 in 13700	Chromosome microdeletion / Paternal uniparental disomy (10%) / Trisomy with paternal duplication / Loss of imprinting at IGF2 / Loss of function mutations in p57^{KIP2} / Balanced chromosome rearrangements	Somatic overgrowth, macroglossia, visceromegaly, susceptibility to childhood tumours	p57^{KIP2}, IGF2	2
Wilms' tumour	11p15		Biallelic expression of IGF2 (70%) / Biallelic expression of H19 (30%) / Loss of heterozygosity		IGF2, H19	
	11p13		Loss of heterozygosity (30%) / Mutations in WT1 (10%)		WT1	

References: 1, Nicholls *et al.* (1999); 2, Feinberg (1994).

Figure 17.8 Scanning electron micrograph of a multiradial configuration from a patient with ICF syndrome. Scale bar = 2 μm. See also Fig. 7.5.

deficiency of the methyltransferase Dnmt3b (Okano et al., 1999).

In Rett syndrome – an X-linked neurodevelopmental disease affecting almost exclusively girls – the deficiency is not in the methylation process but in the MeCP2 protein that binds to methylated DNA (Willard & Hendrich, 1999). Protein MeCP2 is a global transcriptional repressor and Rett syndrome may simply be the result of too much transcriptional 'noise'. Severity of the disease is related not only to the type of mutation, but also to whether the pattern of X inactivation is skewed (Shahbazian & Zoghbi, 2001). Both in humans and in mouse models there is a delay before symptoms appear; possibly MeCP2 is required to stabilize brain function rather than for its development (Guy et al., 2001).

In ATR-X (alpha-thalassaemia, mental retardation, X-linked) there are both increases and decreases in methylation of various repeated DNA sequences, including ribosomal DNA, compared with normal individuals (Gibbons et al., 2000). The ATRX gene encodes an SWI/SNF-like protein (Section 4.2.5) that may be involved in chromatin remodelling associated with the MBD/HDAC protein complex, which modulates methylation and histone acetylation.

Down's syndrome is the result of trisomy for chromosome 21 (Section 17.2.1), so that many of the genes identified on chromosome 21 are expressed at trisomic levels. However, the CpG island of the h2-calponin gene is specifically methylated so that it is expressed at the normal diploid level (Kuromitsu et al., 1997). The intriguing suggestion has been made that the downregulation by methylation of certain genes in Down's syndrome patients might be necessary to allow their survival.

17.8 Telomeres and disease

Decrease in telomere length is associated with senescence (Section 13.6.1), and cancer is associated with reactivation of telomerase and stabilization of telomere length (Section 13.6.2), but there are other conditions in which telomere shortening is associated with disease. In the premature ageing disease Werner's syndrome (Section 17.4) the rate of telomere shortening is greater than in normal people (Tahara et al., 1997), although most of the symptoms of Werner's syndrome are due to defective DNA repair. Accelerated shortening of telomeres also occurs in another chromosome breakage syndrome, ataxia telangiectasia (Metcalfe et al., 1996; Section 17.4), and in Down's syndrome (Vaziri et al., 1993). In these cases, however, the shortening of the telomeres is probably an accompaniment to the general progression of the syndrome rather than a causative factor. Nevertheless, there are at least two diseases in which telomere shortening itself is an important factor. In dyskeratosis congenita (DKC) there are defects in tissues such as skin and bone marrow that have a high rate of cell division, and there is chromosome instability and a tendency to develop certain types of malignancies. Dyskeratosis congenita is caused by mutations in the dyskerin gene. Dyskerin is involved in ribosomal RNA processing, but is also a component of the human telomerase complex (Mitchell *et al.*, 1999). The DKC cells consequently have reduced telomerase activity and are defective in telomere maintenance. Highly proliferative tissues such as skin and bone marrow require an efficient system of telomere maintenance, and these are precisely the tissues in which the principal defects are found in DKC. As DKC progresses, there is an increase in chromosome rearrangements in skin and bone marrow, although DKC cells have normal sensitivity to DNA damaging agents, unlike the chromosome breakage syndromes described in Section 17.4.

Telomere loss is also responsible for end-stage organ failure in cirrhosis of the liver (Rudolph *et al.*, 2000), a disease that affects several hundred million people throughout the world. A variety of hepatotoxic agents can destroy hepatocytes, so that excessive hepatocyte regeneration is stimulated. In people with cirrhosis, there is significantly increased shortening of the telomeres in the hepatocytes. When end-stage liver failure is reached, hepatocyte proliferation ceases. In mice with experimentally induced cirrhosis, the telomeres also become shortened and the failure of hepatocyte regeneration can be reversed by adenoviral delivery of telomerase.

17.9 Cancer – anything and everything can go wrong with chromosomes

Cancer is a collection of diseases with multiple causes. Chromosomal changes are characteristic of cancers and, although many are undoubtedly secondary events, it is clear that some rearrangements, by activating oncogenes, are important in initiating certain cancers. Many of the chromosomal diseases mentioned in this chapter, such as the chromosome breakage syndromes (Section 17.4) and the imprinting diseases (Section 17.6), are also associated with an increased risk of cancer, and reactivation of telomerase is usually required for immortalization and transformation

of cells (Section 13.6.2). Both increases and decreases of DNA methylation are seen in cancers. Cancer cells, therefore, are cells in which a great variety of chromosomal changes have occurred.

17.9.1 Structural and numerical chromosomal changes in cancer

A wide variety of chromosomal changes are found in cancers, although a few retain a diploid karyotype, and display microsatellite instability rather than chromosomal instability (Atkin, 2001). The best known example of a cancer with microsatellite instability is hereditary non-polyposis colorectal cancer (HNPCC), which is a result of a defect in mismatch repair (Section 3.6.1) (Lengauer *et al.*, 1998). Chromosomal alterations in cancers include changes in number, translocations and other rearrangements, amplifications (Section 17.9.2) and deletions (Lengauer *et al.*, 1998), many of which are associated with genes that are directly responsible for causing cancers. Other alterations appear to be epiphenomena that occur as a result of the cancer, rather than being a cause. The number and variety of chromosomal changes observed in cancers is now extremely large, and they have been catalogued in book form (Mitelman *et al.*, 1994) and on CD-ROM (Mitelman *et al.*, 1998). Specific breakpoints are associated with specific types of cancer (Mitelman *et al.*, 1997). More restricted listings have been given in various review articles (e.g. Rabbitts, 1994; Sánchez-García, 1997; Cobaleda *et al.*, 1998), and these are particularly useful because they concentrate on alterations that are characteristic of specific cancers and involve genes that are known to be responsible for the development of the cancers. Because of the vast amount of data now available, no attempt will be made to tabulate it here, and just a few examples will be given as illustrations of general principles. A disproportionate amount of information has been obtained from leukaemias and other haematopoietic cancers; this is because of the ease of obtaining good chromosome preparations from blood cells and the difficulty of obtaining them from solid tissues. There is, however, no good reason to suppose that changes in solid tumours, or the effects of such changes, differ in any fundamental way between leukaemias and solid tumours.

Losses and gains of whole chromosomes can result in the loss of a tumour-suppressor gene or the gain of a mutant oncogene, respectively (Lengauer et al., 1998). Thus in glioblastomas chromosome 10 is often lost, resulting in a loss of the tumour suppressor gene *PTEN*. In papillary renal carcinoma, on the other hand, there can be a gain of chromosome 7, so that the dose of a mutant oncogene, *MET*, is doubled.

Specific chromosome translocations are characteristic of very many tumours, and these can affect specific genes in various ways (Cobaleda et al., 1998; Lengauer et al., 1998). In many leukaemias and lymphomas, a translocation juxtaposes an oncogene with an immunoglobulin or T-cell receptor gene, so that the oncogene comes under the control of the immunoglobulin or T-cell receptor gene (Rabbitts, 1994; Klein, 2000). Because the latter are transcribed at high rates in the appropriate types of cells, the oncogene is also transcribed at a high rate. Such rearrangements have been identified in Burkitt's lymphoma, acute lymphocytic leukaemia (ALL), chronic lymphocytic leukaemia (CLL), non-Hodgkin's lymphoma (NHL) and a few other diseases. Obviously this mode of gene activation is restricted to cells in which the genome undergoes rearrangement as part of the process of maturation to form immunoglobulin-producing cells or T cells, and in these cases the rearrangement involves a different chromosome in error.

More generally, specific chromosome rearrangements in cancers involve breakpoints that occur in the introns of genes on different chromosomes, which then fuse to form a new, chimeric protein. Such rearrangements have been reported from numerous haematopoietic and solid tumours (Cobaleda et al., 1998; Rowley, 1998; Klein, 2000). The prototype of such proteins was the *BCR-ABL* oncogene produced by the 9;22 translocation that gives rise to the Philadelphia (Ph[1]) chromosome. Depending on where the breakpoint is in the *BCR* gene, alternative chimeric proteins can be produced that are

characteristic of either chronic myelogenous leukaemia (CML) or of acute lymphoblastic leukaemia (ALL). Many such proteins have turned out to be chimeric transcription factors (Sánchez-García, 1997; Cobaleda et al., 1998).

What determines the location of the breakpoints for rearrangements in cancer? In those rearrangements that involve immunoglobulin or T-cell receptor genes, the occurrence of V(D)J recombination at the site of such genes is a predisposing factor for rearrangement. It seems quite plausible, however, that the sites of many breakpoints could be random, and that only those rearrangements that gave the cells a selective advantage would persist. The 9;22 translocation that gives rise to the Ph[1] chromosome appears to be favoured because these chromosomes lie close to each other in the interphase nucleus (Kozubek et al., 1999).

17.9.2 Chromosome amplifications, HSRs and double minutes

Amplifications and deletions of segments of chromosomes occur commonly in cancers, and can be detected readily using comparative genomic hybridization (CGH) (Box 17.1). In some cases the degree of amplification is sufficient to produce morphologically distinct structures, which are of two kinds: HSRs (Fig. 17.9a) and double minutes (DMs) (Fig. 17.9b). The HSRs were originally discovered in methotrexate-resistant Chinese hamster cells, and appeared in G-banded chromosomes as extended regions with an intermediate level of uniform staining, which were named 'homogeneously staining regions' (Biedler & Spengler, 1976); they can occur in a high proportion of patients with certain types of cancer (e.g. Bernardino et al., 1998). In fact, HSRs are not always homogeneous, but may have regularly repeated patterns of banding that depend on the banding technique used.

Double minutes (DMs) are small chromatin bodies that vary in size from below the resolution limit of the light microscope (<250 kb) to small visible dots up to 2 μm in diameter (>7000 kb) (Hahn, 1993). There may be up to

Box 17.1 Comparative genome hybridization

In comparative genome hybridization (CGH), the object is to compare normal and abnormal sets of chromosomes (Forozan *et al.*, 1997). Fluorescence *in situ* hybridization probes derived from whole genomes are used, one from each of the genomes that is to be compared. One probe, say the normal genome probe, might be labelled with a green fluorochrome, while the probe from the abnormal genome might be labelled with a red fluorochrome. The two probes are hybridized simultaneously to normal chromosome spreads, and in general the colour of the fluorescence will be a mixture of that from the two probes: yellow. If, however, a chromosomal segment has been deleted from the abnormal karyotype, only green fluorescence will be seen on that segment of the normal chromosomes after hybridization. Conversely, if there is a duplication, either of a whole chromosome or just a

chromosomal segment, in the abnormal karyotype, then the fluorescence of the duplicated regions will tend to be red after hybridization, because there is more of the red probe for those regions in the hybridization mixture.

The CGH technique can be used to detect very small deletions and translocations that are not readily detected by banding techniques (Kirchhoff *et al.*, 2000). With suitable technique, deletions as small as 3 Mbp can be detected, compared with about 10 Mbp for a typical chromosome band. What makes this technique so powerful is that the DNA probes can be derived from interphase nuclei; this is particularly important when studying cancer cells, which are often difficult to get good chromosomes from. Also, the hybridization can be carried out on normal chromosomes (which are easily available) to give information about abnormal chromosomes.

several hundred in a cell (Rattner & Lin, 1984). They lack both centromeres and telomeres, but nevertheless are transmitted effectively to daughter cells, perhaps to some extent by random segregation, but also as a result of a tendency to associate with normal chromosomes. The absence of telomeres is because DMs are formed from circular DNA molecules.

Both DMs and HSRs contain a number of copies of the amplified gene, usually an oncogene (Schwab, 1998), but in methotrexate-resistant cells the dihydrofolate reductase gene is amplified. As well as the genes of interest, other chromatin is amplified, sometimes including heterochromatin, centromeres and nucleolus organizer regions (NORs) (Holden *et al.*, 1985). The unit of amplification can vary in size between 110 kb and 10 Mb, and there can be between 5 and 5000 copies in an HSR; DMs contain one or a few copies of the unit of amplification (Hahn, 1993; Schwab, 1998). The amplified oncogenes in both HSRs and DMs are normal, not mutated, and their effects are presumably the result of the much greater quantity of protein

made by these amplified genes (Schwab, 1998). In the case of methotrexate-resistant cells with multiple copies of dihydrofolate reductase, the cells are resistant to chemotherapy, and the degree of resistance is directly related to the size of the HSR (Biedler *et al.*, 1980).

There has been a lot of interest in possible mechanisms of amplification to form HSRs and DMs, not merely because of their intrinsic interest, but also because knowledge of such mechanisms should help in designing therapy for cancers that produce HSRs and DMs. Various mechanisms have been proposed, and indeed all may be used at different stages in the amplification process, or in different situations (Stark *et al.*, 1989). Formation of DMs may be the first stage of amplification in some cases; they might be derived by chromosome breakage, perhaps of chromosome segments that were delayed in replication or condensation (Hahn, 1993). Cells that accumulated more DMs might well have a selective advantage. At some stage, or in certain conditions, DMs seem to integrate into chromosomes and continue to amplify there; although

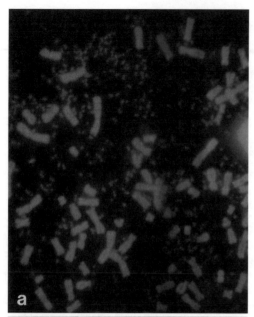

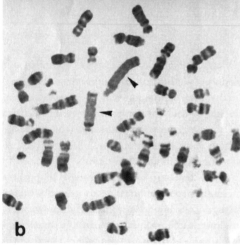

Figure 17.9 Morphological alterations in chromosomes as a result of DNA amplification in tumour cells: (a) double minutes (DMs); (b) homogeneously staining regions (HSRs). Reproduced by permission of Wiley – Liss, Inc., a subsidiary of John Wiley & Sons, Inc., from Schwab (1998) *Bioessays* **20**, 473–479. © John Wiley & Sons.

some HSRs are at the original site of the gene that they amplify, others are on completely different chromosomes, consistent with a process involving reintegration of DMs (Stark *et al.*, 1989). Unequal crossing-over (Section 3.3.1.3) or

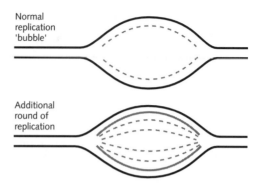

Figure 17.10 Onion-skin amplification. Multiple copies of a DNA sequence are produced by repeated initiation of replication during a single S phase.

sister-chromatid exchange has been proposed as an early, perhaps the initial, amplification event (Stark *et al.*, 1989; Smith *et al.*, 1990). Amplification could also occur by an 'onion-skin' process, in which multiple copies of a segment of DNA could be produced by multiple rounds of replication during the same S phase (Fig. 17.10). Such a mechanism could easily produce either extra-chromosomal circular DNA molecules (DMs) or an intrachromosomal linear array (HSR) (Stark *et al.*, 1989). Rolling circle models (see Fig. 11.4) have also been proposed, but although it has been shown that onion-skin and rolling-circle mechanisms are used for amplification in certain cases, it is not clear whether they are involved in the formation of HSRs or DMs.

17.9.3 Changes in methylation and imprinting in cancers

The genome of cancer cells often shows a lower overall level of methylation than that of normal cells, although paradoxically promoters (CpG islands) are commonly hypermethylated. The latter has a number of consequences at the molecular level: silencing tumour suppressors and DNA repair genes (Jones & Gonzalgo, 1997; Jones, 1999; Jones & Laird, 1999; Baylin & Herman, 2000), and mutation by deamination of 5-methylcytosine to thymine (Laird & Jaenisch, 1994, 1996). Equally, hypomethylation of proto-oncogenes has been found (Laird & Jaenisch,

1994), which would be expected to lead to their activation. Experimental methylation tends to induce tumorigenesis, while inhibition of methylation inhibits the production of tumours.

At the chromosomal level, hypomethylation appears to lead to reduced chromosome stability (Jones & Gonzalgo, 1997; Jones, 1999; Rizwana & Hahn, 1999). Cells with a reduced level of methylation seem to be more liable to undergo loss, gain or rearrangement of chromosomes, although the connection between these events and the methylation level is not clear.

Imprinting diseases have an increased susceptibility to cancer. Imprinting in mammals is, of course, a result of differential methylation, and tumorigenesis can be induced by inappropriate methylation, as for example in Wilms' tumour (Peterson & Sapienza, 1993; Jones & Laird, 1999). Loss of expression of imprinted genes has also been identified as a cause of hepatocarcinomas (Schwienbacher *et al.*, 2000).

17.9.4 Cancer as a result of chromosome instability

As described above (Section 17.9.1), there are numerous chromosomal alterations that are associated with cancer, and to these can be added DNA repair defects (Section 17.4), reactivation of telomerase activity (Section 13.6.2), failure of sister-chromatid separation (Zou *et al.*, 1999) and fragile sites (LeBeau & Rowley, 1984). Is there any common factor involved? It would perhaps be naive to look for a single chromosomal cause for the enormous variety of cancers, yet a feature common to most of them is chromosome instability. Although in many cases specific chromosome rearrangements have been identified as causes, or at least early events, in cancers, further rearrangements, losses and gains develop as the cancer progresses. The variety of mechanisms that can lead to chromosomal changes and instability have been noted earlier in this chapter. The systems for replicating, segregating and maintaining chromosomes are of remarkable complexity, and perhaps it is not surprising that all too often these systems fail, as in cancer, which is a leading cause of death, and in fetal loss, a major proportion of which is caused by chromosomal abnormalities (Section 17.2). Study of situations where things go wrong does, of course, lead to understanding of the processes involved in normal and abnormal cells. Such understanding, in turn, leads to better management and treatment of these chromosomal diseases.

Websites

An excellent and comprehensive site that gives information on a wide range of subjects connected with human chromosomal and genetic diseases is the Atlas of Genetics and Cytogenetics in Oncology and Haematology, which includes short articles on specific diseases and other subjects:

www.infobiogen.fr/services/chromcancer

References to information on chromosomal abnormalities (e.g. trisomies, fragile sites, etc.) in mammals, birds, etc. can be found in the ANGIS (Australian National Genomic Information Service) website:

www.angis.su.oz.au/

Chromosome engineering and artificial chromosomes

18

18.1 Engineering chromosomes – an ancient technique

Chromosome engineering has been practised for a very long time, and indeed could go back thousands of years to the times when the first cereal crops were being domesticated by man, and selection for favourable characteristics could have resulted in chromosomal changes. A recent example of this is the reduction in heterochromatin in maize bred for growing in cooler, more northerly latitudes (Rayburn et al., 1985; Section 16.5). In such cases, however, the chromosomal changes are essentially a side-effect of the breeding and selection processes, which have been carried out without any intention of modifying the chromosomes, and the changes observed lie essentially within the range of normal variation.

Deliberately modified chromosomes were first produced in Drosophila, and have proved invaluable for a wide variety of studies. The standard procedure is to break chromosomes using X-rays, and to select individual flies with the desired chromosomal breaks and rearrangements (Novitski, 1976; Novitski & Childress, 1976; Golic & Golic, 1996). Once flies with modified chromosomes have been produced, they can be crossed to form almost any desired karyotype. Compound chromosomes produced by such methods have been valuable for studying various aspects of recombination and meiosis (Novitski & Childress, 1976; Holm, 1976), the function of heterochromatin (Yamamoto & Miklos, 1978; Section 7.4) and for gene mapping (Lindsley et al., 1972). Although the value of this approach has been immense, and it became possible to construct Drosophila chromosomes having almost any properties that were desired, the manipulations involved in producing the modified chromosomes are quite labour-intensive, and because the action of X-rays on the chromosomes is random, it may require several experiments to produce the required results. Moreover, because X-rays are mutagenic, it can be difficult to know if the observed effects are the result of the chromosome breakage and rearrangement itself, or of associated mutations.

A much more controllable method for producing chromosome rearrangements and deletions, and for integrating DNA into specific regions of chromosomes, is Cre/lox recombination (Box 18.1). When other methods are used for integrating exogenous DNA into chromosomes, multiple copies of the DNA may become integrated, the efficiency is low and the sites of integration can vary, with unpredictable consequences. With Cre/lox recombination, integration is always at a single, specific site, giving reproducible results and the ability to compare different experiments (Mills & Bradley, 2001; Yu & Bradley, 2001). Functional studies of large chromosomal regions, gene clusters, imprinted regions and suchlike require the production of large rearrangements and deletions. The Cre/lox system is ideal for this, and has been developed for use in mice (Mills & Bradley, 2001). The system can also be used to induce loss of heterozygosity (LOH) to create model systems for studying cancers (Zheng *et al.*, 2000). Cre/*lox* recombination has also been used to produce chromosome rearrangements and deletions in plants (Medberry *et al.*, 1995), and in the fission yeast *Schizosaccharomyces pombe* (Qin *et al.*, 1995), and the similar FLP recombinase system has been used to create rearrangements in *Drosophila* chromosomes (Golic & Golic, 1996).

In plants, the passive alteration of karyotypes as a result of breeding and selection has long been complemented by deliberate manipulation

of chromosomes. Interspecific hybridization is used to introduce desirable properties (e.g. disease resistance) from one species to another, and this can be done not merely between species in the same genus, but between species in different families ('wide hybrids') (Gill & Friebe, 1998). Sometimes quite small chromosome fragments are introduced as a result of recombination; because cereal chromosomes (for example) are homologous over a wide range of species and genome size (Section 16.3.6), such recombination can occur quite readily. Microprotoplasts containing only a single chromosome have been used to transfer the chromosome into cells of

Box 18.1   Cre/*lox*   recombination

The Cre recombinase from bacteriophage P1 induces reciprocal recombination at specific sites in DNA (*lox*), each of which is a 34 bp sequence consisting of two 13 bp inverted repeats separated by an asymmetrical 8 bp spacer (Fig. 1). When the Cre recombinase has bound to the inverted repeats of one *lox* site, it binds to a second *lox* site and then cleaves

the spacer DNA in each site to produce strand exchange between the two synapsed *lox* sites (Mills & Bradley, 2001; Yu & Bradley, 2001). Cre recombinase works with high efficiency, and does not require any accessory factors. The budding yeast FLP recombinase works in a similar way (Kilby *et al.*, 1993).

A T A A C T T C G T A T A | A T G T A T G C | T A T A C G A A G T T A T

| Inverted repeat (13 bp) | Asymmetrical spacer (8 bp) | Inverted repeat (13 bp) |

Figure 1 The *lox* 34 bp nucleotide sequence.

Because any specific 34 bp sequence is highly unlikely to occur by chance in a eukaryotic genome, a *lox* site will not occur unless it is deliberately introduced. This can be done by the standard but inefficient process of homologous recombination, using a selectable marker to ensure that only those cells that have incorporated the *lox* site will survive. Once the *lox* site has been introduced, it can be used as a site for targeted integration or deletion of specific sequences, and for the production of large inversions and translocations (Fig. 2).

Because rearrangements produced by the Cre/*lox* system are reversible, it is desirable to have some control over the activity of Cre

recombinase. Instead of being constitutively active in the cells of interest, it may be introduced at a specific time as Cre-containing adenovirus, or by fusing with a Cre-containing transgenic cell line (Akagi *et al.*, 1997). Selectable markers can be incorporated to ensure that only cells survive in which recombination has occurred in the desired direction. Another method uses mutant *lox* sites that favour integration over deletion (Araki *et al.*, 1997). An inducible Cre recombinase has been produced by fusing the recombinase gene with the gene for a mutated ligand-binding domain of the human oestrogen receptor, so that it can be induced by tamoxifen (Feil *et al.*, 1996).

Continued on p. 230

Box 18.1 *Cont.*

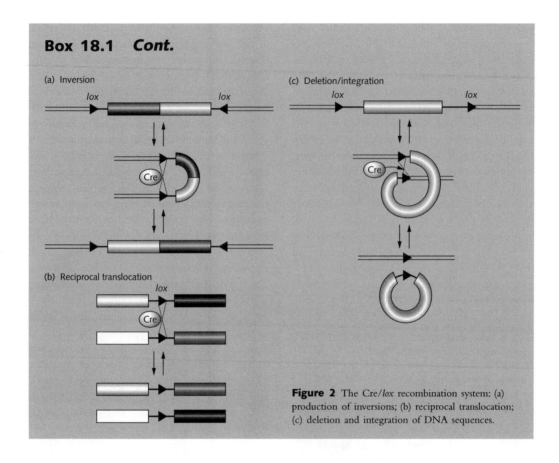

Figure 2 The Cre/*lox* recombination system: (a) production of inversions; (b) reciprocal translocation; (c) deletion and integration of DNA sequences.

another species. Genomic *in situ* hybridization (GISH, see Box 5.3) has proved valuable to monitor the nature and amount of the material transferred, and whether and where it has been integrated into the recipient genome (Leitch *et al.*, 1997; Gill & Friebe, 1998).

Cultured human cell lines have been engineered to become monosomic for specific chromosomes, for the study of gene dosage effects, imprinting and recessive mutants (Clarke *et al.*, 1998). Monosomy is induced by partial inhibition of topoisomerase II at mitosis to induce non-disjunction (Section 2.3.1).

18.2 What is an artificial chromosome?

All the modifications of karyotypes and individual chromosomes described in Section 18.1 involve the modification of existing chromosomes, with variable degrees of knowledge of the composition of the modified chromosomes. The first modified eukaryotic chromosomes to which the epithet 'artificial' was applied were yeast artificial chromosomes (YACs) (Murray & Szostak, 1983; Burke *et al.*, 1987), which have become very valuable for cloning genomic libraries from *Caenorhabditis*, *Drosophila* and mammals. Conceptually they differ from the modified chromosomes described in Section 18.1 in being assembled from known components (Murray & Szostak, 1983). Although yeast centromeres and telomeres must be used, the remaining DNA can come from any source. Thus human genes have been introduced into YACs and cloned, and this technique has been very important for gene mapping (Burke *et al.*, 1987). Yeast artificial chromosomes have also been useful for studying chromosome behaviour during mitosis and

Table 18.1 Properties of artificial chromosomes and other systems for introducing genes into cells.

System	Size of mammalian DNA carried	Pathogenicity	Site in nucleus	Stability
Viral vectors		Possible	Integrated into chromosomes	Low
BACs (bacterial artificial chromosomes)	100–300 kb		Integrated into chromosomes	Low
PACs (P1 bacteriophage artificial chromosomes)	100–300 kb		Integrated into chromosomes	Low
YACs (yeast artificial chromosomes)	1–2 Mb	No	Integrated or free	Stable
Human artificial episomal chromosomes (HAECs)	60–650 kb	Possible	Independent episome	Stable
Mammalian artificial chromosomes (MACs)	2.5–10 Mb	No	Free	Stable

Data from Vos (1997).

meiosis (Murray & Szostak, 1983, 1985) and for establishing the nature of the essential components of a functional chromosome (Clarke, 1990; Monaco & Larin, 1994).

Another potential application of artificial chromosomes is in gene therapy for genetic diseases that result from deficiency of one or a few defined genes. Various methods have been proposed for introducing functional genes into cells and organisms, although so far none have proved wholly satisfactory, and at present we seem to be a long way from a safe and workable system. Such a system needs to be able to carry gene-sized fragments of human DNA, including their promoters and other control elements. It must be able to enter cells efficiently, be maintained stably in the nucleus for an unlimited number of cell divisions, not interfere with the functioning of other genes and function normally in the recipient nucleus. Viral vectors, such as adenovirus and retroviruses, have a limited capacity for extra DNA (Table 18.1), can potentially disrupt normal genes when they integrate into the chromosome and are potentially pathogenic. In practice, they may fail to function when integrated into a chromosome. In fact, most DNA sequences integrated into mammalian chromosomes become rearranged, disrupted or deleted, so viral vectors and bacterial and P1 bacteriophage artificial chromosomes tend not to work

efficiently (Calos, 1996). In other cases, they may be inactivated as a result of position effects (Vos, 1999; Section 7.4.5). One answer to these problems is human artificial episomal chromosomes (HAECs) (Sun et al., 1994; Vos, 1997, 1999), which as independent episomes do not have the problems associated with integrated DNA and appear to be stably transmitted. However, because they include viral genomes, usually that of the potentially pathogenic Epstein–Barr virus, there is still some concern about possible pathogenicity (Vos, 1999).

Yeast artificial chromosomes (Section 18.3.1) can be stably transmitted, the genes in them can be expressed, there are no pathogenicity problems and they can hold a much larger amount of DNA than any of the artificial chromosomes or vectors just mentioned (Table 18.1). They may either remain free in the nucleus or become incorporated into a mammalian chromosome (Allshire et al., 1987; Jakobovits et al., 1993). However, it might be thought that it would be more appropriate to make mammalian artificial chromosomes (MACs) and introduce them into mammalian cells, where they would be expected to behave similarly to the endogenous chromosomes. Various approaches have been used to make MACs (Brown et al., 2000; Sections 18.3.2–4) but, although a lot of progress has been made, MACs are still very much at the experimental stage. Whereas YACs can be assembled from known components, this is not practicable for MACs; although mammalian telomeres are well defined, it is not yet clear what the salient features are of either mammalian centromeres (Section 12.2.3) or mammalian origins of replication (Section 2.2.2.1), so any MAC must include certain poorly defined components.

18.3 How to make artificial chromosomes

This section does not provide instructions on how to make artificial chromosomes, but instead gives brief accounts of the principles of construction of YACs and MACs. Production of YACs is now routine, but several approaches to the construction of MACs are being tried; only time will tell which is best, and indeed some other method may emerge, or perhaps combinations of different methods might eventually prove to be the most effective.

18.3.1 Making YACs

The manufacture of YACs is now a standard procedure of molecular biology. To produce a successful YAC, genes, origins of replication, centromeres and telomeres are all required (Burke et al., 1987), each of which is well defined in yeast. A plasmid that can replicate in *Escherichia coli* is assembled that contains a yeast centromere and origin of replication (autonomously replicating sequence, ARS; Section 2.2.2.1), a small number of yeast genes, including selectable markers, and *Tetrahymena* telomere sequences (which work in yeast cells) (Burke et al., 1987; Fig. 18.1). The plasmid also contains specific restriction enzyme recognition sites. Digestion with the appropriate restriction enzymes cuts the plasmid into left and right arms, each of which has a telomere at one end, and ligation in the presence of (for example) human DNA produces a linear YAC containing a yeast centromere, a human DNA sequence and *Tetrahymena* telomeres. Of course YACs can be made that contain DNA from a wide variety of

other animals, and or that contain plant DNA (e.g. Saji et al., 2001).

A YAC made as described above is reasonably stable at both mitosis and meiosis (Clarke & Carbon, 1980; Murray & Szostak, 1983), although the stability of the YAC depends on its size. A YAC with 55 kb of DNA is quite stable, but one as small as 20 kb is relatively unstable and is often lost at cell division. These sizes are much smaller than those of normal yeast chromosomes, which range from 150 kb to 1000 kb.

Yeast artificial chromosomes can be transferred into cells by lipofection, or by fusion of yeast spheroplasts (Box 18.2). They can be inserted into mouse embryonic stem cells, and cells that have successfully incorporated the YAC DNA can be selected using the selectable markers on the YAC. The embryonic stem cells can then be injected into a mouse blastocyst, where they repopulate various tissues in the developing embryo (Jakobovits et al., 1993), including the germ line, so that the introduced DNA can be transmitted to the offspring. When YACs are introduced into mammalian cells, they often integrate into the mammalian chromosomes (Allshire et al., 1987; Jakobovits et al., 1993; McManus et al., 1994). Integration of YACs into mammalian chromosomes is probably necessary for their long-term survival, as YAC telomeres do not function in mammalian cells (Farr et al., 1995).

18.3.2 Making MACs – the synthetic approach

There are at least three distinct methods of producing MACs (Brown et al., 2000). In principle, they could be assembled from known components using the standard methods of molecular biology, just as YACs are made. This is the 'bottom-up' approach described in this section; according to some (Willard, 1996), only chromosomes that are produced in this way should be referred to as artificial. The other, 'top-down', methods described (telomere-associated fragmentation, Section 18.3.3; and SATACs, Section 18.3.4) depend on the elimination of unwanted

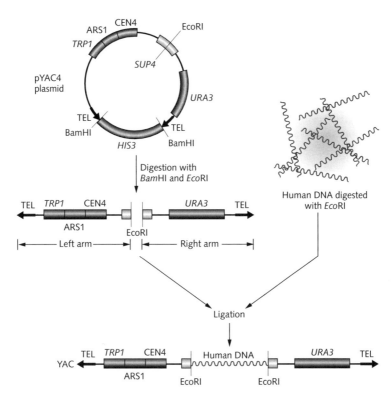

Figure 18.1 Construction of yeast artificial chromosomes (YACs). The plasmid pYAC4 is assembled containing the centromere of yeast chromosome 4 (CEN4), a yeast replication origin (ARS1), four yeast genes (*TRP1, SUP4, URA3* and *HIS3*) restriction enzyme recognition sites (BamHI and EcoRI) and *Tetrahymena* telomeres (TEL). Digestion with *Bam*HI and *Eco*RI produces two fragments ('left arm' and 'right arm'), each with a telomere at one end; the HIS3 gene is eliminated. Ligation in the presence of human DNA digested with *Eco*RI produces a linear YAC that can be cloned in yeast cells and contains human DNA segment; the genes *TRP1* and *URA3* can be used as selectable markers so that cells that do not contain the complete YAC are eliminated.

material from existing chromosomes, and in that sense they are simply normal chromosomes that have undergone extensive deletion, and are sometimes referred to as minichromosomes. However, because the aim in all cases is to produce chromosomes containing defined components, it seems logical as well as convenient to describe them all as artificial chromosomes.

Any attempt to make a MAC using defined components comes up against the problem already mentioned, that although the organization of mammalian telomeres is understood quite well (Chapter 13), the same cannot be said of centromeres or origins of replication. To make human artificial chromosomes, alpha-satellite has been

used as the centromeric component and seems to work quite well, although it is possible that it is neither necessary nor sufficient to form a functional centromere (Sections 12.2.3 and 18.4). Mammalian origins of replication have not yet been defined satisfactorily (Section 2.2.2.1), but it is generally assumed that if a large enough piece of DNA is used an origin is likely to be included.

Two main 'bottom-up' approaches have been tried. One is essentially that used to make YACs (Section 18.3.1), but using mammalian DNA sequences. Telomeres, alpha-satellite and a piece of DNA of interest (normally, at this stage of development, containing a suitable marker gene) are introduced into *Saccharomyces cerevisiae*, and a

Box 18.2   Getting DNA and artificial chromosomes into cells

There are several ways of getting DNA and artificial chromosomes into cells. The most direct is *microinjection*, in which the chromosome is injected directly into the nucleus of the recipient cell. This method is slow, expensive and only small numbers of cells can be injected.

Deoxyribonucleic acid can be transfected into cells using *calcium phosphate precipitation* or *electroporation*. Plasmids and other DNA molecules are readily precipitated by calcium phosphate, and the resulting precipitate, containing the DNA, is readily taken up by cells. *Electroporation* relies on the transient induction of pores in the cell membrane when the cell is placed in an appropriate electric field. The method can be used to get plasmids into cells (Voet *et al.*, 2001).

Other methods of introducing material into cells rely on enclosing the material in a lipid membrane and fusing this with the cell membrane, usually with the help of a substance such as polyethylene glycol or the commercial product Lipofectamine. The material to be introduced can be enclosed in an artificial membrane (lipofection; e.g. Lee & Jaenisch, 1996), or can be in a proper cell; usually it is desirable to engineer a cell containing just a single chromosome for *microcell fusion*. Microcells are produced when cells are subjected to prolonged mitotic arrest, which causes the formation of micronuclei containing one or a few chromosomes. The micronuclei can be extruded from cells by treatment with cytochalasin B followed by centrifugation, resulting in microcells consisting of micronuclei surrounded by a cell membrane (Fig. 1). There is no difficulty about performing cell fusion with animal cells, but with yeasts and plants the cell wall must be removed, to yield *spheroplasts* and *protoplasts*, respectively.

The efficiency of introducing plasmids or artificial chromosomes into cells is usually quite low, so it is usual for the plasmid or chromosome to contain a selectable marker, so that cells that have not incorporated it can be eliminated.

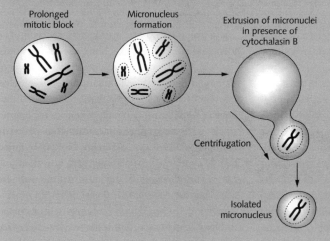

Figure 1 Preparation of microcells containing one or a few chromosomes.

MAC is assembled by homologous recombination (Brown *et al.*, 1996). Ikeno *et al.* (1998) used this method to make MACs of 1–5 Mb in size: they used a recombination-deficient strain of yeast to clone large arrays of alpha-satellite, which would otherwise be broken up and dispersed, and incorporated a selectable marker so that cells that had not formed an artificial chromosome could be eliminated. After lipofection or microinjection (Box 18.2), 18–68% of cells contained a distinct minichromosome; in the others the MAC had integrated into one of the endogenous chromosomes or was not detectable. The minichromosomes were much larger than the YAC, but contained only those DNA sequences that had been derived from the YAC that contained the mammalian sequences. Most cells with the minichromosome had only a single copy of it. More recently another group has used similar methods to produce a human artificial chromosome 1 Mb in size, and showed that it was stable for at least 100 cell generations (Henning *et al.*, 1999). Both of these results show that alpha-satellite can be sufficient for centromere activity, although there is still no information on the requirements for an efficient replication origin.

Using YACs for cloning mammalian DNAs has the advantage that they can handle large pieces of DNA. On the other hand, the stability of genes and alpha-satellite is greater in BACs and PACs (bacterial and P1 bacteriophage artificial chromosomes, respectively). Pieces of DNA cloned in PACs can be isolated into agarose gel and joined together using Cre recombinase (Box 18.1) in a method known as 'in gel site-specific recombination' (IGSSR). The advantages of IGSSR are that no further purification is needed, and the formation of MACs should be highly efficient (Schindelhauer, 1999), but at the time of writing it has yet to be shown how effective this might be in practice.

A different approach was taken by Harrington *et al.* (1997), who transfected mixtures of alpha-satellite containing a selectable marker, telomeric DNA and genomic fragments of undetermined composition into a human cell line, where it was expected that they would be joined together by non-homologous recombination (Fig. 18.2). In many cases the introduced DNA integrated into the endogenous chromosomes, or the telomeric sequences induced truncation of the existing chromosomes (Section 18.3.3), but in rare cases minichromosomes were formed, about 6–10 Mb in size (a fifth to a tenth of the size of the smallest human chromosomes), which were segregated normally for at least six months in culture.

Although this approach achieved some success, it appears to be largely a matter of chance whether any minichromosomes are produced.

18.3.3 Making MACs – telomere-associated fragmentation

The 'bottom-up' approach (Section 18.3.2) has a number of difficulties, and therefore people have been attracted to the 'top-down' approach of truncating an existing chromosome to produce a minimal chromosome. Although the composition of the latter cannot easily be predetermined, this is also true to some extent of the 'bottom-up' approach, because the requirements for an origin of replication are not at all clear (Section 2.2.2.1).

The most popular 'top-down' approach is telomere-associated fragmentation. The principle of this method is that when telomeric sequences are inserted interstitially in a chromosome, they cause breakage at the point of insertion. This process is cell-line specific, the frequency of seeding of new telomeres being high (>60%) in some cell lines, while the phenomenon occurs rarely or not at all in others (Farr, 1996, 1999). A telomere-seeding construct requires a selectable marker as well as telomeric [(TTAGGG)$_n$] sequences, and these are introduced into the

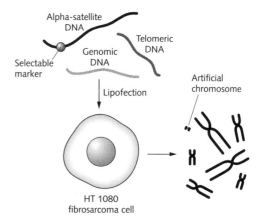

Figure 18.2 Construction of human artificial chromosomes by lipofection of cells with predetermined DNA components, followed by non-homologous recombination (Harrington *et al.*, 1997).

recipient cells by some sort of transfection process, usually lipofection or electroporation (Box 18.2). A simple telomere-seeding construct will seed telomeres randomly in a chromosome, but it is possible to make the process more specific, either by using strong selection to enrich for particular truncated chromosomes, or by using homologous recombination to target a particular chromosomal site (Fig. 18.3). Homologous recombination occurs at only a very low level in most mammalian cell lines, and for this reason the chicken pre-B-cell line DT40, which shows high levels of homologous recombination, has been adopted for targeted truncation experiments. Human chromosomes can be transferred into DT40 cells by microcell-mediated chromosome transfer (Box 18.2).

Using this technique, minichromosomes have been produced from human X and Y chromosomes (Heller *et al.*, 1996; Mills *et al.*, 1999). Such minichromosomes range in size from 2.4 Mb to 10 Mb (between one-sixteenth and a quarter of the size of the smallest normal human chromosomes), and can be maintained stably in cells for at least 100 divisions. As they have not been assembled from defined components, they cannot easily be used to answer questions about, for example, what sequences are required to form a functional centromere, but they obviously have potential for therapeutic purposes.

18.3.4 Making MACs – satellite DNA artificial chromosomes (SATACs)

Integration of foreign DNA into the pericentric heterochromatin of mouse chromosomes resulted

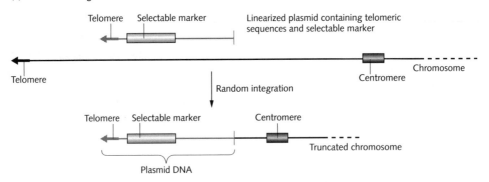

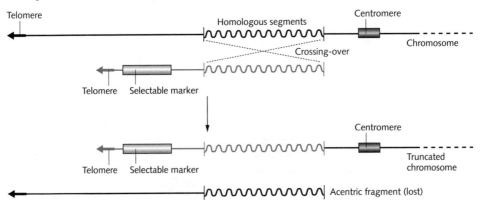

Figure 18.3 Production of artificial chromosomes by telomere-associated fragmentation: (a) random seeding of telomeres; (b) targeted insertion of telomeres by homologous recombination.

in large-scale chromosome amplification to produce mega- and giga-chromosomes (Keresö *et al.*, 1996). At the same time, minichromosomes were formed by breakage in the region where the foreign DNA had integrated. As well as a neo-centromere, this minichromosome contained mouse satellite DNA, and a euchromatic segment that contained the foreign DNA. This 'neo-minichromosome' was 20–30 Mb in size. This procedure is the basis for producing satellite DNA artificial chromosomes (SATACs), and human as well as mouse satellite DNA-based artificial chromosomes have been developed (Csonka *et al.*, 2000). Such SATACs can be transferred to mouse, human and bovine cells by microcell fusion and, using a selectable marker gene (hygromycin resistance) and a reporter gene (β-galactosidase), cells with minichromosomes can be selected (Telenius *et al.*, 1999). The SATACs can be maintained for prolonged periods in the cells into which they have been introduced, with a segregation efficiency of up to >90%, and can be isolated on a large scale by flow cytometry (de Jong *et al.*, 1999). They can be microinjected into pronuclei of murine and bovine zygotes, and in mice they have been maintained into adulthood and passed on to the offspring (Co *et al.*, 2000). Thus SATACs appear to be a very promising tool for a variety of purposes.

18.4 Artificial chromosomes – the future

In the mid-1990s, MACs that were made of known components and maintained for an indefinite number of cell generations were little more than a dream, although the first tentative experiments to produce MACs were already under way. At that time, YACs were already well-established tools, but the lack of knowledge about the composition of essential chromosomal components made the production of MACs much more difficult. Indeed, an important application of MACs was to establish the requirements for a functional mammalian centromere (Brown *et al.*, 2000); minichromosomes could be formed with alpha-satellite that contained the CENP-B

box (Section 12.2.3), but not with alpha-satellite that lacked the CENP-B box (Ikeno *et al.*, 1998; Masumoto *et al.*, 1998). Similarly, in *Drosophila*, Sun *et al.* (1997) used traditional methods of manipulating chromosomes with X-rays to delineate a minimal centromeric sequence. Experiments of this sort have the advantage that they provide a functional assay (in the case of centromeres, efficient mitotic segregation), and no doubt artificial chromosomes will continue to be used to elucidate fundamental aspects of chromosome organization and function. At present, however, the technical difficulties of producing artificial chromosomes that contain precisely defined chromosomal components restricts their value as purely experimental tools.

In fact, two main potential applications of artificial chromosomes can be identified that could justify the labour involved in producing them: gene therapy; and the production of transgenic animals, particularly as sources of proteins that could be used for therapeutic purposes (Brown *et al.*, 2000). There will be many regulatory hurdles to overcome before artificial chromosomes can be used in such ways, because safety must be a prime consideration. We have seen (Section 17.2) that extra chromosomes often have disastrous consequences for the individual, and if there should be any tendency for an artificial chromosome to become integrated into the endogenous chromosomes, or induce rearrangements, there would be concerns that such changes might lead to cancer (Section 17.9.1). However, we do seem to be reaching a stage when artificial chromosomes can be produced that are mitotically stable for an indefinite period and do not integrate into other chromosomes. As an example, human chromosome fragments carrying immunoglobulin genes have been introduced into mice, where they express the immunoglobulins and are maintained stably and transmitted through both male and female germ lines (Tomizuka *et al.*, 2000). Such procedures might have direct therapeutic potential or could be used to produce therapeutic products in bulk in transgenic animals. Given the speed of advance in most scientific fields, it will probably be no more than a few years before satisfactory artificial chromosomes can be produced that are eminently suitable for such applications, although targeted gene integration might turn out to be a more convenient technique (Suraokar & Bradley, 2000). Provided that the safety questions can be answered satisfactorily, which could indeed take several years, there seems no reason why such applications should not proceed.

There remains the question of how to deliver the artificial chromosomes to the required tissues. We have seen that it is possible to microinject MACs into zygotes, and that these MACs are maintained in the adult animal and are passed through the germ line to the offspring (Co et al., 2000). Alternatively, artificial chromosomes can be introduced into embryonic stem cells, which then repopulate various tissues and can also be transmitted to offspring (Jakobovits et al., 1993). Such methods would be practicable for producing transgenic animals for the production of therapeutic proteins, although targeted integration of genes in sheep is currently a more advanced technique (Suraokar & Bradley, 2000). If it were desirable to restrict the expression of the gene to a single tissue, it should be possible to arrange that it would be under the control of a tissue-specific promoter.

Injection of MACs into zygotes would obviously not be practicable for gene therapy. Apart from the technical difficulties, most gene defects would only be discovered at a much later stage. For haematological disorders, it would be possible to introduce MACs into bone marrow cells, and then re-introduce the cells to the marrow. Transfection into cells of other accessible tissues, such as lung (for example, for cystic fibrosis), gut, etc., would be a possibility, but much more development would be needed to make this an efficient method. Other tissues, such as brain, might seem much more inaccessible, yet if it is practicable to introduce cells into the brain to treat Parkinson's disease (for example) (Brundin et al., 2000), it should be possible to introduce MACs into brain cells. The problems of efficient delivery of MACs for gene therapy may seem very great, but there is no fundamental reason to suppose that they should be insuperable. Many technologies that are now routine were once regarded as being very difficult.

The future impact of artificial chromosome technology on biology and medicine is difficult to estimate, but if it does become a practical, successful technology it will be as a result of application of the accumulated knowledge about chromosomes that has been described in this book. In turn, artificial chromosome technology, although ultimately aimed at practical applications, has helped us to define what is required to form a functional chromosome, and has added to our knowledge of chromosome organization.

References

Aagaard, L., Schmid, M., Warburton, P. & Jenuwein, T. (2000) Mitotic phosphorylation of SUV39H1, a novel component of active centromeres, coincides with transient accumulation at mammalian centromeres. *Journal of Cell Science* **113**, 817–829.

Abranches, R., Beven, A.F., Aragón-Alcaide, L. & Shaw, P.J. (1998) Transcription sites are not correlated with chromosome territories in wheat nuclei. *Journal of Cell Biology* **143**, 5–12.

Adams, M.D., Celniker, S.E., Holt, R.A. *et al.* (2000) The genome sequence of *Drosophila melanogaster*. *Science* **287**, 2185–2195.

Adams, R.L.P., Knowler, J.T. & Leader, D.P. (1992) *The Biochemistry of the Nucleic Acids*, 11th edn. Chapman & Hall, London.

Adams, R.R., Carmena, M. & Earnshaw, W.C. (2001) Chromosomal passengers and the (aurora) ABCs of mitosis. *Trends in Cell Biology* **11**, 49–54.

Adams, S.P., Leitch, I.J., Bennett, M.D. & Leitch, A.R. (2000) *Aloe* L. – a second plant family without (TTTAGGG)$_n$ telomeres. *Chromosoma* **109**, 201–205.

Adolph, K.W., Cheng, S.M. & Laemmli, U.K. (1977) Role of nonhistone proteins in metaphase chromosome structure. *Cell* **12**, 805–816.

Aïssani, B., D'Onofrio, G., Mouchiroud, D., Gardiner K., Gautier, C. & Bernardi, G. (1991) The compositional properties of human genes. *Journal of Molecular Evolution* **32**, 493–503.

Akagi, K., Sandig, V., Vooijs, M. *et al.* (1997) Cre-mediated somatic site-specific recombination in mice. *Nucleic Acids Research* **25**, 1766–1773.

Akhtar, A., Zink, D. & Becker, P.B. (2000) Chromodomains are protein–RNA interaction modules. *Nature* **407**, 405–409.

Albertson, D.G. & Thomson, J.N. (1982) The kinetochores of *Caenorhabditis elegans*. *Chromosoma* **86**, 409–428.

Allen, T.D., Cronshaw, J.M., Bagley, S., Kiseleva, E. & Goldberg, M.W. (2000) The nuclear pore complex: mediation of translocation between nucleus and cytoplasm. *Journal of Cell Science* **113**, 1651–1659.

Allshire, R.C., Cranston, G., Gosden, J.R., Maule, J.C., Hastie, N.D. & Fantes, P.A. (1987) A fission yeast chromosome can replicate autonomously in mouse cells. *Cell* **50**, 391–403.

Amati, B.B. & Gasser, S.M. (1988) Chromosomal ARS and CEN elements bind specifically to the yeast nuclear scaffold. *Cell* **54**, 967–978.

Ambros, P. & Sumner, A.T. (1987) Correlation of pachytene chromomeres and metaphase bands of human chromosomes, and distinctive properties of telomeric regions. *Cytogenetics and Cell Genetics* **44**, 223–228.

Ammermann, D. (1987) Giant chromosomes in ciliates. In: *Structure and Function of Eukaryotic Chromosomes* (ed. W. Hennig), pp. 59–67. Springer-Verlag, Berlin.

Amon, A. (1999) The spindle checkpoint. *Current Opinion in Genetics and Development* **9**, 69–75.

Ananiev, E.V., Phillips, R.L. & Rines, H.W. (1998a) Chromosome-specific molecular organization of maize (*Zea mays* L.) centromeric regions. *Proceedings of the National Academy of Sciences of the USA* **95**, 13073–13078.

Ananiev, E.V., Phillips, R.L. & Rines, H.W. (1998b) Complex structure of knob DNA on maize chromosome 9: retrotransposon invasion into heterochromatin. *Genetics* **149**, 2025–2037.

Andersen, S.S.L. (1999) Balanced regulation of microtubule dynamics during the cell cycle: a contemporary view. *Bioessays* **21**, 53–60.

Anderson, L.K., Offenberg, M.M., Verkuijlen, W.H.M.C. & Heyting, C. (1997) RecA-like proteins are components of early meiotic nodules in lily. *Proceedings of the National Academy of Sciences of the USA* **94**, 6868–6873.

Andersson, K., Björkroth, B. & Daneholt, B. (1980) The in situ structure of the active 75S RNA genes in Balbiani rings of *Chironomus tentans*. *Experimental Cell Research* **130**, 313–326.

Andersson, K., Mähr, R., Björkroth, B. & Daneholt, B. (1982) Rapid reformation of the thick chromosome fibre upon completion of RNA synthesis at the Balbiani ring genes in *Chironomus tentans*. *Chromosoma* **87**, 33–48.

Angelier, N., Bonnanfant-Jaïs, M.L., Moreau, N., Gounon, P. & Lavand, A. (1986) DNA methylation and RNA transcriptional activity in amphibian lampbrush chromosomes. *Chromosoma*, **94**, 169–182.

Angell, R.R. (1997) First-meiotic-division nondisjunction in human oocytes. *American Journal of Human Genetics* **61**, 23–32.

Aragón-Alcaide, L., Miller, T., Schwarzacher, T., Reader, S. & Moore, G. (1996) A cereal centromeric sequence. *Chromosoma* **105**, 261–268.

Araki, K., Araki, M. & Yamamura, K-I. (1997) Targeted integration of DNA using mutant *lox* sites in embryonic stem cells. *Nucleic Acids Research* **25**, 868–872.

Arkhipova, I. & Meselson, M. (2000) Transposable elements in sexual and ancient asexual taxa. *Proceedings of the National Academy of Sciences of the USA* **97**, 14473–14477.

Armstrong, J.A. & Emerson, B.M. (1998) Transcription of chromatin: these are complex times. *Current Opinion in Genetics and Development* **8**, 165–172.

Arnheim, N. & Shibata, D. (1997) DNA mismatch repair in mammals: role in disease and meiosis. *Current Opinion in Genetics and Development* **7**, 364–370.

Ascenzi, R. & Gantt, J.S. (1999) Subnuclear distribution of the entire complement of linker histone variants in *Arabidopsis thaliana*. *Chromosoma* **108**, 345–355.

Ashburner, M. (1972) Puffing patterns in *Drosophila melanogaster* and related species. In: *Developmental Studies on Giant Chromosomes* (ed. W. Beermann), pp. 101–151. Springer-Verlag, Berlin.

Ashley, T. (1990) G-bands and chromosomal meiotic behaviour. In: *Chromosomes Today*, Vol. 10 (eds K. Fredga, B.A. Kihlman & M.D. Bennett), pp. 311–320. Unwin Hyman, London.

Atkin, N.B. (2001) Microsatellite instability. *Cytogenetics and Cell Genetics* **92**, 177–181.

Attia, T. & Lelley, T. (1987) Effects of constitutive heterochromatin and genotype on frequency and distribution of chiasmata in the seven individual rye bivalents. *Theoretical and Applied Genetics* **74**, 527–530.

Avery, O.T., Macleod, C.M. & McCarty, M. (1944) Studies on the chemical nature of the substance inducing transformation of pneumococcal types. Induction of transformation by a desoxyribonucleic acid fraction isolated from *Pneumococcus* type III. *Journal of Experimental Medicine* **79**, 137–158.

Avner, P. & Heard, E. (2001) X-chromosome inactivation: counting, choice and initiation. *Nature Reviews Genetics* **2**, 59–67.

Avner, P., Prissette, M., Arnaud, D., Coutier, B., Cecchi, C. & Heard, E. (1998) Molecular correlates of the murine *Xce* locus. *Genetical Research* **72**, 217–224.

Ayoub, N., Richler, C. & Wahrman, J. (1997) *Xist* RNA is associated with the transcriptionally inactive XY body in mammalian male meiosis. *Chromosoma* **106**, 1–10.

Azzalin, C.M., Nergadze, S.G. & Giulotto, E. (2001) Human intrachromosomal telomeric-like repeats: sequence organization and mechanisms of origin. *Chromosoma* **110**, 75–82.

Bachant, J.B. & Elledge, S.J. (1999) Mitotic treasures in the nucleolus. *Nature* **398**, 757–758.

Bailey, J.A., Carrel, L., Chakravarti, A. & Eichler, E.E. (2000) Molecular evidence for a relationship between LINE-1 elements and X chromosome inactivation: the Lyon repeat hypothesis. *Proceedings of the National Academy of Sciences of the USA* **97**, 6634–6639.

Bailey, J.A., Yavor, A.M., Viggiano, L. *et al.* (2002) Human-specific duplication and mosaic transcripts: the recent paralogous structure of chromosome 22. *American Journal of Human Genetics* **70**, 83–100.

Baker, B.S., Gorman, M. & Marín, I. (1994) Dosage compensation in *Drosophila*. *Annual Review of Genetics* **28**, 491–521.

Baldwin, L. & Macgregor, H.C. (1985) Centromeric satellite DNA in the newt *Triturus cristatus karelinii* and related species: its distribution and transcription on lampbrush chromosomes. *Chromosoma* **92**, 100–107.

Balhorn, R. (1982) A model for the structure of chromatin in mammalian sperm. *Journal of Cell Biology* **93**, 298–304.

Balicek, P., Zizka, J. & Skalská, H. (1977) Length of human constitutive heterochromatin in relation to chromosomal contraction. *Human Genetics* **38**, 189–193.

Bannister, A.J., Zegerman, P., Partridge, J.F., Miska, E.A., Thomas, J.O., Allshire, R.C. & Kouzarides, T. (2001) Selective recognition of methylated lysine 9 on histone H3 by the HP1 chromo domain. *Nature* **410**, 120–124.

Barch, M.J., Knutsen, T. & Spurbeck, J.L. (eds) (1997) *The ACT Cytogenetics Laboratory Manual*, 3rd edn. Raven Press, New York.

Barlow, A.L. & Hultén, M.A. (1998) Crossing over analysis at pachytene in man. *European Journal of Human Genetics* **6**, 350–358.

Barlow, A.L., Benson, F.E., West, S.C. & Hultén, M.A. (1997) Distribution of the Rad51 recombinase in human and mouse spermatocytes. *EMBO Journal* **16**, 5207–5215.

Barlow, D.P. (1994) Imprinting: a gamete's point of view. *Trends in Genetics* **10**, 194–199.

Barry, A.E., Howman, E.V., Cancilla, M.R., Saffery, R. & Choo, K.H.A. (1999) Sequence analysis of an 80 kb human neocentromere. *Human Molecular Genetics* **8**, 217–227.

Bartolmé, S., Bermúdez, A. & Daban, J-R. (1994) Internal structure of the 30 nm chromatin fibre. *Journal of Cell Science* **107**, 2983–2992.

Bass, H.W., Marshall, W.F., Sedat, J.W., Agard, D.A. & Cande, W.Z. (1997) Telomeres cluster de novo before the initiation of synapsis: a three-dimensional spatial analysis of telomere positions before and during meiotic prophase. *Journal of Cell Biology* **137**, 5–18.

Bassham, S., Beam, A. & Shampay, J. (1998) Telomere variation in *Xenopus laevis*. *Molecular and Cellular Biology* **18**, 269–275.

Baum, M., Ngan, V.K. & Clarke, L. (1994) The centromeric K-type repeat and the central core are together sufficient to establish a functional *Schizosaccharomyces pombe* centromere. *Molecular Biology of the Cell* **5**, 747–761.

Baumer, A., Balmer, D., Binkert, F. & Schinzel, A. (2000) Parental origin and mechanisms of formation of triploidy: a study of 25 cases. *European Journal of Human Genetics* **8**, 911–917.

Baylin, S.B. & Herman, J.G. (2000) DNA hypermethylation in tumorigenesis. *Trends in Genetics* **16**, 168–174.

Bechert, T., Heck, S., Fleig, U., Diekman, S. & Hegemann, J.H. (1999) All 16 centromere DNAs from *Saccharomyces cerevisiae* show DNA curvature. *Nucleic Acids Research* **27**, 1444–1449.

Bedo, D.G. (1975) C banding in polytene chromosomes of *Simulium ornatipes* and *S. melatum* (Diptera: Simuliidae). *Chromosoma* **51**, 291–300.

Bedo, D.G. (1982) Differential sex chromosome replication and dosage compensation in polytene trichogen cells of *Lucilia cuprina* (Diptera: Calliphoridae). *Chromosoma* **87**, 21–32.

Bedo, D.G. (1994) Differential replication of heterochromatin in polytene chromosomes of the Old World screw-worm fly, *Chrysomya bezziana* (Diptera: Calliphoridae), analysed by fluorescence staining. *Chromosome Research* **2**, 191–199.

Beechey, C.V., Cattanach, B.M. & Blake, A. (2001) MRC Mammalian Genetics Unit, Harwell, Oxfordshire. World Wide Web Site – Genetic and physical imprinting map of the mouse: http://www.mgu.har.mrc.ac.uk/imprinting/all_impmaps.html

Beerman, S. (1977) The diminution of heterochromatic chromosomal segments in *Cyclops* (Crustacea, Copepoda). *Chromosoma* **60**, 297–344.

Beermann, W. (1972) Chromomeres and genes. In: *Developmental Studies on Giant Chromosomes* (ed. W. Beermann), pp. 1–33. Springer-Verlag, Berlin.

Bell, A.C. & Felsenfeld, G. (1999) Stopped at the border: boundaries and insulators. *Current Opinion in Genetics and Development* **9**, 191–198.

Bellini, M. (2000) Coilin, more than a molecular marker of the Cajal (coiled) body. *Bioessays* **22**, 861–867.

Belmont, A.S., Dietzel, S., Nye, A.C., Strukov, Y.G. &

Tumbar, T. (1999) Large-scale chromatin structure and function. *Current Opinion in Cell Biology* **11**, 307–311.

Belyaev, N.D., Houben, A., Baranczewski, P. & Schubert, I. (1997) Histone H4 acetylation in plant heterochromatin is altered during the cell cycle. *Chromosoma* **106**, 193–197.

Belyaeva, E.S., Demakova, O.V., Umbetova, G.H. & Zhimulev, I.F. (1993) Cytogenetic and molecular aspects of position-effect variegation in *Drosophila melanogaster*. V. Heterochromatin-associated protein HP1 appears in euchromatic chromosomal regions that are inactivated as a result of position-effect variegation. *Chromosoma* **102**, 583–590.

Bennett, M.D. (1985) Intraspecific variation in DNA amount and the nucleotypic dimension in plant genetics. In: *Plant Genetics* (ed. M. Freeling), pp. 283–302. Alan R. Liss, New York.

Bennett, M.D. (1995) The development and use of genomic *in situ* hybridization (GISH) as a new tool in plant biosystematics. In: *Kew Chromosome Conference IV* (eds P.E. Brandham & M.D. Bennett), pp. 167–183. Royal Botanic Garden, Kew.

Bennetzen, J.L., SanMiguel, P., Chen, M., Tikhonov, A., Francki, M. & Avramova, Z. (1998) Grass genomes. *Proceedings of the National Academy of Sciences of the USA* **95**, 1975–1978.

Berendes, H.D. (1965) Salivary gland function and chromosomal puffing patterns in *Drosophila hydei*. *Chromosoma* **17**, 35–77.

Berendes, H.D. (1972) The control of puffing in *Drosophila hydei*. In: *Developmental Studies on Giant Chromosomes* (ed. W. Beermann), pp. 181–207. Springer-Verlag, Berlin.

Berezney, R. & Wei, X. (1998) The new paradigm: integrating genomic function and nuclear architecture. *Journal of Cell Biochemistry* **30–31** (suppl.), 238–242.

Beridze, T. (1986) *Satellite DNA*. Springer-Verlag, Berlin.

Bernard, P., Maure, J-F., Partridge, J.F., Genier, S., Javerzat, J-P. & Allshire, R.C. (2001) Requirement of heterochromatin for cohesion at centromeres. *Science* **294**, 2539–2542.

Bernardi, G. (1989) The isochore organization of the human genome. *Annual Review of Genetics* **23**, 637–661.

Bernardi, G. (1993a) The isochore organization of the human genome and its evolutionary history – a review. *Gene* **135**, 57–66.

Bernardi, G. (1993b) The vertebrate genome: isochores and evolution. *Molecular Biology and Evolution* **10**, 186–204.

Bernardino, J., Gerbault-Seureau, M., Zafrani, B. *et al.* (1998) Homogeneously staining regions in 223 breast carcinomas: cytogenetic and clinicopathological correlations. *British Journal of Cancer* **78**, 1214–1218.

Bertoni, L., Attolini, C., Faravelli, M., Simi, S. & Giulotto,

E. (1996) Intrachromosomal telomere-like DNA sequences in Chinese hamster. *Mammalian Genome* **7**, 853–855.

Beukeboom, L.W. (1995) Sex determination in Hymenoptera: a need for genetic and molecular studies. *Bioessays* **17**, 813–817.

Bi, X. & Broach, J.R. (2001) Chromosomal boundaries in *S. cerevisiae*. *Current Opinion in Genetics and Development* **11**, 199–204.

Bianchi, M.E. & Beltrame, M. (2000) Upwardly mobile proteins. *EMBO Reports* **1**, 109–119.

Bickham, J.W. (1981) Two-hundred-million-year-old chromosomes: deceleration of the rate of karyotypic evolution in turtles. *Science* **212**, 1291–1293.

Bickmore, W.A. & Craig, J. (1997) *Chromosome Bands: Patterns in the Genome*. R.G. Landes, Austin, Texas.

Bickmore, W.A. & Oghene, K. (1996) Visualizing the spatial relationships between defined DNA sequences and the axial region of extracted metaphase chromosomes. *Cell* **84**, 95–104.

Bidichandani, S.I., Ashizawa, T. & Patel, P.I. (1998) The GAA triplet-repeat expansion in Friedreich ataxia interferes with transcription and may be associated with an unusual DNA structure. *American Journal of Human Genetics* **62**, 111–121.

Biedler, J.L. & Spengler, B.A. (1976) Metaphase chromosome anomaly: association with drug resistance and cell-specific products. *Science* **191**, 185–187.

Biedler, J.L., Melera, P.W. & Spengler, B.A. (1980) Specifically altered metaphase chromosomes in antifolate-resistant Chinese hamster cells that overproduce dihydrofolate reductase. *Cancer Genetics and Cytogenetics* **2**, 47–60.

Biessmann, H. & Mason, J.M. (1992) Genetics and molecular biology of telomeres. *Advances in Genetics*. **30**, 185–249.

Biessmann, H. & Mason, J.M. (1997) Telomere maintenance without telomerase. *Chromosoma* **106**, 63–69.

Biessmann, H., Donath, J. & Walter, M.F. (1996) Molecular characterization of the *Anopheles gambiae* 2L telomeric region via an integrated transgene. *Insect Molecular Biology* **5**, 11–20.

Biessmann, H., Zurocova, M., Yao, J.G., Lozovskaya, E. & Walter, M.F. (2000) A telomeric satellite in *Drosophila virilis* and its sibling species. *Chromosoma* **109**, 372–380.

Biggins, S. & Murray, A.W. (1999) Sister chromatid cohesion in mitosis. *Current Opinion in Genetics and Development* **9**, 230–236.

Birchler, J.A. (1997) Do these sequences make CENs yet? *Genome Research* **7**, 1035–1037.

Bird, A. (1987) CpG islands as gene markers in the vertebrate nucleus. *Trends in Genetics* **3**, 342–347.

Blackburn, E.H. (1991a) Structure and function of telomeres. *Nature* **350**, 569–573.

Blackburn, E.H. (1991b) Telomeres. *Trends in Biochemical Sciences* **16**, 378–381.

Blackburn, G.M. & Gait, M.J. (1996) *Nucleic Acids in Chemistry and Biology*, 2nd edn. Oxford University Press, Oxford.

Blasco, M.A., Lee, H-W., Hande, M.P. *et al.* (1997) Telomere shortening and telomere formation by mouse cells lacking telomerase RNA. *Cell* **91**, 25–34.

Blasco, M.A., Gasser, S.M. & Lingner, J. (1999) Telomeres and telomerase. *Genes and Development* **13**, 2353–2359.

Bloom, K. (1993) The centromere frontier: kinetochore components, microtubule-based motility, and the CEN-value paradox. *Cell* **73**, 621–624.

Blow, J.J. & Prokhorova, T.A. (1999) Saying a firm 'no' to DNA re-replication. *Nature Cell Biology* **1**, E175–E177.

Blower, M.D. & Karpen, G.H. (2001) The role of *Drosophila* CID in kinetochore formation, cell-cycle progression and heterochromatin interactions. *Nature Cell Biology* **3**, 730–739.

Bobrow, M. (1985) Heterochromatic chromosome variation and reproductive failure. *Experimental and Clinical Immunogenetics* **2**, 97–105.

Bodnar, A.G., Ouellette, M., Frolkis, M. *et al.* (1998) Extension of life-span by introduction of telomerase into normal human cells. *Science* **279**, 349–352.

Bork, P. & Copley, R. (2001) Filling in the gaps. *Nature* **409**, 818–820.

Borsani, G. & Ballabio, A. (1993) X chromosome dosage compensation in female mammals. *Seminars in Developmental Biology* **4**, 129–139.

Borst, P., van der Ploeg, M., van Hoek, J.F.M., Tas, J. & James, J. (1982) On the DNA content and ploidy of trypanosomes. *Molecular and Biochemical Parasitology* **6**, 13–23.

Bostock, C.J. & Sumner, A.T. (1978) *The Eukaryotic Chromosome*. North-Holland, Amsterdam.

Boy de la Tour, E. & Laemmli, U.K. (1988) The metaphase scaffold is helically folded: sister chromatids have predominantly opposite helical handedness. *Cell* **55**, 937–944.

Boyes, J., Byfield, P., Nakatani, Y. & Ogryzko, V. (1998) Regulation of activity of the transcription factor GATA-1 by acetylation. *Nature* **396**, 594–598.

Boyle, A.L., Ballard, S.G. & Ward, D.C. (1990) Differential distribution of long and short interspersed element sequences in the mouse genome: chromosome karyotyping by fluorescence in situ hybridisation. *Proceedings of the National Academy of Sciences of the USA* **87**, 7757–7761.

Boyle, S., Gilchrist, S., Bridger, J.M., Mahy, N.L., Ellis, J.A. & Bickmore, W.A. (2001) The spatial organization of human chromosomes within the nuclei of normal and emerin-mutant cells. *Human Molecular Genetics* **10**, 211–219.

Brady, T. & Clutter, M.E. (1974) Structure and replication of *Phaseolus* polytene chromosomes. *Chromosoma* **45**, 63–79.

Breneman, J.W., Yau, P., Teplitz, R.L. & Bradbury, E.M. (1993) A light microscope study of linker histone distribution in rat metaphase chromosomes and interphase nuclei. *Experimental Cell Research* **206**, 16–26.

Breneman, J.W., Yau, P.M., Swiger, R.R. *et al.* (1996) Activity banding of human chromosomes as shown by histone acetylation. *Chromosoma* **105**, 41–49.

Brewer, B.J., Diller, J.D., Friedman, K.L. *et al.* (1993) The topography of chromosome replication in yeast. *Cold Spring Harbor Symposium on Quantitative Biology* **58**, 425–434.

Bridger, J.M. & Lichter, P. (1999) Analysis of mammalian interphase chromosomes by FISH and immunofluorescence. In: *Chromosome Structural Analysis: a Practical Approach* (ed. W.A. Bickmore), pp. 103–123. Oxford University Press, Oxford.

Bridges, B.A. (1999) DNA repair: polymerases for passing lesions. *Current Biology* **9**, R475–R477.

Britton-Davidian, J., Catalan, J., Ramalhinho, M. da G. *et al.* (2000) Rapid chromosomal evolution in island mice. *Nature* **403**, 158.

Brown, E.J. & Baltimore, D. (2000) *ATR* disruption leads to chromosomal fragmentation and early embryonic lethality. *Genes and Development* **14**, 397–402.

Brown, K.E., Guest, S.S., Smale, S.T., Hahm, K., Merkenschlager, M. & Fisher, A.G. (1997) Association of transcriptionally silent genes with Ikaros complexes at centromeric heterochromatin. *Cell* **91**, 845–854.

Brown, K.E., Baxter, J., Graf, D., Merkenschlager, M. & Fisher, A.G. (1999) Dynamic repositioning of genes in the nucleus of lymphocytes preparing for cell division. *Molecular Cell* **3**, 207–217.

Brown, S.W. (1966) Heterochromatin. *Science* **151**, 417–425.

Brown, S.W. & Nur, U. (1964) Heterochromatic chromosomes in the coccids. *Science* **145**, 130–136.

Brown, W., Heller, R., Loupart, M-L., Shen, M-H. & Chand, A. (1996) Mammalian artificial chromosomes. *Current Opinion in Genetics and Development* **6**, 281–288.

Brown, W.R.A., Mee, P.J. & Shen, M.H. (2000) Artificial chromosomes: ideal vectors? *Trends in Biotechnology* **18**, 218–223.

Brundin, P., Karlsson, J., Emgard, M. *et al.* (2000) Improving the survival of grafted dopaminergic neurons: a review over current approaches. *Cell Transplantation* **9**, 179–195.

Bryan, T.M. & Cech, T.R. (1999) Telomerase and the maintenance of chromosome ends. *Current Opinion in Cell Biology* **11**, 318–324.

Bryan, T.M., Engelzou, A., Dalla-Pozza, L., Dunham, M.A. & Reddel, R.R. (1997) Evidence for an alternative mechanism for maintaining telomere length in human tumors and tumor-derived cell lines. *Nature Medicine* **3**, 1271–1274.

Buchwitz, B.J., Ahmad, K., Moore, L.L., Roth, M.B. & Henikoff, S. (1999) A histone-H3-like protein in *C. elegans*. *Nature* **401**, 547–548.

Buckland, R.A. & Evans, H.J. (1978) Cytogenetic aspects of phylogeny in the Bovidae. I. G-banding. *Cytogenetics and Cell Genetics* **21**, 42–63.

Buermeyer, A.B., Deschênes, S.M., Baker, S.M. & Liskay, R.M. (1999) Mammalian DNA mismatch repair. *Annual Review of Genetics* **33**, 533–564.

Bull, J.J. (1983) *Evolution of Sex Determining Mechanisms.* The Benjamin/Cummings Publishing Co. Inc., Menlo Park, CA.

Bunch, T.D. & Nadler, C.F. (1980) Giemsa-band patterns of the tahr and chromosomal evolution of the tribe Caprini. *Journal of Heredity* **71**, 110–116.

Buongiorno-Nardelli, M., Micheli, G., Carri, M.T. & Marilley, M. (1982) A relationship between replicon size and supercoiled loop domains in the eukaryotic genome. *Nature* **298**, 100–102.

Burke, D.T., Carle, G.F. & Olson, M.V. (1987) Cloning of large segments of exogenous DNA into yeast by means of artificial chromosome vectors. *Science* **236**, 806–812.

Burns, E.M., Christopoulou, L., Corish, P. & Tyler-Smith, C. (1999) Quantitative measurement of mammalian chromosome mitotic loss rates using the green fluorescent protein. *Journal of Cell Science* **112**, 2705–2714.

Busch, H. & Rothblum, L. (eds) (1982) *The Cell Nucleus,* Vol. X. *rDNA*, part A. Academic Press, New York.

Bustin, M. (2001) Revised nomenclature for high mobility group (HMG) proteins. *Trends in Biochemical Sciences* **26**, 152–153.

Bustin, M. & Reeves, R. (1996) High-mobility-group chromosomal proteins: architectural components that facilitate chromatin function. *Progress in Nucleic Acid Research and Molecular Biology* **54**, 35–100.

C. elegans Sequencing Consortium (1998) Genome sequence of the nematode *C. elegans*: a platform for investigating biology. *Science* **282**, 2012–2018.

Caceres, M.E., De Pace, C., Scarascia Mugnozza, G.T., Kotsonis, P., Ceccarelli, M. & Cionini, P.G. (1998) Genome size variations within *Dasypyrum villosum*: correlations with chromosomal traits, environmental factors and plant phenotypic characteristics and behaviour in reproduction. *Theoretical and Applied Genetics* **96**, 559–567.

Callan, H.G. (1982) Lampbrush chromosomes. *Proceedings of the Royal Society of London B* **214**, 417–448.

Callan, H.G. (1986) *Lampbrush Chromosomes.* Springer-Verlag, Berlin.

Callan, H.G., Gall, J.G. & Murphy, C. (1988) The distribution of oocyte 5S, somatic 5S and 18S + 28S rDNA sequences in the lambbrush chromosomes of *Xenopus laevis*. *Chromosoma* **97**, 43–54.

Calos, M.P. (1996) The potential of extrachromosomal replicating vectors for gene therapy. *Trends in Genetics* **12**, 463–466.

Capanna, E. (2000) Chromosomes yesterday: a century of chromosome studies. In: *Chromosomes Today*, Vol. 13 (eds E. Olmo & C.A. Redi), pp. 3–22. Birkhäuser, Basel.

Capy, P. (2000) Is bigger better in cricket? *Science* **287**, 985–986.

Carmena, M., Abad, J.P., Villasante, A. & Gonzalez, C. (1993) The *Drosophila melanogaster dodecasatellite* sequence is closely linked to the centromere and can form connections between sister chromatids during mitosis. *Journal of Cell Science* **105**, 41–50.

Carmi, I., Kopczynski, J.B. & Meyer, B.J. (1998) The nuclear hormone receptor SEX-1 is an X-chromosome signal that determines nematode sex. *Nature* **396**, 168–173.

Carrel, L., Cottle, A.A., Goglin, K.C. & Willard, H.F. (1999) A first-generation X-inactivation profile of the human X chromosome. *Proceedings of the National Academy of Sciences of the USA* **96**, 14440–14444. See also: www.pnas.org/cgi/content/abstract/96/25/14440

Carson, H.L. (1981) Homosequential species of Hawaiian *Drosophila*. In: *Chromosomes Today*, Vol. 7 (eds M.D. Bennett, M. Bobrow & G.M. Hewitt), pp. 150–164. George Allen & Unwin, London.

Carvalho, A.B., Dobo, B.A., Vibranovski, M.D. & Clark, A.G. (2001) Identification of five new genes on the Y chromosome of *Drosophila melanogaster*. *Proceedings of the National Academy of Sciences of the USA* **98**, 13225–13230.

Caspary, T., Cleary, M.A., Perlman, E.J., Zhang, P., Elledge, S.J. & Tilghman, S.M. (1999) Oppositely imprinted genes *p57Kip2* and *Igf2* interact in a mouse model for Beckwith–Wiedemann syndrome. *Genes and Development* **13**, 3115–3124.

Cassagnau (1974) Les chromosomes polytènes de *Neanura monticola* Cassagnau (Collemboles). I. Polymorphisme écologique du chromosome X. *Chromosoma* **46**, 343–363.

Cavalier-Smith, T. (1978) Nuclear volume control by nucleoskeletal DNA, selection for cell volume and cell growth rate, and the solution of the DNA C-value paradox. *Journal of Cell Science* **34**, 247–278.

Cavalier-Smith, T. (1982) Skeletal DNA and the evolution of genome size. *Annual Review of Biophysics and Bioengineering* **11**, 273–302.

Chadwick, B.P. & Willard, H.F. (2001a) A novel chromatin protein, distantly related to histone H2A, is largely excluded from the inactive X chromosome. *Journal of Cell Biology* **152**, 375–384.

Chadwick, B.P. & Willard, H.F. (2001b) Histone H2A vari-

ants and the inactive X chromosome: identification of a second macroH2A variant. *Human Molecular Genetics* **10**, 1101–1113.

Chan, G.K.T., Schaar, B.T. & Yen, T.J. (1998) Characterization of the kinetochore binding domain of CENP-E reveals interactions with the kinetochore proteins CENP-F and hBUBR1. *Journal of Cell Biology* **143**, 49–63.

Chandley, A.C. (1984) Infertility and chromosome abnormality. *Oxford Reviews of Reproductive Biology* **6**, 1–46.

Chandley, A.C. (1986) A model for effective pairing and recombination at meiosis based on early replicating sites (R-bands) along chromosomes. *Human Genetics* **72**, 50–57.

Chandra, H.S. (1994) Proposed role of W chromosome inactivation and the absence of dosage compensation in avian sex determination. *Proceedings of the Royal Society of London B* **258**, 79–82.

Charlesworth, B. & Charlesworth, D. (2000) The degeneration of Y chromosomes. *Philosophical Transactions of the Royal Society of London B* **355**, 1563–1572.

Charlesworth, B., Sniegowski, P. & Stephan, W. (1994) The evolutionary dynamics of repetitive DNA in eukaryotes. *Nature* **371**, 215–220.

Chen, K-S., Manian, P., Koeuth, T. *et al.* (1997) Homologous recombination of a flanking repeat cluster is a mechanism for a common contiguous gene deletion syndrome. *Nature Genetics* **17**, 154–163.

Cheung, P., Allis, C.D. & Sassone-Corsi, P. (2000) Signaling to chromatin through histone modifications. *Cell* **103**, 263–271.

Chong, J.P.J., Thömmes, P. & Blow, J.J. (1996) The role of MCM/P1 proteins in the licensing of DNA replication. *Trends in Biochemical Sciences* **21**, 102–106.

Choo, K.H. (1990) Role of acrocentric cen-pter satellite DNA in Robertsonian translocation and chromosomal non-disjunction. *Molecular Biology and Medicine* **7**, 437–449.

Choo, K.H.A. (1997) Centromere DNA dynamics: latent centromeres and neocentromere formation. *American Journal of Human Genetics* **61**, 1225–1233.

Choo, K.H.A. (2000) Centromerization. *Trends in Cell Biology* **10**, 182–188.

Chowdhary, B.P. & Raudsepp, T. (2000) HSA4 and GGA4: remarkable conservation despite 300-Myr divergence. *Genomics* **64**, 102–105.

Chowdhary, B.P., Raudsepp, T., Frönicke, L. & Scherthan, H. (1998) Emerging patterns of comparative genome organisation in some mammalian species as revealed by Zoo-FISH. *Genome Research* **8**, 577–589.

Chua, P.R. & Roeder, G.S. (1998) Zip2, a meiosis-specific protein required for the initiation of chromosome synapsis. *Cell* **93**, 349–359.

Chung, J., Lee, J-H., Arumuganathan, K., Graef, G.L. &

Specht, J.E. (1998) Relationships between nuclear DNA content and seed and leaf size in soybean. *Theoretical and Applied Genetics* **96**, 1064–1068.

Clarke, D.J. & Giménez-Abián, J.F. (2000) Checkpoints controlling mitosis. *Bioessays* **22**, 351–363.

Clarke, D.J., Giménez-Abián, J.F., Tönnies, H. *et al.* (1998) Creation of monosomic derivatives of human cultured cell lines. *Proceedings of the National Academy of Sciences of the USA* **95**, 167–171.

Clarke, L. (1990) Centromeres of budding and fission yeasts. *Trends in Genetics* **6**, 150–154.

Clarke, L. (1998) Centromeres: proteins, protein complexes, and repeated domains at centromeres of simple eukaryotes. *Current Opinion in Genetics and Development* **8**, 212–218.

Clarke, L. & Carbon, J. (1980) Isolation of a yeast centromere and construction of functional small circular chromosomes. *Nature* **287**, 504–509.

Clemson, C.M., McNeil, J.A., Willard, H.F. & Lawrence, J.B. (1996) XIST RNA paints the inactive X chromosome at interphase: evidence for a novel RNA involved in nuclear/chromosome structure. *Journal of Cell Biology* **132**, 1–17.

Clemson, C.M., Chow, J.C., Brown, C.J. & Lawrence, J.B. (1998) Stabilization and localization of Xist RNA are controlled by separate mechanisms and are not sufficient for X inactivation. *Journal of Cell Biology* **142**, 13–23.

Cline, T.W. (1993) The *Drosophila* sex determination signal: how do flies count to two? *Trends in Genetics* **9**, 385–390.

Cline, T.W. & Meyer, B.J. (1996) Vive la différence: males vs females in flies vs worms. *Annual Review of Genetics* **30**, 637–702.

Co, D.O., Borowski, A.H., Leung, J.D. *et al.* (2000) Generation of transgenic mice and germline transmission of a mammalian artificial chromosome introduced into embryos by pronuclear microinjection. *Chromosome Research* **8**, 183–191.

Cobaleda, C., Pérez-Losada, J. & Sánchez-García, I. (1998) Chromosomal abnormalities and tumor development: from genes to therapeutic mechanisms. *Bioessays* **20**, 922–930.

Cockell, M. & Gasser, S.M. (1999a) Nuclear compartments and gene regulation. *Current Opinion in Genetics and Development* **9**, 199–205.

Cockell, M.M. & Gasser, S.M. (1999b) Nucleolar space for RENT. *Current Biology* **9**, R575–R576.

Coelho, P.S.R., Monesi, N., De Almeida, J.C., Toledo, F., Buttin, G. & Paçó-Larson, M.L. (1993) DNA puff C4 of *Bradysia hygida* (Diptera: Sciaridae) contains genes unequally amplified and differentially expressed during development. *Chromosome Research* **1**, 121–126.

Collins, K. (1999) Ciliate telomerase biochemistry. *Annual Review of Biochemistry* **68**, 187–218.

Collins, K. (2000) Mammalian telomeres and telomerase. *Current Opinion in Cell Biology* **12**, 378–383.

Colot, V. & Rossignol, J-L. (1999) Eukaryotic DNA methylation as an evolutionary device. *Bioessays* **21**, 402–411.

Compton, D.A. (1998) Focusing on spindle poles. *Journal of Cell Science* **111**, 1477–1481.

Compton, D.A. & Cleveland, D.W. (1994) NuMA, a nuclear protein involved in mitosis and nuclear reformation. *Current Opinion in Cell Biology* **6**, 343–346.

Constância, M., Pickard, B., Kelsey, G. & Reik, W. (1998) Imprinting mechanisms. *Genome Research* **8**, 881–900.

Cook, P.R. (1995) A chromomeric model for nuclear and chromosome structure. *Journal of Cell Science* **108**, 2927–2935.

Cook, P.R. (1999) The organization of replication and transcription. *Science* **284**, 1790–1795.

Cooke, C.A., Heck, M.M.S. & Earnshaw, W.C. (1987) The inner centromere protein (INCENP) antigens: movement from inner centromere to midbody during mitosis. *Journal of Cell Biology* **105**, 2053–2067.

Cooke, C.A., Bazett-Jones, D.P., Earnshaw, W.C. & Rattner, J.B. (1993) Mapping DNA within the mammalian kinetochore. *Journal of Cell Biology* **120**, 1083–1091.

Cooke, C.A., Schaar, B., Yen, T.J. & Earnshaw, W.C. (1997) Localization of CENP-E in the fibrous corona and outer plate of mammalian kinetochores from prometaphase through anaphase. *Chromosoma* **106**, 446–455.

Costanzi, C. & Pehrson, J.R. (1998) Histone macroH2A1 is concentrated in the inactive X chromosome of female mammals. *Nature* **393**, 599–601.

Coyne, R.S., Chalker, D.L. & Yao, M-C. (1996) Genome downsizing during ciliate development: nuclear division of labour through chromosome restructuring. *Annual Review of Genetics* **30**, 557–578.

Craig, J. (1999) Isolation of vertebrate metaphase chromosomes and their analysis by FISH. In: *Chromosome Structural Analysis: a Practical Approach* (ed. W.A. Bickmore), pp. 59–80. Oxford University Press, Oxford.

Craig, J.M. & Bickmore, W.A. (1993) Chromosome bands – flavours to savour. *Bioessays*, **15**, 349–354.

Craig, J.M., Boyle, S., Perry, P. & Bickmore, W.A. (1997) Scaffold attachments within the human genome. *Journal of Cell Science* **110**, 2673–2682.

Cremer, M., von Hase, J., Volm, T., Brero, A., Kreth, G., Walter, J., Fischer, C., Solovei, I., Cremer, C. & Cremer, T. (2001) Non-random radial higher-order chromatin arrangements in nuclei of diploid human cells. *Chromosome Research* **9**, 541–567.

Cremer, T. & Cremer, C. (2001) Chromosome territories, nuclear architecture and gene regulation in mammalian cells. *Nature Reviews Genetics* **2**, 292–301.

Crosland, M.W.J & Crozier, R.H. (1986) *Mymecia pilosula*, an ant with only one pair of chromosomes. *Science* **231**, 1278.

Crouse, H. (1960) The controlling element in sex chromosome behavior in *Sciara*. *Genetics* **45**, 1425–1443.

Crouse, H.V. & Keyl, H-G. (1968) Extra replications in the 'DNA puffs' of *Sciara coprophila*. *Chromosoma* **25**, 357–364.

Cryderman, D.E., Morris, E.J., Biessmann, H., Elgin, S.C.R. & Wallrath, L.L. (1999) Silencing at *Drosophila* telomeres: nuclear organization and chromatin structure play critical roles. *EMBO Journal* **18**, 3724–3735.

CSHL/WUGSC/PEB *Arabidopsis* Sequencing Consortium (2000) The complete sequence of a heterochromatic island from a higher eukaryote. *Cell* **100**, 377–386.

Csink, A.K. & Henikoff, S. (1998) Something from nothing: the evolution and utility of satellite repeats. *Trends in Genetics* **14**, 200–204.

Csonka, E., Cserpán, I., Fodor, K. *et al.* (2000) Novel generation of human satellite DNA-based artificial chromosomes in mammalian cells. *Journal of Cell Science* **113**, 3207–3216.

Cummings, C.J. & Zoghbi, H.Y. (2000) Fourteen and counting: unravelling trinucleotide repeat diseases. *Human Molecular Genetics* **9**, 909–916.

Cutts, S.M., Fowler, K.J., Kile, B.T. *et al.* (1999) Defective chromosome segregation, microtubule bundling and nuclear bridging in inner centromere protein gene (*Incenp*)-disrupted mice. *Human Molecular Genetics* **8**, 1145–1155.

Czepulkowski, B. (2001) *Analyzing Chromosomes*. Bios Scientific Publishers, Oxford.

Damm, K., Hemmann, U., Garin-Chesa, P. *et al.* (2001) A highly selective telomerase inhibitor limiting human cancer cell proliferation. *EMBO Journal* **20**, 6958–6968.

Daneholt, B. (1999) Pre-mRNP particles: from gene to nuclear pore. *Current Biology* **9**, R412–R415.

Darlington, C.D. (1966) The chromosomes as *we* see them. In: *Chromosomes Today*, Vol. 1 (eds C.D. Darlington & K.R. Lewis), pp. 1–6. Oliver & Boyd, Edinburgh.

Davie, J.R. (1998) Covalent modifications of histones: expression from chromatin templates. *Current Opinion in Genetics and Development* **8**, 173–178.

Dawe, R.K., Reed, L.M., Yu, H-G., Muszynski, M.G. & Hiatt, E.N. (1999) A maize homolog of mammalian CENPC is a constitutive component of the inner kinetochore. *Plant Cell* **11**, 1227–1238.

Dawes, H.E., Berlin, D.S., Lapidus, D.M., Nusbaum, C., Davis, T.L. & Meyer, B.J. (1999) Dosage compensation proteins targeted to X chromosomes by a determinant of hermaphrodite fate. *Science* **284**, 1800–1804.

de Belle, I., Cai, S. & Kohwi-Shigematsu, T. (1998) The genomic sequences bound to special AT-rich sequence-binding protein 1 (SATB1) in vivo in Jurkat T cells are tightly associated with the nuclear matrix at the bases of the chromatin loops. *Journal of Cell Biology* **141**, 335–348.

de Grouchy, J. & Turleau, C. (1986) Microcytogenetics 1984. *Experientia* **42**, 1090–1097.

de Jong, G., Telenius, A., Telenius, H., Perez, C.F., Drayer, J.I. & Hadlaczky, Gy. (1999) Mammalian artificial chromosome pilot production facility: large-scale isolation of functional satellite DNA-based artificial chromosomes. *Cytometry* **35**, 129–133.

de la Torre, J., Herrero, P., García de la Vega, C., Sumner, A.T. & Gosálvez, J. (1996) Patterns of DNase I sensitivity in the chromosomes of the grasshopper *Chorthippus parallelus* (Orthoptera). *Chromosome Research* **4**, 56–60.

de Laat, W.L., Jaspers, N.G.J. & Hoeijmakers, J.H.J. (1999) Molecular mechanism of nucleotide excision repair. *Genes and Development* **13**, 768–785.

de Lange, T. & Jacks, T. (1999) For better or worse? Telomerase inhibition and cancer. *Cell* **98**, 273–275.

de Vries, B.B.A., Wiegers, A.M., de Graaff, E. *et al.* (1993) Mental status and fragile X expression in relation to FMR-1 gene mutation. *European Journal of Human Genetics* **1**, 72–79.

de Vries, B.B.A., Halley, D.J.J., Oostra, B.A. & Niermeijer, M.F. (1998) The fragile X syndrome. *Journal of Medical Genetics* **35**, 579–589.

Dean, W. & Ferguson-Smith, A. (2001) Genomic imprinting: mother maintains methylation marks. *Current Biology* **11**, R527–R530.

Deeming, D.C. & Ferguson, M.W.J. (1988) Environmental regulation of sex determination in reptiles. *Philosophical Transactions of the Royal Society of London B* **322**, 19–39.

Deloukas, P., Schuler, G.D., Gyapay, G. *et al.* (1998) A physical map of 30000 human genes. *Science* **282**, 744–746.

Demple, B. & Harrison, L. (1994) Repair of oxidative damage to DNA: enzymology and biology. *Annual Review of Biochemistry* **63**, 915–948.

den Boer, B.G.W. & Murray, J.A.H. (2000) Triggering the cell cycle in plants. *Trends in Cell Biology* **10**, 245–250.

DePamphilis, M.L. (1999) Replication origins in metaphase chromosomes: fact or fiction? *Bioessays* **21**, 5–16.

Derenzini, M., Sirri, V. & Trerè, D. (1994) Nucleolar organizer regions in tumor cells. *Cancer Journal* **7**, 71–77.

Derenzini, M., Trerè, D., Pession, A., Govoni, M., Sirri, V. & Chieco, P. (2000) Nucleolar size indicates the rapidity of cell proliferation in cancer tissues. *Journal of Pathology* **191**, 181–186.

Dernburg, A.F., Sedat, J.W. & Hawley, R.S. (1996) Direct evidence of a role for heterochromatin in meiotic chromosome segregation. *Cell* **86**, 135–146.

Diede, S.J. & Gottschling, D.E. (1999) Telomerase-mediated telomere addition in vivo requires DNA primase and DNA polymerases α and δ. *Cell* **99**, 723–733.

Dimitri, P. & Junakovic, N. (1999) Revising the selfish DNA hypothesis: new evidence on accumulation of transposable elements in heterochromatin. *Trends in Genetics* **15**, 123–124.

Dimitrova, D.S. & Gilbert, D.M. (1999) The spatial position and replication timing of chromosomal domains are both established in early G1 phase. *Molecular Cell* **4**, 983–993.

Dimitrova, D.S. & Gilbert, D.M. (2000) Temporally coordinated assembly and disassembly of replication factories in the absence of DNA synthesis. *Nature Cell Biology* **2**, 686–694.

Disney, J.E., Johnson, K.R., Magnuson, N.S., Sylvester, S.R. & Reeves, R. (1989) High-mobility group protein HMG-I localizes to G/Q- and C-bands of human and mouse chromosomes. *Journal of Cell Biology* **109**, 1975–1982.

Disteche, C.M. (1995) Escape from X inactivation in human and mouse. *Trends in Genetics* **11**, 17–22.

Disteche, C.M. (1999) Escapees on the X chromosome. *Proceedings of the National Academy of Sciences of the USA* **96**, 14180–14182.

Djian, P., Hancock, J.M. & Chana, H.S. (1996) Codon repeats in genes associated with human diseases: fewer repeats in the genes of nonhuman primates and nucleotide substitutions concentrated at the sites of reiteration. *Proceedings of the National Academy of Sciences of the USA* **93**, 417–421.

Dobie, K.W., Hari, K.L., Maggert, K.A. & Karpen, G.H. (1999) Centromere proteins and chromosome inheritance: a complex affair. *Current Opinion in Genetics and Development* **9**, 206–217.

Dobson, S.L. & Tanouye, M.A. (1998) Evidence for a genomic imprinting sex determination mechanism in *Nasonia vitripennis* (Hymenoptera; Chalcidoidea). *Genetics* **149**, 233–242.

Donaldson, A.D. & Blow J.J. (1999) The regulation of replication origin activation. *Current Opinion in Genetics and Development* **9**, 62–68.

Dong, F. & Jiang, J. (1998) Non-Rabl patterns of centromere and telomere distribution in the interphase nuclei of plant cells. *Chromosome Research* **6**, 551–558.

Dong, F., Miller, J.T., Jackson, S.A., Wang, G-L., Ronald, P.C. & Jiang, J. (1998) Rice (*Oryza sativa*) centromeric regions consist of complex DNA. *Proceedings of the National Academy of Sciences of the USA* **95**, 8135–8140.

Doshi, P., Kaushal, S., Benyajati, C. & Wu, C-I. (1991) Molecular analysis of the Responder satellite DNA in *Drosophila melanogaster:* DNA bending, nucleosome structure, and Rsp-binding proteins. *Molecular Biology and Evolution* **8**, 721–741.

Drouin, R., Lemieux, N. & Richer, C-L. (1990) Analysis of DNA replication during S-phase by means of dynamic chromosome banding at high resolution. *Chromosoma* **99**, 273–280.

Drouin, R., Holmquist, G.P. & Richer, C-L. (1994) High-resolution replication bands compared with morphologic G- and R-bands. *Advances in Human Genetics* **22**, 47–115.

Dundr, M., Misteli, T. & Olson, M.O.J. (2000) The dynamics of postmitotic reassembly of the nucleolus. *Journal of Cell Biology* **150**, 433–446.

Dunham, I., Shimizu, N., Roe, B.A. *et al.* (1999) The DNA sequence of human chromosome 22. *Nature* **402**, 489–495. See also: http://www.sanger.ac.uk/HGP/Chr22 and http://www.genome.ou.edu/Chr22.html

Dunham, M.A., Neumann, A.A., Fasching, C.L. & Reddel, R.R. (2000) Telomere maintenance by recombination in human cells. *Nature Genetics* **26**, 447–450.

DuPraw, E.J. (1970) *DNA and Chromosomes.* Holt, Rinehart & Winston, New York.

Earle, E., Saxena, A., MacDonald, A. *et al.* (2000) Poly(ADP-ribose) polymerase at active centromeres and neocentromeres at metaphase. *Human Molecular Genetics* **9**, 187–194.

Earnshaw, W.C. & Heck, M.M.S. (1985) Localization of topoisomerase II in mitotic chromosomes. *Journal of Cell Biology* **100**, 1716–1725.

Earnshaw, W.C. & Bernat, R.L. (1991) Chromosomal passengers: toward an integrated view of mitosis. *Chromosoma* **100**, 139–146.

Eastman, E.M., Goodman, R.M., Erlanger, B.F. & Miller, O.J. (1980) The organization of DNA in the mitotic and polytene chromosomes of *Sciara coprophila*. *Chromosoma* **79**, 293–314.

Efstratiadis, A. (1994) Parental imprinting of autosomal mammalian genes. *Current Opinion in Genetics and Development* **4**, 265–280.

Eichler, E.E. (1999) Repetitive conundrums of centromere structure and function. *Human Molecular Genetics* **8**, 151–155.

Eichler, E.E. (2001) Recent duplication, domain accretion and the dynamic mutation of the human genome. *Trends in Genetics* **17**, 661–669.

Eils, R., Dietzel, S., Bertin, E. *et al.* (1996) Three-dimensional reconstruction of painted human interphase chromosomes: active and inactive X chromosome territories have similar volumes but differ in shape and surface structure. *Journal of Cell Biology* **135**, 1427–1440.

Eissenberg, J.C. & Elgin, S.C.R. (2000) The HP1 protein family: getting a grip on chromatin. *Current Opinion in Genetics and Development* **10**, 204–210.

Eissenberg, J.C., Ge, Y-W. & Hartnett, T. (1994) Increased phosphorylation of HP1, a heterochromatin-associated protein of *Drosophila*, is correlated with heterochromatin assembly. *Journal of Biological Chemistry* **269**, 21315–21321.

Ekwall, K., Olsson, T., Turner, B.M., Cranston, G. & Allshire, R.C. (1997) Transient inhibition of histone deacetylation alters the structural and functional imprint at fission yeast centromeres. *Cell* **91**, 1021–1032.

El-Alfy, M. & Leblond, C.P. (1988) Visualization of chromosome assembly during the S and G2 stages of the cycle and chromosome disassembly during the G1 stage in semithin sections of mouse duodenal crypt cells and other cells. *American Journal of Anatomy* **183**, 45–56.

El-Alfy, M. & Leblond, C.P. (1989) An electron microscopic study of mitosis in mouse duodenal crypt cells confirms that the prophasic condensation of chromatin begins during the DNA-synthesizing (S) stage of the cycle. *American Journal of Anatomy* **186**, 69–84.

El-Alfy, M., Turner, J.P., Nadler, N.J., Liu, D.F. & Leblond, C.P. (1994) Subdivision of the mitotic cycle into eleven stages, on the basis of the chromosomal changes observed in mouse duodenal crypt cells stained by the DNA-specific Feulgen reaction. *Anatomical Record* **238**, 289–296.

Elder, F.F.B. & Robinson, T.J. (1989) Rodent common fragile sites: are they conserved? Evidence from mouse and rat. *Chromosoma* **97**, 459–464.

Elgar, G. (1996) Quality not quantity: the pufferfish genome. *Human Molecular Genetics* **5**, 1437–1442.

Elgin, S.C.R. (1996) Heterochromatin and gene regulation in *Drosophila*. *Current Opinion in Genetics and Development* **6**, 193–202.

Elledge, S.J. (1996) Cell cycle checkpoints: preventing an identity crisis. *Science* **274**, 1664–1672.

Elledge, S.J. (1998) Mitotic arrest: Mad2 prevents sleepy waking up the APC. *Science* **279**, 999–1000.

Ellegren, H. (2002) Dosage compensation: do birds do it as well? *Trends in Genetics* **18**, 25–28.

Ericsson, C., Mehlin, H., Björkroth, B., Lamb, M.M. & Daneholt, B. (1989) The ultrastructure of upstream and downstream regions of an active Balbiani ring gene. *Cell* **56**, 631–639.

Evans, E.P. (1989) Standard normal chromosomes. In: *Genetic Variants and Strains of the Laboratory Mouse* (eds M.F. Lyon & A.G. Searle), pp. 576–581. Oxford University Press, Oxford.

Evans, H.J., Buckland, R.A. & Sumner, A.T. (1973) Chromosome homology and heterochromatin in goat, sheep and ox studied by banding techniques. *Chromosoma* **42**, 383–402.

Eyre-Walker, A. & Hurst, L.D. (2001) The evolution of isochores. *Nature Reviews Genetics* **2**, 549–555.

Fakan, S. (1994) Perichromatin fibrils are *in situ* forms of nascent transcripts. *Trends in Cell Biology* **4**, 86–90.

Falciola, L., Spada, F., Calogero, S. *et al.* (1997) High mobility group 1 protein is not stably associated with the chromosomes of somatic cells. *Journal of Cell Biology* **137**, 19–26.

Fanti, L., Dorer, D.R., Berloco, M., Henikoff, S. & Pimpinelli, S. (1998) Heterochromatin protein 1 binds transgene arrays. *Chromosoma* **107**, 286–292.

Farr, C.J. (1996) Mammalian telomeres and chromosome fragmentation. *Seminars in Cell & Developmental Biology* **7**, 41–48.

Farr, C.J. (1999) Chromosome fragmentation in vertebrate cell lines. In: *Chromosome Structural Analysis: a Practical Approach* (ed. W.A. Bickmore), pp. 183–198. Oxford University Press, Oxford.

Farr, C.J., Bayne, R.A.L., Kipling, D., Mills, W., Critcher, R. & Cooke, H.J. (1995) Generation of a human X-derived minichromosome using telomere-associated chromosome fragmentation. *EMBO Journal* **14**, 5444–5454.

Feil, R. & Kelsey, G. (1997) Genomic imprinting: a chromatin connection. *American Journal of Human Genetics* **61**, 1213–1219.

Feil, R. & Khosla, S. (1999) Genomic imprinting in mammals: an interplay between chromatin and DNA methylation? *Trends in Genetics* **15**, 431–435.

Feil, R., Brocard, J., Macrez, B., LeMeur, M., Metzger, D. & Chambon, P. (1996) Ligand-activated site-specific recombination in mice. *Proceedings of the National Academy of Sciences of the USA* **93**, 10887–10890.

Feinberg, A.P. (1994) A developmental context for multiple genetic alterations in Wilms' tumour. *Journal of Cell Science* **18** (suppl.), 7–12.

Felsenfeld, G. (1996) Chromatin unfolds. *Cell* **86**, 13–19.

Felsenfeld, G., Boyes, J., Chung, J., Clark, D. & Studitsky, V. (1996) Chromatin structure and gene expression. *Proceedings of the National Academy of Sciences of the USA* **93**, 9384–9388.

Ferraro, M. & Prantera, G. (1988) Human NORs show correlation between transcriptional activity, DNase I sensitivity, and hypomethylation. *Cytogenetics and Cell Genetics* **47**, 58–61.

Ferraro, M., Buglia, G.L. & Romano, F. (2001) Involvement of histone H4 acetylation in the epigenetic inheritance of different activity states of maternally and paternally derived genomes in the mealybug *Planococcus citri*. *Chromosoma* **110**, 93–101.

Ferrier, D.E.K. & Holland, P.W.H. (2001) Ancient origin of the Hox gene cluster. *Nature Reviews Genetics* **2**, 33–38.

Feulgen, R. & Rossenbeck, H. (1924) Mikroskopisch-chemischer Nachweis einer Nukleinsäure vom Typus der Thymonukleinsäure und die darauf beruhende elektive Färbung von Zellkernen in mikroskopischen Präparaten. *Hoppe-Seyler's Zeitschrift für Physiologische Chemie* **135**, 203–248.

Filatov, D.A., Monéger, F., Negrutiu, I. & Charlesworth, D.

(2000) Low variability in a Y-linked plant gene and its implications for Y-chromosome evolution. *Nature* **404**, 388–390.

Finch, J.T. & Klug, A. (1976) Solenoidal model for superstructure in chromatin. *Proceedings of the National Academy of Sciences of the USA* **73**, 1897–1901.

Finnerty, J.R. & Martindale, M.Q. (1998) The evolution of the Hox cluster: insights from outgroups. *Current Opinion in Genetics and Development* **8**, 681–687.

Flaus, A. & Owen-Hughes, T. (2001) Mechanisms for ATP-dependent chromatin remodelling. *Current Opinion in Genetics and Development* **11**, 148–154.

Flemming, W. (1965) Contributions to the knowledge of the cell and its vital processes (English translation of original paper of 1880). *Journal of Cell Biology* **25**, 1–69.

Follette, P.J., Duronio, R.J. & O'Farrell, P.H. (1998) Fluctuations in cyclin E levels are required for multiple rounds of endocycle S phase in *Drosophila*. *Current Biology* **8**, 235–238.

Forozan, F., Karhu, R., Kononen, J., Kallioniemi, A. & Kallioniemi, O-P. (1997) Genome screening by comparative genomic hybridization. *Trends in Genetics* **13**, 405–409.

Francke, U. (1994) Digitized and differentially shaded human chromosome ideograms for genomic applications. *Cytogenetics and Cell Genetics* **65**, 206–219.

Franke, A. & Baker, B.S. (2000) Dosage compensation rox! *Current Opinion in Cell Biology* **12**, 351–354.

Fredga, K. (1994) Bizarre mammalian sex-determining mechanisms. In: *The Differences Between the Sexes* (eds R.V. Short & E. Balaban), pp. 419–431. Cambridge University Press, Cambridge.

Fretter, V. & Graham, A. (1962) *British Prosobranch Molluscs*. Ray Society, London.

Fridolfsson, A-K., Cheng, H., Copeland, N.G. *et al.* (1998) Evolution of the avian sex chromosomes from an ancestral pair of autosomes. *Proceedings of the National Academy of Sciences of the USA* **95**, 8147–8152.

Frönicke, L. & Scherthan, H. (1997) Zoo-fluorescence *in situ* hybridisation analysis of human and Indian muntjac karyotypes (*Muntiacus muntjak vaginalis*) reveals satellite DNA clusters at the margins of conserved syntenic segments. *Chromosome Research* **5**, 254–261.

Fry, K. & Salser, W. (1977) Nucleotide sequences of HS-α satellite DNA from kangaroo rat *Dipodomys ordii* and characterization of similar sequences in other rodents. *Cell* **12**, 1069–1084.

Fuchs, J., Strehl, S., Brandes, A., Schweizer, D. & Schubert, I. (1998) Molecular-cytogenetic characterization of the *Vicia faba* genome – heterochromatin differentiation, replication patterns and sequence localization. *Chromosome Research* **6**, 219–230.

Fukagawa, T., Pendon, C., Morris, J. & Brown, W. (1999) CENP-C is necessary but not sufficient to induce formation of a functional centromere. *EMBO Journal* **18**, 4196–4209.

Fukagawa, T., Mikami, Y., Nishihashi, A. *et al.* (2001) CENP-H, a constitutive centromere component, is required for centromere targeting of CENP-C in vertebrate cells. *EMBO Journal* **20**, 4603–4617.

Fussell, C.P. (1975) The position of interphase chromosomes and late replicating DNA in centromere and telomere regions of *Allium cepa* L. *Chromosoma* **50**, 201–210.

Gacy, A.M. & McMurray, C.T. (1998) Influence of hairpins on template reannealing at trinucleotide repeat duplexes: a model for slipped DNA. *Biochemistry* **37**, 9426–9434.

Gacy, A.M., Goellner, G.M., Spiro, C. *et al.* (1998) GAA instability in Friedreich's ataxia shares a common, DNA-directed and intraallelic mechanism with other trinucleotide diseases. *Molecular Cell* **1**, 583–593.

Gall, J.G. (1996) *Views of the Cell*. American Society for Cell Biology, Bethesda, MD.

Gall, J.G. (2000) Cajal bodies: the first 100 years. *Annual Review of Cell and Developmental Biology* **16**, 273–300.

Gall, J.G. (2001) A role for Cajal bodies in assembly of the nuclear transcription machinery. *FEBS Letters* **498**, 164–167.

Gall, J.G., Cohen, E.H. & Polan, M.L. (1971) Repetitive DNA sequences in *Drosophila*. *Chromosoma* **33**, 319–344.

Gallardo, M.H., Bickham, J.W., Honeycutt, R.L., Ojeda, R.A. & Kohler, N. (1999) Discovery of tetraploidy in a mammal. *Nature* **401**, 341.

Galy, V., Olivo-Marin, J.C., Scherthan, H., Doye, V., Rascalou, N. & Nehrbass, U. (2000) Nuclear pore complexes in the organization of silent telomeric chromatin. *Nature* **403**, 108–112.

Gant, T.M. & Wilson, K.L. (1997) Nuclear assembly. *Annual Review of Cell and Developmental Biology* **13**, 669–695.

Garagna, S., Redi, C.A., Capanna, E. *et al.* (1993) Genome distribution, chromosomal allocation, and organization of the major and minor satellite DNAs in 11 species and subspecies of the genus *Mus*. *Cytogenetics and Cell Genetics* **64**, 247–255.

Gardiner, K. (1996) Base composition and gene distribution: critical patterns in mammalian genome organization. *Trends in Genetics* **12**, 519–524.

Gardiner, K., Aïssani, B. & Bernardi, G. (1990) A compositional map of human chromosome 21. *EMBO Journal* **9**, 1853–1858.

Gardiner-Garden, M. & Frommer, M. (1992) Significant CpG-rich regions in angiosperm genes. *Journal of Molecular Evolution* **34**, 231–245.

Gardiner-Garden, M., Sved, J.A. & Frommer, M. (1992) Methylation sites in angiosperm genes. *Journal of Molecular Evolution* **34**, 219–230.

Gardner, R.D. & Burke, D.J. (2000) The spindle checkpoint: two transitions, two pathways. *Trends in Cell Biology* **10**, 154–158.

Garrett, M.D. & Fattaey, A. (1999) CDK inhibition and cancer therapy. *Current Opinion in Genetics and Development* **9**, 104–111.

Gasser, S.M. (2000) A sense of the end. *Science* **288**, 1377–1379.

Gasser, S.M., Amati, B.B., Cardenas, M.E. & Hofmann, J.F.-X. (1989) Studies on scaffold attachment sites and their relation to genome function. *International Review of Cytology* **119**, 57–96.

Gatti, M. & Pimpinelli, S. (1992) Functional elements in *Drosophila melanogaster* heterochromatin. *Annual Review of Genetics* **26**, 239–275.

Ghosh, S. & Paweletz, N. (1993) Mitosis: dissociability of its events. *International Review of Cytology* **144**, 217–257.

Gibbons, R.J., McDowell, T.L., Raman, S. *et al.* (2000) Mutations in ATRX, encoding a SWI/SNF-like protein, cause diverse changes in the pattern of DNA methylation. *Nature Genetics* **24**, 368–371.

Gilbert, D.M. (1998) Replication origins in yeast versus metazoa: separation of the haves and the have nots. *Current Opinion in Genetics and Development* **8**, 194–199.

Gilbert, D.M. (2001) Making sense of eukaryotic DNA replication origins. *Science* **294**, 96–100.

Gill, B.S. & Friebe, B. (1998) Plant cytogenetics at the dawn of the 21st century. *Current Opinion in Plant Biology* **1**, 109–115.

Gill, K.S., Gill, B.S. & Endo, T.R. (1993) A chromosome region-specific mapping strategy reveals gene-rich telomeric ends in wheat. *Chromosoma* **102**, 374–381.

Gilson, E, Laroche, T. & Gasser, S.M. (1993) Telomeres and the functional architecture of the nucleus. *Trends in Cell Biology* **3**, 128–134.

Giménez-Abián, J.F., Clarke, D.J., Mullinger, A.M., Downes, C.S. & Johnson, R.T. (1995) A postprophase topoisomerase II-dependent chromatid core separation step in the formation of metaphase chromosomes. *Journal of Cell Biology* **131**, 7–17.

Giménez-Abián, J.F., Clarke, D.J., Devlin, J. *et al.* (2000) Premitotic chromosome individualization in mammalian cells depends on topoisomerase II activity. *Chromosoma* **109**, 235–244.

Ginisty, H., Sicard, H., Roger, B. & Bouvet, P. (1999) Structure and functions of nucleolin. *Journal of Cell Science* **112**, 761–772.

Glaser, R.L., Karpen, G.H. & Spradling, A.C. (1992) Replication forks are not found in a *Drosophila* minichromosome demonstrating a gradient of polytenization. *Chromosoma* **102**, 15–19.

Glover, D.M., Zaha, A., Stocker, A.J. *et al.* (1982) Gene amplification in *Rhynchosciara* salivary gland chromosomes. *Proceedings of the National Academy of Sciences of the USA* **79**, 2947–2951.

Goldberg, I.G., Sawhney, H., Pluta, A.F., Warburton, P.E. & Earnshaw, W.C. (1996) Surprising deficiency of CENP-B binding sites in African green monkey α-satellite DNA: implications for CENP-B function at centromeres. *Molecular and Cellular Biology* **16**, 5156–5168.

Golic, K.G. & Golic, M.M. (1996) Engineering the *Drosophila* genome: chromosome rearrangements by design. *Genetics* **144**, 1693–1711.

Golic, K.G., Golic, M.M. & Pimpinelli, S. (1998) Imprinted control of gene activity in *Drosophila*. *Current Biology* **8**, 1273–1276.

Goodrich, J. (1998) Plant development: Medea's maternal instinct. *Current Biology* **8**, R480–R484.

Görlich, D. (1998) Transport into and out of the nucleus. *EMBO Journal* **17**, 2721–2727.

Gorman, M. & Baker, B.S. (1994) How flies make one equal two: dosage compensation in *Drosophila*. *Trends in Genetics* **10**, 376–380.

Gosden, J.R. (ed.) (1994) *Chromosome Analysis Protocols*. Humana Press, Totowa, NJ.

Gottschling, D.E. & Berg, B.L. (1998) Chromosome dynamics: yeast pulls it apart. *Current Biology* **8**, R76–R79.

Gowher, H., Leismann, O. & Jeltsch, A. (2000) DNA of *Drosophila melanogaster* contains 5-methylcytosine. *EMBO Journal* **19**, 6918–6923.

Grafi, G. & Larkins, B.A. (1995) Endoreduplication in maize endosperm: involvement of M phase-promoting factor inhibition and induction of S-phase-related kinases. *Science* **269**, 1262–1264.

Graves, J.A.M. (1994) Mammalian sex-determining genes. In: *The Differences Between the Sexes* (eds R.V. Short & E. Balaban), pp. 397–418. Cambridge University Press, Cambridge.

Graves, J.A.M. (1996) Mammals that break the rules: genetics of marsupials and monotremes. *Annual Review of Genetics* **30**, 233–260.

Graves, J.A.M. (2001) The X – a sexy chromosome. *Bioessays* **23**, 1091–1094.

Graves, J.A.M., Wakefield, M.J. & Toder, R. (1998) The origin and evolution of the pseudoautosomal regions of human sex chromosomes. *Human Molecular Genetics* **7**, 1991–1996.

Gregory, P.D. & Hörz, W. (1998) Life with nucleosomes: chromatin remodelling in gene regulation. *Current Opinion in Cell Biology* **10**, 339–345.

Gregory, T.R. & Hebert, P.D.N. (1999) The modulation of DNA content: proximate causes and ultimate consequences. *Genome Research* **9**, 317–324.

Greider, C.W. (1996) Telomere length regulation. *Annual Review of Biochemistry* **65**, 337–365.

Greider, C.W. (1998) Telomeres and senescence: the history, the experiment, the future. *Current Biology* **8**, R178–R181.

Greider, C.W. (1999) Telomerase activation: one step on the road to cancer? *Trends in Genetics* **15**, 109–112.

Griffith, J.D., Comeau, L., Rosenfield, S. *et al.* (1999) Mammalian telomeres end in a large duplex loop. *Cell* **97**, 503–514.

Grosschedl, R., Giese, K. & Pagel, J. (1994) HMG domain proteins: architectural elements in the assembly of nucleoprotein structures. *Trends in Genetics* **10**, 94–100.

Grunstein, M. (1998) Yeast heterochromatin: regulation of its assembly and inheritance by histones. *Cell* **93**, 325–328.

Guo, X.W., Th'ng, J.P.H., Swank, R.A. *et al.* (1995) Chromosome condensation induced by fostriecin does not require p34^{cdc2} kinase activity and histone H1 phosphorylation, but is associated with enhanced histone H2A and H3 phosphorylation. *EMBO Journal* **14**, 976–985.

Guy, J., Hendrich, B., Holmes, M., Martin, J.E. & Bird, A. (2001) A mouse Mecp2-null mutation causes neurological symptoms that mimic Rett syndrome. *Nature Genetics* **27**, 322–326.

Haaf, T. & Willard, H.F. (1997) Chromosome-specific α-satellite DNA from the centromere of chimpanzee chromosome 4. *Chromosoma* **106**, 226–232.

Haber, J.E. (1997) A super new twist on the initiation of meiotic recombination. *Cell* **89**, 163–166.

Habermann, F.A., Cremer, M., Walter, J., Kreth, G., von Hase, J., Bauer, K., Wienberg, J., Cremer, C., Cremer, T. & Solovei, I. (2001) Arrangements of macro- and microchromosomes in chicken cells. *Chromosome Research* **9**, 569–584.

Hackstein, J.H.P. & Hochstenbach, R. (1995) The elusive fertility genes of *Drosophila*: the ultimate haven for selfish genetic elements. *Trends in Genetics* **11**, 195–200.

Hägele, K. (1977) Differential staining of polytene chromosome bands in *Chironomus* by Giemsa banding methods. *Chromosoma* **59**, 207–216.

Hahn, P.J. (1993) Molecular biology of double-minute chromosomes. *Bioessays* **15**, 477–484.

Hahn, W.C., Stewart, S.A., Brooks, M.W. *et al.* (1999) Inhibition of telomerase limits the growth of human cancer cells. *Nature Medicine* **5**, 1164–1170.

Haig, D. (1999) Multiple paternity and genomic imprinting. *Genetics* **151**, 1229–1231.

Halleck, M.S. & Schlegel, R.A. (1983) C-banding of *Peromyscus* constitutive heterochromatin persists following histone hyperacetylation. *Experimental Cell Research* **147**, 269–279.

Hanawalt, P.C. (2000) The bases for Cockayne syndrome. *Nature* **405**, 415–416.

Hancock, J.M. (1996) Simple sequences and the expanding genome. *Bioessays* **18**, 421–425.

Hancock, J.M. (1999) Microsatellites and other simple sequences: genomic context and mutational mechanisms. In: *Microsatellites* (eds D.B. Goldstein & C. Schlötterer), pp. 1–8. Oxford University Press: Oxford.

Hancock, R. (2000) A new look at the nuclear matrix. *Chromosoma* **109**, 219–225.

Hao, S., Jiao, M. & Huang, B. (1990) Chromosome organization revealed upon the decondensation of telophase chromosomes in *Allium*. *Chromosoma* **99**, 371–378.

Harbour, J.W. & Dean, D.C. (2000) Rb function in cell-cycle regulation and apoptosis. *Nature Cell Biology* **2**, E65–E67.

Harley, C.B. & Villeponteau, B. (1995) Telomeres and telomerase in aging and cancer. *Current Opinion in Genetics and Development* **5**, 249–255.

Harrington, J.J., Van Bokkelen, G., Mays, R.W., Gustashaw, K. & Willard, H.F. (1997) Formation of *de novo* centromeres and construction of first-generation human artificial microchromosomes. *Nature Genetics* **15**, 345–355.

Hartl, D.L. (1996) The most unkindest cut of all. *Nature Genetics* **12**, 227–229.

Hartley, S.E. (1987) The chromosomes of salmonid fishes. *Biological Reviews* **62**, 197–214.

Hartman, T.P.V. & Southern, D.I. (1995) Genome reorganization from polyteny to polyploidy in the nurse cells found in onion fly (*Delia antiqua*) and cabbage root fly (*Delia radicum*) ovaries (Diptera, Anthomyiidae). *Chromosome Research* **3**, 271–280.

Hassold, T. & Hunt, P. (2001) To err (meiotically) is human: the genesis of human aneuploidy. *Nature Reviews Genetics* **2**, 280–291.

Hattori, H., Fujiyama, A., Taylor, T.D. *et al.* (2000) The DNA sequence of human chromosome 21. *Nature* **405**, 311–319. See also: http://chr21.r2-berlin.mpg.de

Hauffe, H.C. & Searle, J.B. (1998) Chromosomal heterozygosity and fertility in house mice (*Mus musculus domesticus*) from Northern Italy. *Genetics* **150**, 1143–1154.

Hawley, R.S. (1997) Unresolvable endings: defective telomeres and failed separation. *Science* **275**, 1441–1443.

Hawley, R.S. & Therkauf, W.E. (1993) Requiem for distributive segregation: achiasmate segregation in *Drosophila* females. *Trends in Genetics* **9**, 310–317.

He, D., Zeng, C., Woods, K. *et al.* (1998) CENP-G: a new centromeric protein that is associated with the α-1 satellite DNA subfamily. *Chromosoma* **107**, 189–197.

He, H., Liao, C. & Edström, J.E. (1998) Centromere 3 specific tandem repeat from *Chironomus pallidivittatus*. *Chromosoma* **107**, 304–310.

Heald, R., Tournebize, R., Blank, T. *et al.* (1996) Self-organization of microtubules into bipolar spindles around artificial chromosomes in *Xenopus* egg extracts. *Nature* **382**, 420–425.

Heatwole, V.M. & Haynes, S.R. (1996) Association of RB97D, an RRM protein required for male fertility, with a Y chromosome lampbrush loop in *Drosophila* spermatocytes. *Chromosoma* **105**, 285–292.

Hecht, F., Ramesh, K.H. & Lockwood, D.H. (1990) A guide to fragile sites on human chromosomes. *Cancer Genetics and Cytogenetics* **44**, 37–45.

Heck, M.M.S. (1997) Condensins, cohesins, and chromosome architecture: how to make and break a mitotic chromosome. *Cell* **91**, 5–8.

Hegemann, J.H. & Fleig, U.N. (1993) The centromere of budding yeast. *Bioessays* **15**, 451–460.

Heller, R., Brown, K.E., Burgtorf, C. & Brown, W.R.A. (1996) Mini-chromosomes derived from the human Y chromosome by telomere directed chromosome breakage. *Proceedings of the National Academy of Sciences of the USA* **93**, 7125–7130.

Hemann, M.T. & Greider, C.W. (2000) Wild-derived inbred mouse strains have short telomeres. *Nucleic Acids Research* **28**, 4474–4478.

Hemann, M.T., Strong, M.A., Hao, L.-Y. & Greider, C.W. (2001) The shortest telomere, not average telomere length, is critical for cell viability and chromosome stability. *Cell* **107**, 67–77.

Hemerley, A.S., Ferreira, P.C.G., Van Montagu, M. & Inzé, D. (1999) Cell cycle control and plant morphogenesis: is there an essential link? *Bioessays* **21**, 29–37.

Henderson, S., Allsopp, R., Spector, D., Wang, S.-S. & Harley, C. (1996) *In situ* analysis of changes in telomere size during replicative aging and cell transformation. *Journal of Cell Biology* **134**, 1–12.

Hendrich, B. (2000) Human genetics: methylation moves into medicine. *Current Biology* **10**, R60–R63.

Hendzel, M.J., Wei, Y., Mancini, M.A. *et al.* (1997) Mitosis-specific phosphorylation of histone H3 initiates primarily within pericentric heterochromatin during G2 and spreads in an ordered fashion coincident with mitotic chromosome condensation. *Chromosoma* **106**, 348–360.

Henikoff, S. (1998) Conspiracy of silence among repeated transgenes. *Bioessays* **20**, 532–535.

Hennig, W. (1985) Y chromosome function and spermatogenesis in *Drosophila hydei*. *Advances in Genetics* **23**, 179–234.

Hennig, W. & Meer, B. (1971) Reduced polyteny of ribosomal RNA cistrons in giant chromosomes of *Drosophila hydei*. *Nature New Biology* **233**, 70–71.

Henning, K.A., Novotny, E.A., Compton, S.T., Guan, X.-Y., Liu, P.P. & Ashlock, M.A. (1999) Human artificial chromosomes generated by a modification of a yeast artificial chromosome containing both human alpha satellite and single-copy DNA sequences. *Proceedings of the National Academy of Sciences of the USA* **96**, 592–597.

Heppich, S. (1978) Hybridogenesis in *Rana esculenta*: C-band karyotypes of *Rana ridibunda*, *Rana lessonae* and *Rana esculenta*. *Zeitschrift für Zoologische Systematik und Evolutionsforschung* **16**, 27–39.

Heppich, S., Tunner, H.G. & Greilhuber, J. (1982) Premeiotic chromosome doubling after genome elimination during spermatogenesis of the species hybrid *Rana esculenta*. *Theoretical and Applied Genetics* **61**, 101–104.

Herbert, B.-S., Pitts, A.E., Baker, S.I. *et al.* (1999) Inhibition of human telomerase in immortal human cells leads to progressive telomere shortening and cell death. *Proceedings of the National Academy of Sciences of the USA* **96**, 14276–14281.

Hernandez-Verdun, D. & Gautier, T. (1994) The chromosome periphery during mitosis. *Bioessays* **16**, 179–185.

Herrera, E., Samper, E., Martin-Caballero, J., Flores, J.M., Lee, H.-W. & Blasco, M.A. (1999) Disease states associated with telomerase deficiency appear earlier in mice with short telomeres. *EMBO Journal* **18**, 2950–2960.

Herrero, P., de la Torre, J., Arano, B., Gosálvez, J. & Sumner, A.T. (1995) Patterns of replication and DNase sensitivity in the chromosomes of *Rana perezi* (Amphibia: Anura). *Genome* **38**, 339–343.

Heslop-Harrison, J.S., Leitch, A.R. & Schwarzacher, T. (1993) The physical organization of interphase nuclei. In: *The Chromosome* (eds J.S. Heslop-Harrison & R.B. Flavell), pp. 221–232. Bios Scientific Publishers, Oxford.

Hewett, D.R., Handt, O., Hobson, L. *et al.* (1998) *FRA10B* structure reveals common elements in repeat expansion and chromosomal fragile site genesis. *Molecular Cell* **1**, 773–781.

Heyting, C. (1996) Synaptonemal complexes: structure and function. *Current Opinion in Cell Biology* **8**, 389–396.

Hirano, T. (1998) SMC protein complexes and higher-order chromosome dynamics. *Current Opinion in Cell Biology* **10**, 317–322.

Hirano, T. (2000) Chromosome cohesion, condensation, and separation. *Annual Review of Biochemistry* **69**, 115–144.

Hirano, T., Kobayashi, R. & Hirano, M. (1997) Condensins, chromosome condensation protein complexes containing XCAP-C, XCAP-E and a *Xenopus* homolog of the *Drosophila* barren protein. *Cell* **89**, 511–521.

Hock, R., Carl, M., Lieb, B., Gebauer, D. & Scheer, U. (1996) A monoclonal antibody against DNA topoisomerase II labels the axial granules of *Pleurodeles* lampbrush chromosomes. *Chromosoma* **104**, 358–366.

Hock, R., Scheer, U. & Bustin, M. (1998) Chromosomal proteins HMG-14 and HMG-17 are released from mitotic chromosomes and imported into the nucleus by active transport. *Journal of Cell Biology* **143**, 1427–1436.

Hodgkin, J. (1987) Primary sex determination in the nematode *C. elegans*. *Development* **101** (suppl.), 5–16.

Holden, J.J.A., Reimer, D.L., Higgins, M.J., Roder, J.C. & White, B.N. (1985) Amplified sequences from chromosome 15, including centromeres, nucleolar organizer regions, and centromeric heterochromatin, in homogeneously staining regions in the human melanoma cell line MeWo. *Cancer Genetics and Cytogenetics* **14**, 131–146.

Holm, D.G. (1976) Compound autosomes. In: *The Genetics and Biology of Drosophila*, Vol. 1b (eds M. Ashburner & E. Novitski), pp. 529–561. Academic Press, London.

Holmes, S.E., O'Hearn, E.E., McInnis, M.G. *et al.* (1999) Expansion of a novel CAG trinucleotide repeat in the 5′ region of *PPP2R2B* is associated with SCA12. *Nature Genetics* **23**, 391–392.

Holmes, V.F. & Cozzarelli, N.R. (2000) Closing the ring: links between SMC proteins and chromosome partitioning, condensation, and supercoiling. *Proceedings of the National Academy of Sciences of the USA* **97**, 1322–1324.

Holmquist, G. (1979) The mechanism of C-banding: depurination and β-elimination. *Chromosoma* **72**, 203–224.

Holmquist, G.P. (1992) Chromosome bands, their chromatin flavors, and their functional features. *American Journal of Human Genetics* **51**, 17–37.

Holmquist, G., Gray, M., Porter, T. & Jordan, J. (1982) Characterization of Giemsa dark- and light-band DNA. *Cell* **31**, 121–129.

Horsthemke, B., Dittrich, B. & Buiting, K. (1997) Imprinting mutations on human chromosome 15. *Human Mutation* **10**, 329–337.

Houben, A., Wako, T., Furushima-Shimogawara, R. *et al.* (1999) The cell cycle dependent phosphorylation of histone H3 is correlated with the condensation of plant mitotic chromosomes. *The Plant Journal* **18**, 675–679.

Howe, M., McDonald, K.L., Albertson, D.G. & Meyer, B.J. (2001) HIM-10 is required for kinetochore structure and function on *Caenorhabditis elegans* holocentric chromosomes. *Journal of Cell Biology* **153**, 1227–1238.

Howell, W.M. (1982) Selective staining of nucleolus organizer regions (NORs). In: *The Cell Nucleus*, Vol. XI (eds H. Busch & L. Rothblum), pp. 89–142. Academic Press, New York.

Howman, E.V., Fowler, K.J., Newson, A.J. *et al.* (2000) Early disruption of centromeric chromatin organization in centromere protein A (*Cenpa*) null mice. *Proceedings of the National Academy of Sciences of the USA* **97**, 1148–1153.

Hsieh, J. & Fire, A. (2000) Recognition and silencing of repeated DNA. *Annual Reviews of Genetics* **34**, 187–204.

Hsu, L.Y.F., Benn, P.A., Tannenbaum, H.L., Perlis, T.E. & Carlson, A.D. (1987) Chromosomal polymorphisms of 1, 9, 16 and Y in 4 major ethnic groups: a large prenatal study. *American Journal of Medical Genetics* **26**, 95–101.

Huang, S., Li, B., Gray, M.D., Oshima, J., Mian, I.S. & Campisi, J. (1998) The premature ageing syndrome protein, WRN, is a 3′→5′ exonuclease. *Nature Genetics* **20**, 114–116.

Hudson, D.F., Fowler, K.J., Earle, E. *et al.* (1998) Centromere protein B null mice are mitotically and meiotically normal but have lower body and testis weights. *Journal of Cell Biology* **141**, 309–319.

Humphreys, D., Eggan, K., Akutsu, H. *et al.* (2001) Epigenetic instability in ES cells and cloned mice. *Science* **293**, 95–97.

Hunter, N., Börner, G.V., Lichten, M. & Kleckner, N. (2001) γ-H2AX illuminates meiosis. *Nature Genetics* **27**, 236–238.

Hurst, L.D. & McVean, G.T. (1998) Do we understand the evolution of genomic imprinting? *Current Opinion in Genetics and Development* **8**, 701–708.

Huskins, C.L. (1941) The coiling of chromonemata. *Cold Spring Harbor Symposium on Quantitative Biology* **9**, 13–18.

IHGSC (International Human Genome Sequencing Consortium) (2001) Initial sequencing and analysis of the human genome. *Nature* **409**, 860–921.

Ikeno, M., Grimes, B., Okazaki, T. *et al.* (1998) Construction of YAC-based mammalian artificial chromosomes. *Nature Biotechnology* **16**, 431–439.

Ina, S., Sasaki, T., Yokota, Y. & Shinomiya, T. (2001) A broad replication origin of *Drosophila melanogaster*, *oriD*α, consists of AT-rich multiple discrete initiation sites. *Chromosoma* **109**, 551–564.

Irick, H. (1994) A new function for heterochromatin. *Chromosoma* **103**, 1–3.

ISCN (1995) *An International System for Human Cytogenetic Nomenclature*. Karger, New York.

Jablonka, E. & Lamb, M.J. (1990) The evolution of heteromorphic sex chromosomes. *Biological Reviews* **65**, 249–276.

Jablonski, S.A., Chan, G.K.T., Cooke, C.A., Earnshaw, W.C. & Yen, T.J. (1998) The hBUB1 and hBUBR1 kinases sequentially assemble onto kinetochores during prophase with hBUBR1 concentrating at the kinetochore plates in mitosis. *Chromosoma* **187**, 386–396.

Jackson, D.A., Hassan, A.B., Errington, R.J. & Cook, P.R. (1993) Visualization of focal sites of transcription within human nuclei. *EMBO Journal* **12**, 1059–1065.

Jacobs, P.A. & Hassold, T.J. (1995) The origin of numerical chromosome abnormalities. *Advances in Genetics* **33**, 101–133.

Jacq, C., Alt-Mörbe, J., Andre, B. *et al.* (1997) The nucleotide sequence of *Saccharomyces cerevisiae* chromosome IV. *Nature* **387** (suppl.), 75–78.

Jaenisch, R. (1997) DNA methylation and imprinting: why bother? *Trends in Genetics* **13**, 323–329.

Jakobovits, A., Moore, A.L., Green, L.L. *et al.* (1993) Germline transmission and expression of a human-derived yeast artificial chromosome. *Nature* **362**, 255–258.

James, T.C., Eissenberg, J.C., Craig, C., Dietrich, V., Hobson, A. & Elgin, S.C.R. (1989) Distribution patterns of HP1, a heterochromatin-associated nonhistone chromosomal protein of *Drosophila*. *European Journal of Cell Biology* **50**, 170–180.

Jantsch, M., Hamilton, B., Mayr, B. & Schweizer, D. (1990) Meiotic chromosome behaviour reflects levels of sequence divergence in *Sus scrofa domestica* satellite DNA. *Chromosoma* **99**, 330–335.

Jeffreys, A.J., Murray, J. & Neumann, R. (1998) High-resolution mapping of crossovers in human sperm defines a minisatellite-associated recombination hotspot. *Molecular Cell* **2**, 267–273.

Jenuwein, T. & Allis, C.D. (2001) Translating the histone code. *Science* **293**, 1074–1080.

Jeppesen, P. (1997) Histone acetylation: a possible mechanism for the inheritance of cell memory at mitosis. *Bioessays* **19**, 67–74.

Jeppesen, P., Mitchell, A., Turner, B. & Perry, P. (1992) Antibodies to defined histone epitopes reveal variations in chromatin conformation and underacetylation of centric heterochromatin in human metaphase chromosomes. *Chromosoma* **101**, 322–332.

Jiang, Y.-H., Tsai, T.-F., Bressler, J. & Beaudet, A.L. (1998) Imprinting in Angelman and Prader–Willi syndrome. *Current Opinion in Genetics and Development* **8**, 334–342.

Joenje, H. & Patel, K.J. (2001) The emerging genetic and molecular basis of Fanconi anaemia. *Nature Reviews Genetics* **2**, 446–457.

John, B. (1988) The biology of heterochromatin. In: *Heterochromatin, Molecular and Structural Aspects* (ed. R.S. Verma), pp. 1–147. Cambridge University Press, Cambridge.

John, B. (1990) *Meiosis*. Cambridge University Press, Cambridge.

John, R.M. & Surani, M.A. (2000) Genomic imprinting, mammalian evolution, and the mystery of egg-laying mammals. *Cell* **101**, 585–588.

Johnson, K.R., Wright, J.E., Jr & May, B. (1987) Linkage relationships reflecting ancestral tetraploidy in salmonid fish. *Genetics* **116**, 579–591.

Johnson, K.R., Lehn, D.A., Elton, T.S., Barr, P.J. & Reeves, R. (1988) Complete murine cDNA sequence, genomic structure, and tissue expression of the high mobility group protein HMG-I(Y). *Journal of Biological Chemistry* **263**, 18338–18342.

Jones, C., Müllenbach, R., Grossfeld, P. *et al.* (2000) Co-localisation of CCG repeats and chromosome deletion breakpoints in Jacobsen syndrome: evidence for a common mechanism of chromosome breakage. *Human Molecular Genetics* **9**, 1201–1208.

Jones, C.J., Soley, A., Skinner, J.W. *et al.* (1998) Dissociation of telomere dynamics from telomerase activity in human thyroid cancer cells. *Experimental Cell Research* **240**, 333–339.

Jones, K.W. & Singh, L. (1985) Snakes and the evolution of sex chromosomes. *Trends in Genetics* **1**, 55–61.

Jones, P.A. (1999) The DNA methylation paradox. *Trends in Genetics* **15**, 34–37.

Jones, P.A. & Gonzalgo, M.L. (1997) Altered DNA methylation and genome instability: a new pathway to cancer? *Proceedings of the National Academy of Sciences of the USA* **94**, 2103–2105.

Jones, P.A. & Laird, P.W. (1999) Cancer epigenetics comes of age. *Nature Genetics* **21**, 163–167.

Jones, P.A. & Takai, D. (2001) The role of DNA methylation in mammalian epigenetics. *Science* **293**, 1068–1070.

Judd, B.H., Shen, M.W. & Kaufmann, Z.C. (1972) The anatomy and function of a segment of the X chromosome of *Drosophila melanogaster*. *Genetics* **71**, 139–156.

Kalish, W.-E., Whitmore, T. & Schwitalla, G. (1985) Electron-microscopic map of surface-spread polytene chromosomes of *Drosophila hydei*. *Chromosoma*, **92**, 265–272.

Kalitsis, P., Fowler, K.J., Earle, E., Hill, J. & Choo, K.H.A. (1998) Targeted disruption of mouse centromere protein C gene leads to mitotic disarray and early embryo death. *Proceedings of the National Academy of Sciences of the USA* **95**, 1136–1141.

Kamnert, I., López, C.C., Rosén, M. & Edström, J.-E. (1997) Telomeres terminating with long complex tandem repeats. *Hereditas* **127**, 175–180.

Kanaar, R., Hoeijmakers, J.H.J. & van Gent, D.C. (1998) Molecular mechanisms of DNA double-strand break repair. *Trends in Cell Biology* **8**, 483–489.

Karpen, G.H. (1994) Position-effect variegation and the new biology of heterochromatin. *Current Opinion in Genetics and Development* **4**, 281–291.

Karran, P. (2000) DNA double strand break repair in mammalian cells. *Current Opinion in Genetics and Development* **10**, 144–150.

Karube, T. & Watanabe, S. (1988) Analysis of the chromosomal DNA replication pattern using the bromodeoxyuridine labelling method. *Cancer Research* **48**, 219–222.

Kashi, Y. & Soller, M. (1999) Functional roles of microsatellites and minisatellites. In: *Microsatellites* (eds D.B. Goldstein & C. Schlötterer), pp. 9–22. Oxford University Press: Oxford.

Kaytor, M.D. & Orr, H.T. (2001) RNA targets of the fragile X protein. *Cell* **107**, 555–557.

Kazazian, H.H. (1998) Mobile elements and disease. *Current Opinion in Genetics and Development* **8**, 343–350.

Kearns, L.P. & Sigee, D.C. (1980) The occurrence of period IV elements in dinoflagellate chromatin: an X-ray microanalytical study. *Journal of Cell Science* **46**, 113–127.

Keegan, K.S., Holtzman, D.A., Plug, A.W. *et al.* (1996) The

Atr and Atm protein kinases associate with different sites along meiotically pairing chromosomes. *Genes and Development* **10**, 2423–2437.

Keeney, S., Giroux, C.N. & Kleckner, N. (1997) Meiosis-specific DNA double-strand breaks are catalyzed by Spo11, a member of a widely conserved protein family. *Cell* **88**, 375–384.

Kelley, R.L., Meller, V.H., Gordadze, P.R., Roman, G., Davis, R.L. & Kuroda, M.I. (1999) Epigenetic spreading of the *Drosophila* dosage compensation complex from *roX* RNA genes into flanking chromatin. *Cell* **98**, 513–522.

Kennedy, B.K., Barbie, D.A., Classon, M., Dyson, N. & Harlow, E. (2000) Nuclear organization of DNA replication in primary mammalian cells. *Genes and Development* **14**, 2855–2868.

Keppel, F. (1986) Transcribed human ribosomal RNA genes are attached to the nuclear matrix. *Journal of Molecular Biology* **187**, 15–21.

Keresö, J., Praznovszky, T., Cserpán, I. *et al.* (1996) *De novo* chromosome formations by large-scale amplification of the centromeric region of mouse chromosomes. *Chromosome Research* **4**, 226–239.

Khochbin, S., Verdel, A., Lemercier, C. & Seigneurin-Berny, D. (2001) Functional significance of histone deacetylase diversity. *Current Opinion in Genetics and Development* **11**, 162–166.

Khosla, S., Augustus, M. & Brahmachari, V. (1999) Sex-specific organisation of middle repetitive DNA sequences in the mealybug *Planococcus lilacinus*. *Nucleic Acids Research* **27**, 3745–3751.

Kilby, N.J., Snaith, M.R. & Murray, J.A.H. (1993) Site-specific recombinases: tools for genome engineering. *Trends in Genetics* **9**, 413–421.

Kill, I.R., Bridger, J.M., Campbell, K.H., Maldonado-Codina, G. & Hutchison, C.J. (1991) The timing of the formation and usage of replicase clusters in S-phase nuclei of human diploid fibroblasts. *Journal of Cell Science* **100**, 869–876.

Kim, N.W., Piatyszek, M.A., Prowse, K.P. *et al.* (1994) Specific association of human telomerase with immortal cells and cancer. *Science* **266**, 2011–2015.

Kim, S.-H., Kaminker, P. & Campisi, J. (1999) TIN2, a new regulator of telomere length in human cells. *Nature Genetics* **23**, 405–412.

Kimura, H. & Cook, P.R. (2001) Kinetics of core histones in living human cells: little exchange of H3 and H4 and some rapid exchange of H2B. *Journal of Cell Biology* **153**, 1341–1353.

Kimura, K., Hirano, M., Kobayashi, R. & Hirano, T. (1998) Phosphorylation and activation of 13S condensin by Cdc2 in vitro. *Science* **282**, 487–490.

King, M. (1980) C-banding studies on Australian hybrid frogs: secondary constriction structure and the concept of euchromatin transformation. *Chromosoma* **80**, 191–217.

King, M. (1993) *Species Evolution: the Role of Chromosome Change.* Cambridge University Press, Cambridge.

Kipling, D. & Warburton, P.E. (1997) Centromeres, CENP-B and *Tigger* too. *Trends in Genetics* **13**, 141–145.

Kipling, D., Ackford, H.E., Taylor, B.A. & Cooke, H.J. (1991) Mouse minor satellite DNA genetically maps to the centromere and is physically linked to the proximal telomere. *Genomics* **11**, 235–241.

Kirchhoff, M., Rose, H., Maahr, J. *et al.* (2000) High resolution comparative genomic hybridisation analysis reveals imbalances in dyschromosomal patients with normal or apparently balanced conventional karyotypes. *European Journal of Human Genetics* **8**, 661–668.

Klein, F., Laroche, T., Cardenas, M.E., Hofmann, J.F.-X., Schweizer, D. & Gasser, S.M. (1992) Localization of RAP1 and topoisomerase II in nuclei and meiotic chromosomes of yeast. *Journal of Cell Biology* **117**, 935–948.

Klein, G. (2000) Dysregulation of lymphocyte proliferation by chromosomal translocations and sequential genetic changes. *Bioessays* **22**, 414–422.

Kleinjan, D.-J. & van Heyningen, V. (1998) Position effect in human genetic disease. *Human Molecular Genetics* **7**, 1611–1618.

Klobutcher, L.A. & Jahn, C.L. (1991) Developmentally controlled genomic rearrangements in ciliated protozoa. *Current Opinion in Genetics and Development* **1**, 397–403.

Klobutcher, L.A., Gygax, S.E., Podoloff, J.D. *et al.* (1998) Conserved DNA sequences adjacent to chromosome fragmentation and telomere addition sites in *Euplotes crassus*. *Nucleic Acids Research* **26**, 4230–4240.

Kneissel, S., Franke, W.W., Gall, J.G. *et al.* (2001) A novel karyoskeletal protein: characterization of protein NO145, the major component of nucleolar cortical skeleton in *Xenopus* oocytes. *Molecular Biology of the Cell* **12**, 3904–3918.

Koch, J. (2000) Neocentromeres and alpha satellite: a proposed structural code for functional human centromere DNA. *Human Molecular Genetics* **9**, 149–154.

Kolodner, R.D. & Marsischky, G.T. (1999) Eukaryotic DNA mismatch repair. *Current Opinion in Genetics and Development* **9**, 89–96.

Koob, M.D., Moseley, M.L., Schut, L.J. *et al.* (1999) An untranslated CTG expansion causes a novel form of spinocerebellar ataxia (SCA8). *Nature Genetics* **21**, 379–384.

Kornberg, R.D. & Lorch, Y. (1999) Twenty-five years of the nucleosome, fundamental particle of the eukaryote chromosome. *Cell* **98**, 285–294.

Koshland, D. & Hartwell, L.H. (1987) The structure of sister minichromosome DNA before anaphase in *Saccharomyces cerevisiae*. *Science* **238**, 1713–1716.

Kota, R.S. & Runge, K.W. (1999) Tel2p, a regulator of

yeast telomeric length in vivo, binds to single-stranded telomeric DNA in vitro. *Chromosoma* **108**, 278–290.

Kourmouli, N., Theodoropoulos, P.A., Dialynas, G., Bakou, A., Politou, A.S., Cowell, I.G., Singh, P.B. & Georgatos, S.D. (2000) Dynamic associations of heterochromatin protein 1 with the nuclear envelope. *EMBO Journal* **19**, 6558–6568.

Kozubek, S., Lukášová, E., Marecková, A. *et al.* (1999) The topological organization of chromosomes 9 and 22 in cell nuclei has a determinative role in the induction of t(9,22) translocations and in the pathogenesis of t(9,22) leukaemias. *Chromosoma* **108**, 426–435.

Kralewski, M. & Benavente, R. (1997) XY body formation during rat spermatogenesis: an immunocytochemical study using antibodies against XY body-associated proteins. *Chromosoma* **106**, 304–307.

Kress, H. (1993) The salivary gland chromosomes of *Drosophila virilis*: a cytological map, pattern of transcription and aspects of chromosome evolution. *Chromosoma* **102**, 734–742.

Krízenecký, J. (1965) *Fundamenta Genetica*. Czechoslovak Academy of Sciences, Prague.

Kuo, M.-H. & Allis, C.D. (1998) Roles of histone acetyltransferases and deacetylases in gene regulation. *Bioessays* **20**, 615–626.

Kurek, R., Reugels, A.M., Lammermann, U. & Bünemann, H. (2000) Molecular aspects of intron evolution in dynein encoding mega-genes on the heterochromatic Y chromosome of *Drosophila* sp. *Genetica* **109**, 113–123.

Kurnit, D.M. (1979) Satellite DNA and heterochromatin variants: the case for unequal mitotic crossing over. *Human Genetics* **47**, 169–186.

Kuroda, M.I. & Kelley, R.L. (1999) Sex and repression. *Science* **284**, 1787–1788.

Kuromitsu, J., Yamashita, H., Kataoka, H. *et al.* (1997) A unique downregulation of h2-calponin gene expression in Down syndrome: a possible attenuation mechanism for fetal survival by methylation at the CpG island in the trisomic chromosome 21. *Molecular and Cellular Biology* **17**, 707–712.

Kurz, A., Lampel, S. Nickolenko, J.E. *et al.* (1996) Active and inactive genes localize preferentially in the periphery of chromosome territories. *Journal of Cell Biology* **135**, 1195–1205.

Labib, K. & Diffley, J.F.X. (2001) Is the MCM2–7 complex the eukaryotic DNA replication fork helicase? *Current Opinion in Genetics and Development* **10**, 64–70.

Lachner, M., O'Carroll, D., Rea, S., Mechtler, K. & Jenuwein, T. (2001) Methylation of histone H3 lysine 9 creates a binding site for HP1 proteins. *Nature* **410**, 116–120.

Lacroix, L., Liénard, H., Labourier, E. *et al.* (2000) Identification of two human nuclear proteins that recognise the cytosine-rich strand of human telomeres *in vitro*. *Nucleic Acids Research* **28**, 1564–1575.

Laird, C., Wilkinson, L., Johnson, D. & Sandström, C. (1981) Proposed structural principles of polytene chromosomes. In: *Chromosomes Today*, Vol. 7 (eds M.D. Bennett, M., Bobrow & G.M. Hewitt), pp. 74–83. George Allen & Unwin, London.

Laird, P.W. & Jaenisch, R. (1994) DNA methylation and cancer. *Human Molecular Genetics* **3**, 1487–1495.

Laird, P.W. & Jaenisch, R. (1996) The role of DNA methylation in cancer genetics and epigenetics. *Annual Review of Genetics* **30**, 441–464.

Lalioti, M.D., Scott, H.S., Buresi, C. *et al.* (1997) Dodecamer repeat expansion in cystatin B gene in progressive myoclonus epilepsy. *Nature* **386**, 847–851.

Lamb, M.M. & Laird, C.D. (1987) Three euchromatic DNA sequences under-replicated in polytene chromosomes of *Drosophila* are localised in constrictions and ectopic fibres. *Chromosoma* **95**, 227–235.

Lamb, N.E., Feingold, E., Savage, A. *et al.* (1997) Characterization of susceptible chiasma configurations that increase the risk for maternal nondisjunction of chromosome 21. *Human Molecular Genetics* **6**, 1391–1399.

LeBeau, M.M. & Rowley, J.D. (1984) Heritable fragile sites in cancer. *Nature* **308**, 607–608.

Le Beau, M.M., Rassool, F.V., Neilly, M.E. *et al.* (1998) Replication of a common fragile site, FRA3B, occurs late in S phase and is delayed further upon induction: implications for the mechanism of fragile site induction. *Human Molecular Genetics* **7**, 755–761.

Lee, C., Sasi, R. & Lin, C.C. (1993) Interstitial localization of telomeric DNA sequences in the Indian muntjac chromosomes: further evidence for tandem chromosome fusions in the karyotypic evolution of the Asian muntjacs. *Cytogenetics and Cell Genetics* **63**, 156–159.

Lee, C., Li, X., Jabs, E.W., Court, D. & Lin, C.C. (1995) Human gamma X satellite DNA: an X chromosome specific centromeric DNA sequence. *Chromosoma* **104**, 103–112.

Lee, J.T. & Jaenisch, R. (1996) A method for high efficiency YAC lipofection into murine embryonic stem cells. *Nucleic Acids Research* **24**, 5054–5055.

Lee, J.T. & Jaenisch, R. (1997) The (epi)genetic control of mammalian X-chromosome inactivation. *Current Opinion in Genetics and Development* **7**, 274–280.

Lefevre, G. (1976) A photographic representation and interpretation of the polytene chromosomes of *Drosophila melanogaster* salivary glands. In: *The Genetics and Biology of Drosophila*, Vol. 1a (eds M. Ashburner & E. Novitski), pp. 31–66. Academic Press, London.

Leitch, A.R., Schwarzacher, T., Mosgöller, W., Bennett, M.D. & Heslop-Harrison, J.S. (1991) Parental genomes are separated throughout the cell cycle in a plant hybrid. *Chromosoma* **101**, 206–213.

Leitch, I.J., Parokonny, A.S. & Bennett, M.D. (1997) New insights into chromosome evolution in plants from molecular cytogenetics. In: *Chromosomes Today*, Vol. 12 (eds N. Henriques-Gil, J.S. Parker & M.J. Puertas), pp. 333–346. Chapman & Hall, London.

LeMaire-Adkins, R., Radke, K. & Hunt, P.A. (1997) Lack of checkpoint control at the metaphase/anaphase transition: a mechanism of meiotic nondisjunction in mammalian females. *Journal of Cell Biology* **139**, 1611–1619.

Lemon, B.D. & Freedman, L.P. (1999) Nuclear receptor cofactors as chromatin remodelers. *Current Opinion in Genetics and Development* **9**, 499–504.

Lengauer, C., Kinzler, K.W. & Vogelstein, B. (1998) Genetic instabilities in human cancers. *Nature* **396**, 643–649.

León, P. & Kezer, J. (1990) Loop size in newt lampbrush chromosomes. *Chromosoma*, **99**, 83–86.

Leonhardt, H., Rahn, H.-P., Weinzierl, P. *et al.* (2000) Dynamics of DNA replication factories in living cells. *Journal of Cell Biology* **149**, 271–279.

Lewis, C.D. & Laemmli, U.K. (1982) Higher order metaphase chromosome structure: evidence for metalloprotein interactions. *Cell* **29**, 171–181.

Lewis, C.D., Lebkowski, J.S., Daly, A.K. & Laemmli, U.K. (1984) Interphase nuclear matrix and metaphase scaffolding structures. *Journal of Cell Science* **1**(suppl.), 103–122.

Lewis, J.D., Meehan, R.R., Henzel, W.J. *et al.* (1993) Purification, sequence, and cellular localization of a novel chromosomal protein that binds to methylated DNA. *Cell* **69**, 905–914.

Li, B., Oestreich, S. & de Lange, T. (2000) Identification of human Rap1: implications for telomere evolution. *Cell* **101**, 471–483.

Li, W.-H., Gu, Z., Wang, H. & Nekrutenko, A. (2001) Evolutionary analyses of the human genome. *Nature* **409**, 847–849.

Lica, L.M., Narayanswami, S. & Hamkalo, B.A. (1986) Mouse satellite DNA, centromere structure, and sister chromatid pairing. *Journal of Cell Biology* **103**, 1145–1151.

Lichten, M. (2001) Meiotic recombination: breaking the genome to save it. *Current Biology* **11**, R253–R256.

Lin, Q., Sirotkin, A. & Skoultchi, A.I. (2000) Normal spermatogenesis in mice lacking the testis-specific linker histone H1t. *Molecular and Cellular Biology* **20**, 2122–2128.

Lindsley, D.L., Sandler, L., Baker, B.S. *et al.* (1972) Segmental aneuploidy and the genetic gross structure of the *Drosophila* genome. *Genetics* **71**, 157–184.

Lingenfelter, P.A., Adler, D.A., Poslinski, D. *et al.* (1998) Escape from X inactivation of *Smcx* is preceded by silencing during mouse development. *Nature Genetics* **18**, 212–213.

Lingner, J. & Cech, T.R. (1998) Telomerase and chromosome end maintenance. *Current Opinion in Genetics and Development* **8**, 226–232.

Lipps, H.J. & Eder, C. (1996) Macronucleus structure and macronucleus development in hypotrichous ciliates. *International Journal of Developmental Biology* **40**, 141–147.

Lohe, A.R. & Hilliker, A.J. (1995) Return of the H-word (heterochromatin). *Current Opinion in Genetics and Development* **5**, 746–755.

Loidl, J. (1982) Further evidence for a heterochromatin–chiasma correlation in some *Allium* species. *Genetica* **60**, 31–35.

Long, E.O. & Dawid, I.G. (1980) Repeated genes in eukaryotes. *Annual Review of Biochemistry* **49**, 727–764.

López, C.C. & Edström, J.-E. (1998) Interspersed centromeric element with a CENP-B box-like motif in *Chironomus pallidivittatus*. *Nucleic Acids Research* **26**, 4168–4172.

Lopez, J.M., Karpen, G.H. & Orr-Weaver, T.L. (2000) Sister chromatid cohesion via MEI-S332 and kinetochore assembly are separable functions of the *Drosophila* centromere. *Current Biology* **10**, 997–1000.

Lucchesi, J.C. (1994) The evolution of heteromorphic sex chromosomes. *Bioessays* **16**, 81–83.

Lucchesi, J.C. (1998) Dosage compensation in flies and worms: the ups and downs of X-chromosome regulation. *Current Opinion in Genetics and Development* **8**, 179–184.

Luger, K. & Richmond, T.J. (1998) The histone tails of the nucleosome. *Current Opinion in Genetics and Development* **8**, 140–146.

Luger, K., Mäder, A.W., Richmond, R.K., Sargent, D.F. & Richmond, T.J. (1997) Crystal structure of the nucleosome core particle at 2.8 Å resolution. *Nature* **389**, 251–260.

Lupski, J.R. (1998) Genomic disorders: structural features of the genome can lead to DNA rearrangements and human disease traits. *Trends in Genetics* **14**, 417–422.

Lupski, J.R., Roth, J.R. & Weinstock, G.M. (1996) Chromosomal duplications in bacteria, fruit flies, and humans. *American Journal of Human Genetics* **58**, 21–27.

Lustig, A.J. (1999) Crisis intervention: the role of telomerase. *Proceedings of the National Academy of Sciences of the USA* **96**, 3339–3341.

Lyon, M.F. (1993) Epigenetic inheritance in mammals. *Trends in Genetics* **9**, 123–128.

Ma, H., Samarabandu, J., Devdhar, R.S. *et al.* (1998) Spatial and temporal dynamics of DNA replication sites in mammalian cells. *Journal of Cell Biology* **143**, 1415–1425.

Macgregor, H.C. (1980) Recent developments in the study of lampbrush chromosomes. *Heredity* **41**, 3–35.

Macgregor, H.C. (1993) *An Introduction to Animal Cytogenetics*, 1st edn, pp. 139–160. Chapman & Hall, London.

Macgregor, H.C. & Varley, J.M. (1988) *Working with Animal Chromosomes*, 2nd edn. John Wiley, Chichester.

Machado, C. & Andrew, D.J. (2000) D-titin: a giant protein with dual roles in chromosomes and muscles. *Journal of Cell Biology* **151**, 639–651.

Macleod, K. (2000) Tumor suppressor genes. *Current Opinion in Genetics and Development* **10**, 81–93.

Mandel, J.-L. (1997) Breaking the rule of three. *Nature* **386**, 767–769.

Mann, M.R.W. & Bartolomei, M.S. (1999) Towards a molecular understanding of Prader–Willi and Angelman syndromes. *Human Molecular Genetics* **8**, 1867–1873.

Manton, I. (1950) The spiral structure of chromosomes. *Biological Reviews* **25**, 486–507.

Manuelidis, L. & Chen, T. (1990) A unified model of eukaryotic chromosomes. *Cytometry* **11**, 8–25.

Marcand, S., Gasser, S.M. & Gilson, E. (1996) Chromatin: a sticky silence. *Current Biology* **6**, 1222–1225.

Marilley, M. & Gassend-Bonnet, G. (1989) Supercoiled loop organization of genomic DNA: a close relationship between loop domains, expression units, and replicon organization in rDNA from *Xenopus laevis*. *Experimental Cell Research* **180**, 475–489.

Marin, I., Siegal, M.L. & Baker, B.S. (2000) The evolution of dosage-compensation mechanisms. *Bioessays* **22**, 1106–1114.

Marmorstein, R. & Roth, S.Y. (2001) Histone acetyltransferases: function, structure, and catalysis. *Current Opinion in Genetics and Development* **11**, 155–161.

Marsh, J.L., Walker, H., Theisen, H. *et al.* (2000) Expanded polyglutamine peptides alone are intrinsically cytotoxic and cause neurodegeneration in *Drosophila*. *Human Molecular Genetics* **9**, 13–25.

Marshall, W.F. & Rosenbaum, J.L. (1999) Cell division: the renaissance of the centriole. *Current Biology* **9**, R218–R220.

Marshall, W.F., Fung, J.C. & Sedat, J.W. (1997) Deconstructing the nucleus: global architecture from local interactions. *Current Opinion in Genetics and Development* **7**, 259–263.

Martienssen, R. (1998) Chromosomal imprinting in plants. *Current Opinion in Genetics and Development* **8**, 240–244.

Martin, R.H., Balkan, W., Burns, K., Rademaker, A.W., Lin, C.C. & Rudd, N.L. (1983) The chromosome constitution of 1000 human spermatozoa. *Human Genetics* **63**, 305–309.

Martinez-Balbas, A., Rodriguez-Campos, A., Garcia-Ramirez, M. *et al.* (1990) Satellite DNAs contain sequences that induce curvature. *Biochemistry* **29**, 2342–2348.

Mason, J.M. & Biessmann, H. (1995) The unusual telomeres of *Drosophila*. *Trends in Genetics* **11**, 58–62.

Masumoto, H., Masukata, H., Muro, Y., Nozaki, N. & Okazaki, T. (1989) A human centromere antigen (CENP-B) interacts with a short specific sequence in alphoid DNA, a human centromeric satellite. *Journal of Cell Biology* **109**, 1963–1973.

Masumoto, H., Ikeno, M., Nakano, M. *et al.* (1998) Assay of centromere function using a human artificial chromosome. *Chromosoma* **107**, 406–416.

Matera, A.G. (1999) Nuclear bodies: multifaceted subdomains of the interchromatin space. *Trends Cell in Biology* **9**, 302–309.

Matsuura, T., Yamagata, T., Burgess, D.L. *et al.* (2000) Large expansion of the ATTCT pentanucleotide repeat in spinocerebellar ataxia type 10. *Nature Genetics* **26**, 191–194.

Maul, G.G., Negorev, D., Bell, P. & Ishov, A.M. (2000) Properties and assembly mechanisms of ND10, PML bodies, or PODs. *Journal of Structural Biology* **129**, 278–287.

Maxwell, E.S. & Fournier, M.J. (1995) The small nucleolar RNAs. *Annual Review of Biochemistry* **64**, 897–934.

Mayer, W., Niveleau, A., Walter, J., Fundele, R. & Haaf, T. (2000) Demethylation of the zygotic paternal genome. *Nature* **403**, 501–502.

McCullough, A.K., Dodson, M.L. & Lloyd, R.S. (1999) Initiation of base excision repair: glycosylase mechanisms and structures. *Annual Review of Biochemistry* **68**, 255–285.

McEachern, M.J., Krauskopf, A. & Blackburn, E.H. (2000) Telomeres and their control. *Annual Review of Genetics* **34**, 331–358.

McKee, B.D. (1996) The license to pair: identification of meiotic pairing sites in *Drosophila*. *Chromosoma* **105**, 135–141.

McLysaght, A., Enright, A.J., Skrabanek, L. & Wolfe, K.H. (2000) Estimation of synteny conservation and genome compaction between pufferfish (*Fugu*) and human. *Yeast* **17**, 22–36.

McManus, J., Perry, P., Sumner, A.T. *et al.* (1994) Unusual chromosome structure of fission yeast DNA in mouse cells. *Journal of Cell Science* **107**, 469–486.

McMurray, C.T. (1995) Mechanisms of DNA expansion. *Chromosoma* **104**, 2–13.

Méchali, M. (2001) DNA replication origins: from sequence specificity to epigenetics. *Nature Reviews Genetics* **2**, 640–645.

Medberry, S.L., Dale, E., Qin, M. & Ow, D.W. (1995) Intra-chromosomal rearrangements generated by Cre-lox site-specific recombination. *Nucleic Acids Research* **23**, 485–490.

Medina, F.J., Cerdido, A. & de Cárcer, G. (2000) The functional organization of the nucleolus in proliferating plant cells. *European Journal of Histochemistry* **44**, 117–131.

Meller, V.H. (2000) Dosage compensation: making 1X equal 2X. *Trends in Cell Biology* **10**, 54–59.

Mello, J.A. & Almouzni, G. (2001) The ins and outs of nucleosome assembly. *Current Opinion in Genetics and Development* **11**, 136-141.

Messing, J. & Grossniklaus, U. (1999) Genomic imprinting in plants. In: *Genomic Imprinting* (ed. R. Ohlsson), pp. 23–40. Springer-Verlag, Berlin.

Metcalfe, J.A., Parkhill, J., Campbell, L. *et al.* (1996) Accelerated telomere shortening in ataxia telangiectasia. *Nature Genetics* **13**, 350–353.

Meyer, A. & Schartl, M. (1999) Gene and genome duplications in vertebrates: the one-to-four (-to-eight in fish) rule and the evolution of novel gene functions. *Current Opinion in Cell Biology* **11**, 699–704.

Meyer, B.J. (2000) Sex in the worm: counting and compensating X-chromosome dose. *Trends in Genetics* **16**, 247–253.

Meyne, J., Ratliff, R.L. & Moyzis, R.K. (1989) Conservation of the human telomere sequence $(TTAGGG)_n$ among vertebrates. *Proceedings of the National Academy of Sciences of the USA* **89**, 7049–7053.

Meyne, J., Baker, J.R., Hobart, H.H. *et al.* (1990) Distribution of non-telomeric sites of the $(TTAGGG)_n$ telomeric sequence in vertebrate chromosomes. *Chromosoma* **99**, 3–10.

Michael, W.M. (2000) Nucleocytoplasmic shuttling signals: two for the price of one. *Trends in Cell Biology* **10**, 46–50.

Migeon, B.R. (1994) X-chromosome inactivation: molecular mechanisms and genetic consequences. *Trends in Genetics* **10**, 230–235.

Miklos, G.L.G. & Nankivell, R.N. (1976) Telomeric satellite DNA functions in regulating recombination. *Chromosoma* **56**, 143–167.

Miklos, G.L.G. & Gill, A.C. (1982) Nucleotide sequences of highly repeated DNAs; compilation and comments. *Genetical Research* **39**, 1–30.

Miklos, G.L.G. & Cotsell, J.N. (1990) Chromosome structure at interfaces between major chromatin types: alpha- and beta-heterochromatin. *Bioessays* **12**, 1–6.

Miles, C., Elgar, G., Coles, E., Kleinjan, D.-J., van Heyningen, V. & Hastie, N. (1998) Complete sequencing of the *Fugu* WAGR region from WT1 to PAX6: dramatic compaction and conservation of synteny with human chromosome 11p13. *Proceedings of the National Academy of Sciences of the USA* **95**, 13068–13072.

Miller, J.R., Hindkjaer, J. & Thomsen, P.D. (1993) A chromosomal basis for the differential organization of a porcine centromere-specific repeat. *Cytogenetics and Cell Genetics* **62**, 37–41.

Miller, O.L. (1981) The nucleolus, chromosomes, and visualization of genetic activity. *Journal of Cell Biology* **91**, 15s–27s.

Mills, A.A. & Bradley, A. (2001) From mouse to man: generating megabase chromosome rearrangements. *Trends in Genetics* **17**, 331–339.

Mills, W., Critcher, R., Lee, C. & Farr, C.J. (1999) Generation of an approximately 2.4 Mb human X centromere-based minichromosome by targeted telomere-associated chromosome fragmentation in DT40. *Human Molecular Genetics* **8**, 751–761.

Minc, E., Courvalin, J.C. & Buenida, B. (2000) HP1gamma associates with euchromatin and heterochromatin in mammalian nuclei and chromosomes. *Cytogenetics and Cell Genetics* **90**, 279–284.

Miniou, P., Jeanpierre, M., Blanquet, V. *et al.* (1994) Abnormal methylation pattern in constitutive and facultative (X inactive chromosome) heterochromatin of ICF patients. *Human Molecular Genetics* **3**, 2093–2102.

Miniou, P., Bourc'his, D., Molina Gomes, D., Jeanpierre, M. & Viegas-Péquignot, E. (1997) Undermethylation of *Alu* sequences in ICF syndrome: molecular and in situ analysis. *Cytogenetics and Cell Genetics* **77**, 308–313.

Mintz, P.J., Patterson, S.D., Neuwald, A.F., Spahr, C.S. & Spector, D.L. (1999) Purification and biochemical characterization of interchromatin granule clusters. *EMBO Journal* **18**, 4308–4320.

Mirkovitch, J., Gasser, S.M. & Laemmli, U.K. (1988) Scaffold attachment of DNA loops in metaphase chromosomes. *Journal of Molecular Biology* **200**, 101–110.

Misteli, T. (2000) Cell biology of transcription and pre-mRNA splicing: nuclear architecture meets nuclear function. *Journal of Cell Science* **113**, 1841–1849.

Misteli, T. & Spector, D.L. (1998) The cellular organization of gene expression. *Current Opinion in Cell Biology* **10**, 323–331.

Misteli, T., Gunjan, A., Hock, R., Bustin, M. & Brown, D.T. (2000) Dynamic binding of histone H1 to chromatin in living cells. *Nature* **408**, 877-881.

Mitas, M. (1997) Trinucleotide repeats associated with human disease. *Nucleic Acids Research* **25**, 2245–2253.

Mitchell, A. (2001) Synapsis spoilt. *Nature Reviews Molecular Cell Biology* **2**, 5.

Mitchell, A.R., Nicol, L., Malloy, P. & Kipling, D. (1993) Novel structural organisation of a *Mus musculus* DBA/2 chromosome shows a fixed position for the centromere. *Journal of Cell Science* **106**, 79–85.

Mitchell, A.R., Jeppesen, P., Nicol, L., Morrison, H. & Kipling, D. (1996) Epigenetic control of mammalian centromere protein binding: does DNA methylation have a role? *Journal of Cell Science* **109**, 2199–2206.

Mitchell, J.R., Wood, E. & Collins, K. (1999) A telomerase component is defective in the human disease dyskeratosis congenita. *Nature* **402**, 551–555.

Mitelman, F., Johanssen, B. & Mertens, F. (1994) *Catalog of Chromosome Aberrations in Cancer*, 5th edn. Wiley, New York.

Mitelman, F., Mertens, F. & Johansson, B. (1997) A break-point map of recurrent chromosomal rearrangements in human neoplasia. *Nature Genetics* **15**, 417–474.

Mitelman, F., Johanssen, B. & Mertens, F. (1998) *Catalog of Chromosome Aberrations in Cancer 98*, version 1 (CD-ROM). Wiley, New York.

Mlynarczyk, S.K. & Panning, B. (2000) X inactivation: *Tsix* and *Xist* as yin and yang. *Current Biology* **10**, R899–R903.

Moens, P.B. (1994) Molecular perspectives of chromosome pairing at meiosis. *Bioessays* **16**, 101–106.

Moens, P.B. & Pearlman, R.E. (1990) Telomere and centromere DNA are associated with the cores of meiotic prophase chromosomes. *Chromosoma* **100**, 8–14.

Moens, P.B. & Spyropoulos, B. (1995) Immunocytology of chiasmata and chromosomal disjunction at mouse meiosis. *Chromosoma* **104**, 175–182.

Moens, P.B., Chen, D.J., Shen, Z., *et al.* (1997) Rad51 immunocytology in rat and mouse spermatocytes and oocytes. *Chromosoma* **106**, 207–215.

Moens, P.B., Freire, R., Tarsounas, M., Spyropoulos, B. & Jackson, S.P. (2000) Expression and nuclear localization of BLM, a chromosome stability protein mutated in Bloom's syndrome, suggest a role in recombination during meiotic prophase. *Journal of Cell Science* **113**, 663–672.

Monaco, A.P. & Larin, Z. (1994) YACs, BACs, PACs and MACs: artificial chromosomes as research tools. *Trends in Biotechnology* **12**, 280–286.

Montefalcone, G., Tempesta, S., Rocchi, M. & Archidiacono, N. (1999) Centromere repositioning. *Genome Research* **9**, 1184–1188.

Moore, G. (1995) Cereal genome evolution: pastoral pursuits with 'Lego' genomes. *Current Opinion in Genetics and Development* **5**, 717–724.

Moore, G., Abbo, S., Cheung, W. *et al.* (1993) Key features of cereal genome organization as revealed by the use of cytosine methylation-sensitive restriction endonucleases. *Genomics* **15**, 472–482.

Moore, G., Foote, T., Helentjaris, T., Devos, K., Kurata, N. & Gale, M. (1995) Was there a single ancestral cereal chromosome? *Trends in Genetics* **11**, 81–82.

Moore, L.L. & Roth, M.B. (2001) HCP-4, a CENP-C-like protein in *Caenorhabditis elegans*, is required for resolution of sister centromeres. *Journal of Cell Biology* **153**, 1199–1207.

Mora-Garcia, S. & Goodrich, J. (2000) Genomic imprinting: seeds of conflict. *Current Biology* **10**, R71–R74.

Morales, C.P., Holt, S.E., Oullette, M. *et al.* (1999) Absence of cancer-associated changes in human fibroblasts immortalized with telomerase. *Nature Genetics* **21**, 115–118.

Moreau, N., Angelier, N., Bonnanfant-Jaïs, M.-L., Gounon, P. & Kubisz, P. (1986). Association of nucleo-plasmin with transcription products as revealed by immunolocalization in the amphibian oocyte. *Journal of Cell Biology* **103**, 683–690.

Morgan, D.O. (1997) Cyclin-dependent kinases. *Annual Review of Cell Biology* **13**, 261–291.

Morgan, G.T. (2002) Lampbrush chromosomes and associated bodies: new insights into principles of nuclear structure and function. *Chromosome Research* **10**, 177–200.

Morison, I.M., Paton, C.J. & Cleverley, S.D. (2001) The imprinted gene and parent-of-origin effect database. *Nucleic Acids Research* **29**, 275–276.

Müller, F., Bernard, V. & Tobler, H. (1996) Chromatin diminution in nematodes. *Bioessays* **18**, 133–138.

Murphy, T.D. & Karpen, G.H. (1998) Centromeres take flight: alpha satellite and the quest for the human centromere. *Cell* **93**, 317–320.

Murray, A.W. & Szostak, J.W. (1983) Construction of artificial chromosomes in yeast. *Nature* **305**, 189–193.

Murray, A.W. & Szostak, J.W. (1985) Chromosome segregation in mitosis and meiosis. *Annual Review of Cell Biology* **1**, 289–315.

Nabetani, A., Koujin, T., Tsutsumi, C., Haraguchi, T. & Hiraoka, Y. (2001) A conserved protein, Nuf2, is implicated in connecting the centromere to the spindle during chromosome segregation: a link between the kinetochore function and the spindle checkpoint. *Chromosoma* **110**, 322–334.

Nachman, M.W. & Searle, J.B. (1995) Why is the house mouse karyotype so variable? *Trends in Ecology and Evolution* **10**, 397–402.

Nagele, R.G., Freeman, T., McMorrow, L., Thomson, Z., Kitson-Wind, K. & Lee, H.-Y. (1999) Chromosomes exhibit preferential positioning in nuclei of quiescent human cells. *Journal of Cell Science* **112**, 525–535.

Nagl, W. (1978) *Endopolyploidy and Polyteny in Differentiation and Evolution*. North-Holland, Amsterdam.

Nagl, W. & Schmitt, H.-P. (1985) Transcription of repetitive DNA in condensed plant chromatin. *Molecular Biology Reports* **10**, 143–146.

Nakai, Y., Kubota, S., Goto, Y., Ishibashi, T., Davison, W. & Kohno, S.-I. (1995) Chromosome elimination in three Baltic, south Pacific and north-east Pacific hagfish species. *Chromosome Research* **3**, 321–330.

Nakielny, S. & Dreyfuss, G. (1999) Transport of proteins and RNAs in and out of the nucleus. *Cell* **99**, 677–690.

Nanda, I., Shan, Z., Schartl, M. *et al.* (1999) 300 million years of conserved synteny between chicken Z and human chromosome 9. *Nature Genetics* **21**, 258–259.

Nasmyth, K. (2001) Disseminating the genome: joining, resolving, and separating sister chromatids during mitosis and meiosis. *Annual Review of Genetics* **35**, 673–745.

Neff, N.F., Ellis, N.A., Ye, T.Z. *et al.* (1999) The DNA helicase activity of BLM is necessary for the correction of the genomic instability of Bloom syndrome cells. *Molecular Biology of the Cell* **10**, 665–676.

Nelson, W.G., Pienta, K.J., Barrack, E.R. & Coffey, D.S. (1986) The role of the nuclear matrix in the organization and function of DNA. *Annual Review of Biophysics and Biophysical Chemistry* **15**, 457–475.

Neumann, B. & Barlow, D.P. (1996) Multiple roles for DNA methylation in gametic imprinting. *Current Opinion in Genetics and Development* **6**, 159–163.

Nicholls, R.D., Saitoh, S. & Horsthemke, B. (1998) Imprinting in Prader–Willi and Angelman syndromes. *Trends in Genetics* **14**, 194–200.

Nicholls, R.D., Ohta, T. & Gray, T.A. (1999) Genetic abnormalities in Prader–Willi syndrome and lessons from mouse models. *Acta Paediatrica* **88** (suppl. 433), 99–104.

Nicklas, R.B. (1997) How cells get the right chromosomes. *Science* **275**, 632–637.

Nicklas, R.B, Campbell, M.S, Ward, S.C. & Gorbsky, G.J. (1998) Tension-sensitive kinetochore phosphorylation in vitro. *Journal of Cell Science* **111**, 3189–3196.

Nicoll, M., Akerib, C.C. & Meyer, B.J. (1997) X-chromosome-counting mechanisms that determine nematode sex. *Nature* **388**, 200–204.

Nigg, E.A. (1998) Polo-like kinases: positive regulators of cell division from start to finish. *Current Opinion in Cell Biology* **10**, 776–783.

Nokkala, S. & Nokkala, C. (1985) Spiral structures of meiotic chromosomes in plants. *Hereditas* **103**, 187–194.

Noma, K.-I., Allis, C.D. & Grewal, S.I.S. (2001) Transitions in distinct histone H3 methylation patterns at the heterochromatin domain boundaries. *Science* **293**, 1150–1155.

Nöthiger, R. & Steinmann-Zwicky, M. (1987) Genetics of sex determination: what can we learn from *Drosophila*? *Development* **101** (suppl.), 17–24.

Novitski, E. (1976) The construction of an entire compound two chromosome. In: *The Genetics and Biology of* Drosophila, Vol. 1b (eds M. Ashburner & E. Novitski), pp. 562–568. Academic Press, London.

Novitski, E. & Childress, D. (1976) Compound chromosomes involving the X and the Y chromosomes. In: *The Genetics and Biology of* Drosophila, Vol. 1b (eds M. Ashburner & E. Novitski), pp. 487–503. Academic Press, London.

Nur, U. (1990) Heterochromatization and euchromatization of whole genomes in scale insects (Coccoidea: Homoptera). *Development* **104** (suppl.), 29–34.

O'Brien, S.J., Wienberg, J. & Lyons, L.A. (1997) Comparative genomics: lessons from cats. *Trends in Genetics* **13**, 393–399.

O'Brien, S.J., Menotti-Raymond, M., Murphy, W.J. *et al.* (1999) The promise of comparative genomics. *Science*

286, 458–462 & 479–481, and pull-out supplement *Genome Maps 10*. (http://www.sciencemag.org/feature/data/1044631.shl)

O'Connell, M.J., Walworth, N.C. & Carr, A.M. (2000) The G2-phase DNA-damage checkpoint. *Trends in Cell Biology* **10**, 296–303.

O'Neill, R.J.W., O'Neill, M.J. & Graves, J.A.M. (1998) Undermethylation associated with retroelement activation and chromosome remodelling in an interspecific mammalian hybrid. *Nature* **393**, 68–72.

Ofir, R., Wong, A.C.C., McDermid, H.E., Skorecki, K.L. & Selig, S. (1999) Position effect of human telomeric repeats on replication timing. *Proceedings of the National Academy of Sciences of the USA* **96**, 11434–11439.

Ogawa, A., Murata, K. & Mizuno, S. (1998) The location of Z- and W-linked marker genes and sequence on the homomorphic sex chromosomes of the ostrich and the emu. *Proceedings of the National Academy of Sciences of the USA* **95**, 4415–4418.

Ohi, R. & Gould, K.L. (1999) Regulating the onset of mitosis. *Current Opinion in Cell Biology* **11**, 267–273.

Ohnuki, Y. (1965) Demonstration of the spiral structure of human chromosomes. *Nature* **208**, 916–917.

Okano, M., Bell, D.W., Haber, D.A. & Li, E. (1999) DNA methyltransferases Dnmt3a and Dnmt3b are essential for de novo methylation and mammalian development. *Cell* **99**, 247–257.

Okazaki, S., Tsuchida, K., Maekawa, H., Ishikawa, H. & Fujiwara, H. (1993) Identification of a pentanucleotide telomeric sequence, $(TTAGG)_n$, in the silkworm *Bombyx mori* and in other insects. *Molecular and Cellular Biology* **13**, 1424–1432.

O'Keefe, R.T., Henderson, S.C. & Spector, D.L. (1992) Dynamic organization of DNA replication in mammalian cell nuclei: spatially and temporally defined replication of chromosome-specific α-satellite DNA sequences. *Journal of Cell Biology* **116**, 1095–1110.

Olson, M.O.J., Dundr, M. & Szebeni, A. (2000) The nucleolus: an old factory with unexpected capabilities. *Trends in Cell Biology* **10**, 189–196.

Oostra, B.A. & Willems, P.J. (1995) A fragile gene. *Bioessays* **17**, 941–947.

Orphanides, G., LeRoy, G., Chang, C.-H., Luse, D.S. & Reinberg, D. (1998) FACT, a factor that facilitates transcript elongation through nucleosomes. *Cell* **92**, 105–116.

Orr-Weaver, T.L. (1994) Developmental modification of the *Drosophila* cell cycle. *Trends in Genetics* **10**, 321–327.

Otsuki, T., Furukawa, Y., Ikeda, K. *et al.* (2001) Fanconi anaemia protein, FANCA, associates with BRG1, a component of the human SWI/SNF complex. *Human Molecular Genetics* **10**, 2651–2660.

Otto, S.P. & Whitton, J. (2000) Polyploid incidence and evolution. *Annual Review of Genetics* **34**, 401–437.

Page, A.M. & Hieter, P. (1999) The anaphase-promoting complex: new subunits and regulators. *Annual Review of Biochemistry* **68**, 583–609.

Page, A.W. & Orr-Weaver, T.L. (1997) Stopping and starting the meiotic cell cycle. *Current Opinion in Genetics and Development* **7**, 23–31.

Palomeque, T., Garcia, M.F. & Lorite, P. (1998) Patterns of DNase I sensitivity in the chromosomes of the ant *Tapinoma nigerrimum* (Hymenoptera, Formicidae). *Cytogenetics and Cell Genetics* **81**, 112.

Pandita, T.K., Westphal, C.H., Anger, M. *et al.* (1999) *Atm* inactivation results in aberrant telomere clustering during meiotic prophase. *Molecular and Cellular Biology* **19**, 5096–5105.

Panning, B. & Jaenisch, R. (1998) RNA and the epigenetic regulation of X chromosome inactivation. *Cell* **93**, 305–308.

Pardue, M.-L. (1995) *Drosophila* telomeres: another way to end it all. In: *Telomeres* (eds E.H. Blackburn & C.W. Greider), pp. 339–370. Cold Spring Harbor Laboratory Press, New York.

Pardue, M.-L. & DeBaryshe, P.G. (1999) Telomeres and telomerase: more than the end of the line. *Chromosoma* **108**, 73–82.

Pardue, M.L., Danilevskaya, O.N., Lowenhaupt, K., Slot, F. & Traverse, K.L. (1996) *Drosophila* telomeres: new views on chromosome evolution. *Trends in Genetics* **12**, 48–52.

Parkhurst, S.M. & Meneely, P.M. (1994) Sex determination and dosage compensation: lessons from flies and worms. *Science* **264**, 924–932.

Partridge, J.F., Borgstrøm, B. & Allshire, R.C. (2000) Distinct protein interaction domains and protein spreading in a complex centromere. *Genes and Development* **14**, 783–791.

Pasero, P. & Schwob, E. (2000) Think global, act local – how to regulate S phase from individual replication origins. *Current Opinion in Genetics and Development* **10**, 178–186.

Pathak, S. & Wurster-Hill, D.H. (1977) Distribution of constitutive heterochromatin in carnivores. *Cytogenetics and Cell Genetics* **18**, 245–254.

Paule, M.R. & White, R.J. (2000) Transcription by RNA polymerases I and III. *Nucleic Acids Research* **28**, 1283–1298.

Paulson, J.R. & Laemmli, U.K. (1977) The structure of histone-depleted metaphase chromosomes. *Cell* **12**, 817–828.

Pavan, C. & da Cunha, A.B. (1969) Chromosomal activities in *Rhynchosciara* and other Sciaridae. *Annual Review of Genetics* **3**, 425–450.

Paweletz, N. (2001) Walther Flemming: pioneer of mitosis research. *Nature Reviews Molecular Cell Biology* **2**, 72–75.

Pederson, T. (2000) Half a century of 'the nuclear matrix'. *Molecular Biology of the Cell* **11**, 799–805.

Pederson, T. & Politz, J.C. (2000) The nucleolus and the four ribonucleoproteins of translation. *Journal of Cell Biology* **148**, 1091–1095.

Pelling, C. (1972) Transcription in giant chromosomal puffs. In: *Developmental Studies on Giant Chromosomes* (ed. W. Beermann), pp. 87–99. Springer-Verlag, Berlin.

Peterson, C.L. & Workman, J.L. (2000) Promoter targeting and chromatin remodelling by the SWI/SNF complex. *Current Opinion in Genetics and Development* **10**, 187–192.

Peterson, K. & Sapienza, C. (1993) Imprinting the genome: imprinted genes, imprinting genes, and a hypothesis for their interaction. *Annual Review of Genetics* **27**, 7–31.

Petrov, D.A. (2001) Evolution of genome size: new approaches to an old problem. *Trends in Genetics* **17**, 23–28.

Petrov, D.A., Lozovskaya, E.R. & Hartl, D.L. (1996) High intrinsic rate of DNA loss in *Drosophila*. *Nature* **384**, 346–349.

Petrov, D.A., Sangster, T.A., Johnston, J.S., Hartl, D.L. & Shaw, K.L. (2000) Evidence for DNA loss as a determinant of genome size. *Science* **287**, 1060–1062.

Phair, R.D. & Misteli, T. (2000) High mobility of proteins in the mammalian cell nucleus. *Nature* **404**, 604–609.

Picard, A., Galas, S., Peaucellier, G. & Dorée, M. (1996) Newly assembled cyclin B-cdc2 kinase is required to suppress DNA replication between meiosis I and meiosis II in starfish oocytes. *EMBO Journal* **15**, 3590–3598.

Pich, U., Fuchs, J. & Schubert, I. (1996) How do Alliaceae stabilize their chromosome ends in the absence of TTTAGGG sequences? *Chromosome Research* **4**, 207–213.

Pidoux, A.L. & Allshire, R.C. (2000) Centromeres: getting a grip of chromosomes. *Current Opinion in Cell Biology* **12**, 308–319.

Pienta, K.J. & Coffey, D.S. (1984) A structural analysis of the role of the nuclear matrix and DNA loops in the organization of the nucleus and chromosome. *Journal of Cell Science* **1** (suppl.), 123–135.

Pikaard, C.S. (2000) The epigenetics of nucleolar dominance. *Trends in Genetics* **16**, 495–500.

Pikaard, C.S. (2001) Genomic change and gene silencing in polyploids. *Trends in Genetics* **17**, 675–677.

Pimpinelli, S. & Goday, C. (1989) Unusual kinetochores and chromatin diminution in *Parascaris*. *Trends in Genetics* **5**, 310–315.

Pimpinelli, S., Bonaccorsi, S., Gatti, M. & Sandler, L. (1986) The peculiar genetic organization of *Drosophila* heterochromatin. *Trends in Genetics* **2**, 17–20.

Pines, J. (1999) Four-dimensional control of the cell cycle. *Nature Cell Biology* **1**, E73–E79.

Piñol-Roma, S., Swanson, M.S., Gall, J.G. & Dreyfuss, G. (1989) A novel hnRNP protein with a unique distribution on nascent transcripts. *Journal of Cell Biology* **109**, 2575–2587.

Pirotta, V. (1997) PcG complexes and chromatin silencing. *Current Opinion in Genetics and Development* **7**, 249–258.

Pittman, D.L., Cobb, J., Schimenti, K.J. *et al.* (1998) Meiotic prophase arrest with failure of chromosome synapsis in mice deficient for *Dmc1*, a germline-specific RecA homolog. *Molecular Cell* **1**, 697–705.

Planas-Silva, M.D. & Weinberg, R.A. (1997) The restriction point and control of cell proliferation. *Current Opinion in Cell Biology* **9**, 768–772.

Plasterk, R.H.A., Izsvák, Z. & Ivics, Z. (1999) Resident aliens: the Tc1/mariner superfamily of transposable elements. *Trends in Genetics* **15**, 326–332.

Plug, A.W., Peters, A.H.F.M., Keegan, K.S., Hoekstra, M.F., de Boer, P. & Ashley, T. (1998) Changes in protein composition of meiotic nodules during mammalian meiosis. *Journal of Cell Science* **111**, 413–423.

Poirié, M., Périguet, G. & Beukeboom, L. (1993) The hymenopteran way of determining sex. *Seminars in Developmental Biology* **3**, 357–361.

Polak, J.M. & van Noorden, S. (1997) *Introduction to Immunocytochemistry*, 2nd edn. Bios Scientific Publishers, Oxford.

Politz, J.C., Tuft, R.A., Pederson, T. & Singer, R.H. (1999) Movement of nuclear poly(A) RNA throughout the interchromatin space in living cells. *Current Biology* **9**, 285–291.

Polymenis, M. & Schmidt, E.V. (1999) Coordination of cell growth with cell division. *Current Opinion in Genetics and Development* **9**, 76–80.

Pombo, A., Jackson, D.A., Hollinshead, M., Wang, Z., Roeder, R.G. & Cook, P.R. (1999) Regional specialization in human nuclei: visualization of discrete sites of transcription by RNA polymerase III. *EMBO Journal* **18**, 2241–2253.

Potocki, L., Chen, K.-S., Park, S.-S. *et al.* (2000) Molecular mechanism for duplication 17p11.2 – the homologous recombination reciprocal of the Smith–Magenis microdeletion. *Nature Genetics* **24**, 84–87.

Prades, C., Laurent, A.-M., Puechberty, J., Yurov, Y. & Roizès, G. (1996) SINE and LINE within human centromeres. *Journal of Molecular Evolution* **42**, 37–43.

Prescott, D.M. (1992) Cutting, splicing, reordering, and elimination of DNA sequences in hypotrichous ciliates. *Bioessays* **14**, 317–324.

Prescott, D.M. (2000) Genome gymnastics: unique modes of DNA evolution and processing in ciliates. *Nature Reviews Genetics* **1**, 191–198.

Price, C. (1999a) Capping off the ends. *Nature* **397**, 213–214.

Price, C.M. (1999b) Telomeres and telomerase: broad effects on cell growth. *Current Opinion in Genetics and Development* **9**, 218–224.

Pryde, F.E., Gorham, H.C. & Louis, E.J. (1997) Chromosome ends: all the same under their caps. *Current Opinion in Genetics and Development* **7**, 822–828.

Qin, M., Lee, E., Zankel, T. & Ow, D.W. (1995) Site-specific cleavage of chromosomes in vitro through Cre-lox recombination. *Nucleic Acids Research* **23**, 1923–1927.

Rabbitts, T.H. (1994) Chromosomal translocations in human cancer. *Nature* **372**, 143–149.

Ramkissoon, U. & Goodfellow, P. (1996) Early steps in mammalian sex determination. *Current Opinion in Genetics and Development* **6**, 316–321.

Rasch, E.M. (1970) DNA cytophotometry of salivary gland nuclei and other tissue systems in Dipteran larvae. In: *Introduction to Quantitative Cytochemistry*, Vol. II (eds G.L. Wied & G.F. Bahr), pp. 357–397. Academic Press, New York.

Rasch, E.M. (1985) DNA 'standards' and the range of accurate DNA estimates by Feulgen absorption microspectrophotometry. *Advances in Microscopy*, 137–166.

Raška, I., Dundr, M., Koberna, K., Melcák, I., Risueño, M.-C. & Török, I. (1995) Does the synthesis of ribosomal RNA take place within nucleolar fibrillar centres or dense fibrillar components? A critical appraisal. *Journal of Structural Biology* **114**, 1–22.

Rattner, J.B. (1991) The structure of the mammalian centromere. *Bioessays* **13**, 51–56.

Rattner, J.B. (1992) Integrating chromosome structure with function. *Chromosoma* **101**, 259–264.

Rattner, J.B. & Lin, C.C. (1984) Ultrastructural organization of double minute chromosomes and HSR regions in human colon carcinoma cells. *Cytogenetics and Cell Genetics* **38**, 176–181.

Rattner, J.B. & Lin, C.C. (1985) Radial loops and helical coils coexist in metaphase chromosomes. *Cell* **42**, 291–296.

Rattner, J.B. & Lin, C.C. (1987) The higher order structure of the centromere. *Genome* **29**, 588–593.

Rattner, J.B., Kingwell, B.G. & Fritzler, M.J. (1988) Detection of distinct structural domains within the primary constriction using autoantibodies. *Chromosoma* **96**, 360–367.

Raukas, E. & Mikelsaar, R.-H. (1999) Are there molecules of nucleoprotamine? *Bioessays* **21**, 440–448.

Ray, S., Jahn, C., Tebeau, C.M., Larson, M.N. & Price, C.M. (1999) Differential expression of linker histone variants in *Euplotes crassus*. *Gene* **231**, 15–20.

Rayburn, A.L., Price, H.J., Smith, J.D. & Gold, J.R. (1985)

C-band heterochromatin and DNA content in *Zea mays*. *American Journal of Botany* **72**, 1610–1617.

Razin, S.V. & Gromova, I.I. (1995) The channels model of nuclear matrix structure. *Bioessays* **17**, 443–450.

Rea, S., Eisenhaber, F., O'Carroll, D., Strahl, B.D., Sun, Z.-W. & Schmid, S.M. (2000) Regulation of chromatin structure by site-specific histone H3 methyltransferases. *Nature* **406**, 593–599.

Reddy, P.S. & Housman, D.E. (1997) The complex pathology of trinucleotide repeats. *Current Opinion in Cell Biology* **9**, 364–372.

Redi, C.A. & Capanna, E. (1988) Robertsonian heterozygotes in the house mouse and the fate of their germ cells. In: *The Cytogenetics of Mammalian Autosomal Rearrangements* (ed. A. Daniel), pp. 315–359. Alan R. Liss, New York.

Reik, W. & Maher, E.R. (1997) Imprinting in clusters: lessons from the Beckwith–Wiedemann syndrome. *Trends in Genetics* **13**, 330–334.

Reik, W. & Walter, J. (1998) Imprinting mechanisms in mammals. *Current Opinion in Genetics and Development* **8**, 154–164.

Reik, W. & Murrell, A. (2000) Silence across the border. *Nature* **405**, 408–409.

Reik, W. & Walter, J. (2001a) Genomic imprinting: parental influence on the genome. *Nature Reviews Genetics* **2**, 21–32.

Reik, W. & Walter, J. (2001b) Evolution of imprinting mechanisms: the battle of the sexes begins in the zygote. *Nature Genetics* **27**, 255–256.

Reik, W., Brown, K.W., Schneid, H., Le Bouc, Y., Bickmore, W. & Maher, E.R. (1995) Imprinting mutations in the Beckwith–Wiedemann syndrome suggested by an altered imprinting pattern in the *IGF2-H19* domain. *Human Molecular Genetics* **4**, 2379–2385.

Reik, W., Kelsey, G. & Walter, J. (1999) Dissecting *de novo* methylation. *Nature Genetics* **23**, 380–382.

Reik, W., Dean, W. & Walter, J. (2001) Epigenetic reprogramming in mammalian development. *Science* **293**, 1089–1093.

Reugels, A.M., Kurek, R., Lammermann, U. & Bünemann, H. (2000) Mega-introns in the dynein gene *DhDhc7(Y)* on the heterochromatic Y chromosome give rise to the giant *Threads* loops in primary spermatocytes of *Drosophila hydei*. *Genetics* **154**, 759–769.

Reuter, G. & Spierer, P. (1992) Position effect variegation and chromatin proteins. *Bioessays* **14**, 605–612.

Rhodes, D. (1997) The nucleosome core all wrapped up. *Nature* **389**, 231–233.

Ribbert, D. (1972) Relation of puffing to bristle and footpad differentiation in *Calliphora* and *Sarcophaga*. In: *Developmental Studies on Giant Chromosomes* (ed. W. Beermann), pp. 153–179. Springer-Verlag, Berlin.

Ribbert, D. (1979) Chromomeres and puffing in experi-mentally induced polytene chromosomes of *Calliphora erythrocephala*. *Chromosoma* **74**, 269–298.

Richards, R.I. & Sutherland, G.R. (1992) Fragile X syndrome: the molecular picture comes into focus. *Trends in Genetics* **8**, 249–255.

Richardson, C., Moynahan, M.E. & Jasin, M. (1998) Double-strand break repair by interchromosomal recombination: suppression of chromosomal translocations. *Genes and Development* **12**, 3831–3842.

Rideout, W.M. III., Eggan, K. & Jaenisch, R. (2001) Nuclear cloning and epigenetic reprogramming of the genome. *Science* **293**, 1093–1098.

Ried, T., Schröck, E., Ning, Y. & Wienberg, J. (1998) Chromosome painting: a useful art. *Human Molecular Genetics* **7**, 1619–1626.

Rieder, C.L. & Salmon, E.D. (1998) The vertebrate cell kinetochore and its roles during mitosis. *Trends in Cell Biology* **8**, 310–317.

Rieder, C.L., Schultz, A., Cole, R. & Sluder, G. (1994) Anaphase onset in vertebrate somatic cells is controlled by a checkpoint that monitors sister kinetochore attachment to the spindle. *Journal of Cell Biology* **127**, 1301–1310.

Riesselmann, L. & Haaf, T. (1999) Preferential S-phase pairing of the imprinted region on distal mouse chromosome 7. *Cytogenetics and Cell Genetics* **86**, 39–42.

Ris, H. & Witt, P.L. (1981) Structure of the mammalian kinetochore. *Chromosoma* **82**, 153–170.

Rizwana, R. & Hahn, P.J. (1999) CpG methylation reduces genomic instability. *Journal of Cell Science* **112**, 4513–4519.

Roberge, M. (1992) Checkpoint controls that couple mitosis to completion of DNA replication. *Trends in Cell Biology* **2**, 277–281.

Robinson, W.P. (2000) Mechanisms leading to uniparental disomy and their clinical consequences. *Bioessays* **22**, 452–459.

Rocco, J.W. & Sidransky, D. (2001) p16(MTS-1/CDKN2/INK4a) in cancer progression. *Experimental Cell Research* **264**, 42–55.

Roeder, G.S. (1997) Meiotic chromosomes: it takes two to tango. *Genes and Development* **11**, 2600–2621.

Rooney, D.E. (2001) *Human Cytogenetics: Constitutional Analysis – a Practical Approach*, 3rd edn. Oxford University Press, Oxford.

Roth, M. & Prescott, D.M. (1985) DNA intermediates and telomere addition during genome reorganisation in *Euplotes crassus*. *Cell* **41**, 411–417.

Roth, M.B., Zahler, A.M. & Stolk, J.A. (1991) A conserved family of nuclear phosphoproteins localized to sites of polymerase II transcription. *Journal of Cell Biology* **115**, 587–596.

Round, E.K., Flowers, S.K. & Richards, E.J. (1997) *Arabidopsis thaliana* centromere regions: genetic map positions and repetitive DNA structure. *Genome Research* **7**, 1045–1053.

Roussel, M.R. (2000) An autonomous cell-cycle oscillator involved in the coordination of G1 events. *Bioessays* **22**, 3–5.

Roussel, P. & Hernandez-Verdun, D. (1994) Identification of Ag-NOR proteins, markers of proliferation related to ribosomal gene activity. *Experimental Cell Research* **214**, 465–472.

Rowles, A. & Blow, J.J. (1997) Chromatin proteins involved in the initiation of DNA replication. *Current Opinion in Genetics and Development.* **7**, 152–157.

Rowley, J.D. (1998) The critical role of chromosome translocations in human leukaemias. *Annual Review of Genetics* **32**, 495–519.

Rubinsztein, D.C., Wyttenbach, A. & Rankin, J. (1999) Intracellular inclusions, pathological markers in diseases caused by expanded polyglutamine tracts? *Journal of Medical Genetics* **36**, 265–270.

Rudolph, K.L., Chang, S., Millard, M., Schreiber-Agus, N. & DePinho, R.A. (2000) Inhibition of experimental liver cirrhosis by telomerase gene delivery. *Science* **287**, 1253–1258.

Russell, L.B. (1983) X–autosome translocations in the mouse: their characterization and use as tools to investigate inactivation and gene action. In: *Cytogenetics of the Mammalian X Chromosome*, part A (ed. A.A. Sandberg), pp. 205–250. Alan R. Liss, New York.

Ryan, K.J. & Wente, S.R. (2000) The nuclear pore complex: a protein machine bridging the nucleus and cytoplasm. *Current Opinion in Cell Biology* **12**, 361–371.

Sadoni, N., Langer, S., Fauth, C. *et al.* (1999) Nuclear organization of mammalian genomes: polar chromosome territories build up functionally distinct higher order compartments. *Journal of Cell Biology* **146**, 1211–1226.

Saffery, R., Irvine, D.V., Griffiths, B., Kalitsis, P., Wordeman, L. & Choo, K.H.A. (2000) Human centromeres and neocentromeres show identical distribution patterns of >20 functionally important kinetochore-associated proteins. *Human Molecular Genetics* **9**, 175–185.

Sahara, K., Marec, F. & Traut, W. (1999) TTAGG telomeric repeats in chromosomes of some insects and other arthropods. *Chromosome Research* **7**, 449–460.

Saitoh, N., Goldberg, I.G., Wood, E.R. & Earnshaw, W.C. (1994) ScII: an abundant chromosome scaffold protein is a member of a family of putative ATPases with an unusual predicted tertiary structure. *Journal of Cell Biology* **127**, 303–318.

Saji, S., Umehara, Y., Antonio, B.A. *et al.* (2001) A physical map with yeast artificial chromosome (YAC) clones covering 63% of the 12 rice chromosomes. *Genome* **44**, 32–37.

Samols, D. & Swift, H. (1979) Genomic organization in

the flesh fly *Sarcophaga bullata*. *Chromosoma* **75**, 129–143.

Sancar, A. (1999) Excision repair invades the territory of mismatch repair. *Nature Genetics* **21**, 247–249.

Sánchez-García, I. (1997) Consequences of chromosomal abnormalities in tumor development. *Annual Review of Genetics* **31**, 429–453.

Sauer, K., Knoblich, J.A., Richardson, H. & Lehner, C.F. (1995) Distinct modes of cyclin E/cdc2c kinase regulation and S-phase control in mitotic and endoreduplication cycles of *Drosophila* embryogenesis. *Genes and Development* **9**, 1327–1339.

Saunders, K. & Jones, D. (2001) New complementary techniques. In: *Analysing Chromosomes* (ed. B. Czepulkowski), pp. 159–181. Bios Scientific Publishers, Oxford.

Sauvé, D.M., Anderson, H.J., Ray, J.M., James, W.M. & Roberge, M. (1999) Phosphorylation-induced rearrangement of the histone H3 NH$_2$-terminal domain during mitotic chromosome condensation. *Journal of Cell Biology* **145**, 225–235.

Schaar, B.T., Chan, G.K.T., Maddox, P., Salmon, E.D. & Yen, T.J. (1997) CENP-E function at kinetochores is essential for chromosome alignment. *Journal of Cell Biology* **139**, 1373–1382.

Scheer, U. & Hock, R. (1999) Structure and function of the nucleolus. *Current Opinion in Cell Biology* **11**, 385–390.

Scheer, U., Bangying, X., Merkert, H. & Weissenberger, D. (1997) Looking at Christmas trees in the nucleolus. *Chromosoma* **105**, 470–480.

Scherthan, H. (2001) A bouquet makes ends meet. *Nature Reviews Molecular Cell Biology* **2**, 621–627.

Schindelhauer, D. (1999) Construction of mammalian artificial chromosomes: prospects for defining an optimal centromere. *Bioessays* **21**, 76–83.

Schmekel, K., Skoglund, U. & Daneholt, B. (1993a) The three-dimensional structure of the central region in a synaptonemal complex: a comparison between rat and two insect species, *Drosophila melanogaster* and *Blaps cribosa*. *Chromosoma* **102**, 682–692.

Schmekel, K., Wahrman, J., Skoglund, U. & Daneholt, B. (1993b) The central region of the synaptonemal complex in *Blaps cribosa* studied by electron microscope tomography. *Chromosoma* **102**, 669–681.

Schmid, M., Löser, C., Schmidtke, J. & Engel, W. (1982) Evolutionary conservation of a common pattern of activity of nucleolus organizers during spermatogenesis in vertebrates. *Chromosoma* **86**, 149–179.

Schmid, M., Grunert, D., Haaf, T. & Engel, W. (1983a) A direct demonstration of somatically paired heterochromatin of human chromosomes. *Cytogenetics and Cell Genetics* **36**, 554–561.

Schmid, M., Müller, H., Stasch, S. & Engel, W. (1983b) Silver staining of nucleolus organizer regions during human spermatogenesis. *Human Genetics* **64**, 363–370.

Schmidt, J.V., Levorse, J.M. & Tilghman, S.M. (1999) Enhancer competition between *H19* and *Igf2* does not mediate their imprinting. *Proceedings of the National Academy of Sciences of the USA* **96**, 9733–9738.

Schubert, I. & Oud, J.L. (1997) There is an upper limit of chromosome size for normal development of an organism. *Cell* **88**, 515–520.

Schubert, I. (2001) Alteration of chromosome numbers by generation of minichromosomes – is there a lower limit of chromosome size for stable segregation? *Cytogenetics and Cell Genetics* **93**, 175–181.

Schuffenhauer, S., Bartsch, O., Stumm, M. *et al.* (1995) DNA, FISH and complementation studies in ICF syndrome: DNA hypomethylation of repetitive and single copy loci and evidence for a trans acting factor. *Human Genetics* **96**, 562–571.

Schulz-Schaeffer, J. (1976) A short history of cytogenetics. *Biologische Zentralblatt* **95**, 193–221.

Schwab, M. (1998) Amplification of oncogenes in human cancer cells. *Bioessays* **20**, 473–479.

Schwarz, S., Hess, D. & Jost, J.-P. (1997) The methylated DNA binding protein-2-H1 (MDBP-2-H1) consists of histone H1 subtypes which are truncated at the C-terminus. *Nucleic Acids Research* **25**, 5052–5056.

Schwarzacher, H.G. & Wachtler, F. (1983) Nucleolus organizer regions and nucleoli. *Human Genetics* **63**, 89–99.

Schwarzacher, T. & Heslop-Harrison, P. (2000) *Practical in situ Hybridization.* Bios Scientific Publishers, Oxford.

Schweizer, D. (1976) Giemsa and fluorochrome banding of polytene chromosomes in *Phaseolus vulgaris* and *P. coccineus.* In: *Current Chromosome Research* (eds K. Jones & P.E. Brandham), pp. 51–56. North-Holland, Amsterdam.

Schweizer, D. & Ambros, P. (1979) Analysis of nucleolus organizer regions (NORs) in mitotic and polytene chromosomes of *Phaseolus coccineus* by silver staining and Giemsa C-banding. *Plant Systematics and Evolution* **132**, 27–51.

Schwienbacher, C., Gramantieri, L., Scelfo, R. *et al.* (2000) Gain of imprinting at chromosome 11p15, a pathogenetic mechanism identified in human hepatocarcinomas. *Proceedings of the National Academy of Sciences of the USA* **97**, 5445–5449.

Searle, J.B. (1998) Speciation, chromosomes, and genomes. *Genome Research* **8**, 1–3.

Searle, J.B. & Wójcik, J.M. (1998) Chromosomal evolution: the case of *Sorex araneus.* In: *Evolution of Shrews* (eds J.M. Wójcik & M. Wolsan), pp. 219–268. Mammal Research Institute, Polish Academy of Sciences, Bialowieza, Poland.

Seeberg, E., Eide, L. & Bjørås, M. (1995) The base excision repair pathway. *Trends in Biochemical Sciences* **20**, 391–397.

Shahbazian, M.D. & Zoghbi, H.Y. (2001) Molecular genetics of Rett syndrome and clinical spectrum of MECP2 mutations. *Current Opinion in Neurology* **14**, 171–176.

Shaikh, T.H., Kurahashi, H. & Saitta, S.C. (2000) Chromosome 22-specific low copy repeats and the 22q11.2 deletion syndrome: genomic organization and deletion endpoint analysis. *Human Molecular Genetics* **9**, 489–501.

Shapiro, D.Y. (1994) Sex change in fishes – how and why? In: *The Differences Between the Sexes* (eds R.V. Short & E. Balaban), pp. 105–130, Cambridge University Press, Cambridge.

Sharma, H.W., Sokoloski, J.A., Perez, J.R. *et al.* (1995) Differentiation of immortal cells inhibits telomerase activity. *Proceedings of the National Academy of Sciences of the USA* **92**, 12343–12346.

Shay, J.W. (1999) At the end of the millennium, a view of the end. *Nature Genetics* **23**, 382–383.

Shelby, R.D., Hahn, K.M. & Sullivan, K.F. (1996) Dynamic elastic behaviour of α-satellite DNA domains visualized in situ in living human cells. *Journal of Cell Biology* **135**, 545–557.

Shelby, R.D., Vafa, O. & Sullivan, K.F. (1997) Assembly of CENP-A into centromeric chromatin requires a cooperative array of nucleosomal DNA contact sites. *Journal of Cell Biology* **136**, 501–513.

Shen, J.-C. & Loeb, L.A. (2000) The Werner syndrome gene: the molecular basis of RecQ helicase-deficiency diseases. *Trends in Genetics* **16**, 213–220.

Shen, X., Yu, L., Weir, J.W. & Gorovsky, M.A. (1995) Linker histones are not essential and affect chromatin condensation *in vivo. Cell* **82**, 47–56.

Sherr, C.J. & Roberts, J.M. (1999) CDK inhibitors: positive and negative regulators of G1-phase progression. *Genes and Development* **13**, 1501–1512.

Shibata, F., Hizume, M. & Kuroki, Y. (2000) Differentiation and the polymorphic nature of the Y chromosomes revealed by repetitive sequences in the dioecious plant, *Rumex acetosa. Chromosome Research* **8**, 229–236.

Shiels, P.G., Kind, A.J., Campbell, K.H.S. *et al.* (1999) Analysis of telomere lengths in cloned sheep. *Nature* **399**, 316–317.

Shiloh, Y. (1997) Ataxia-telangiectasia and the Nijmegen breakage syndrome: related disorders but genes apart. *Annual Review of Genetics* **31**, 635–662.

Shinagawa, H. & Iwasaki, H. (1996) Processing the Holliday junction in homologous recombination. *Trends in Biochemical Science* **21**, 107–111.

Shore, D. (2001) Telomeric chromatin: replicating and wrapping up chromosome ends. *Current Opinion in Genetics and Development* **11**, 189–198.

Sims, S.H., Macgregor, H.C., Pellatt, P.S. & Horner, H.A. (1984) Chromosome 1 in crested and marbled newts

(*Triturus*): an extraordinary case of heteromorphism and independent chromosome evolution. *Chromosoma* **89**, 169–185.

Sleeman, J. & Lamond, A.I. (1999) Nuclear organization of pre-mRNA splicing factors. *Current Opinion in Cell Biology* **11**, 372–377.

Smeets, D.F.C.M. & van de Klundert, F.A.J.M. (1990) Common fragile sites in man and three closely related primate species. *Cytogenetics and Cell Genetics* **53**, 8–14.

Smeets, D.F.C.M., Moog, U., Weemaes, C.M.R. *et al.* (1994) ICF syndrome: a new case and review of the literature. *Human Genetics* **94**, 240–246.

Smerdon, M.J. & Conconi, A. (1999) Modulation of DNA damage and DNA repair in chromatin. *Progress in Nucleic Acid Research and Molecular Biology* **62**, 227–255.

Smilenov, L.B., Dhar, S. & Pandita, T.K. (1999) Altered telomere nuclear matrix interactions and nucleosomal periodicity in ataxia telangiectasia cells before and after ionizing radiation treatment. *Molecular and Cellular Biology* **19**, 6963–6971.

Smit, A.F.A. (1999) Interspersed repeats and other mementos of transposable elements in mammalian genomes. *Current Opinion in Genetics and Development* **9**, 657–663.

Smith, A.V. & Orr-Weaver, T.L. (1991) The regulation of the cell cycle during *Drosophila* embryogenesis: the transition to polyteny. *Development* **112**, 997–1008.

Smith, A.V. & Roeder, G.S. (1997) The yeast Red1 protein localizes to the cores of meiotic chromosomes. *Journal of Cell Biology* **136**, 957–967.

Smith, C.M. & Steitz, J.A. (1997) Sno storm in the nucleolus: new roles for myriad small RNPs. *Cell* **89**, 669–672.

Smith, J., Bruley, C.K., Paton, I.R. *et al.* (2000) Differences in gene density on chicken macrochromosomes and microchromosomes. *Animal Genetics* **31**, 96–103.

Smith, K.A., Gorman, P.A., Stark, M.B., Groves, R.P. & Stark, G.R. (1990) Distinctive chromosomal structures are formed very early in amplification of CAD genes in Syrian hamster cells. *Cell* **63**, 1219–1227.

Smith, K.N. & Nicolas, A. (1998) Recombination at work for meiosis. *Current Opinion in Genetics and Development.* **8**, 200–211.

Smith, S. & de Lange, T. (2000) Tankyrase promotes telomere elongation in human cells. *Current Biology* **10**, 1299–1302.

Solovei, I., Gaginskaya, E., Hutchison, N. & Macgregor, H. (1993) Avian sex chromosomes in the lampbrush form: the ZW lampbrush bivalents from six species of bird. *Chromosome Research* **1**, 153–166.

Solovei, I., Gaginskaya, E.R. & Macgregor, H.C. (1994) The arrangement and transcription of telomere DNA sequences at the ends of lampbrush chromosomes of birds. *Chromosome Research* **2**, 460–470.

Solovei, I.V., Joffe, B.I., Gaginskaya, E.R. & Macgregor, H.C. (1996) Transcription on lampbrush chromosomes of a centromerically localized highly repeated DNA in pigeon (*Columba*) relates to sequence arrangement. *Chromosome Research* **4**, 588–603.

Sommerville, J., Crichton, C. & Malcolm, D. (1978) Immunofluorescent localization of transcriptional activity on lampbrush chromosomes. *Chromosoma*, **66**, 99–114.

Sommerville, J., Baird, J. & Turner, B.M. (1993) Histone H4 acetylation and transcription in amphibian chromatin. *Journal of Cell Biology* **120**, 277–290.

Sorsa, V. (1982) Volume of chromatin fibers in interbands and bands of polytene chromosomes. *Hereditas* **97**, 103–113.

Sorsa, V., Saura, A.O. & Heino, T.I. (1984) Electron microscopic map of divisions 61, 62 and 63 of the salivary gland 3L chromosomes in *Drosophila melanogaster*. *Chromosoma* **90**, 177–184.

Spada, F., Brunet, A., Mercier, Y., Renard, J.-P., Bianchi, M.E. & Thompson, E.M. (1998) High mobility group 1 (HMG1) protein in mouse preimplantation embryos. *Mechanisms of Development* **76**, 57–66.

Spencer, H.G. (2000) Population genetics and evolution of genomic imprinting. *Annual Reviews of Genetics* **34**, 457–477.

Sperling, K., Kalscheuer, V. & Neitzel, H. (1987) Transcriptional activity of constitutive heterochromatin in the mammal *Microtus agrestis* (Rodentia, Cricetidae). *Experimental Cell Research* **173**, 463–472.

Spiro, C., Pelletier, R., Rolfsmeier, M.L. *et al.* (1999) Inhibition of FEN-1 processing by DNA secondary structure at trinucleotide repeats. *Molecular Cell* **4**, 1079–1085.

Spradling, A. & Orr-Weaver, T. (1987) Regulation of DNA replication during *Drosophila* development. *Annual Review of Genetics* **21**, 373–403.

Spring, H., Scheer, U., Franke, W.W. & Trendelenburg, M.F. (1975) Lampbrush-type chromosomes in the primary nucleus of the green alga *Acetabularia mediterranea*. *Chromosoma* **50**, 25–43.

Stacey, M., Bennett, M.S. & Hulten, M. (1995) FISH analysis on spontaneously arising micronuclei in the ICF syndrome. *Journal of Medical Genetics* **32**, 502–508.

Stack, S.M. & Anderson, L.K. (2001) A model for chromosome structure during the mitotic and meiotic cell cycles. *Chromosome Research* **9**, 175–198.

Staiber, W. & Behnke, E. (1985) Developmental puffing activity in the salivary gland and Malpighian tubule chromosomes of *Acricotopus lucidus* (Diptera, Chironomidae). *Chromosoma* **93**, 1–16.

Standiford, D.M. (1989) The effects of chromatin diminution on the pattern of C-banding in the chromosomes of *Acanthocyclops vernalis* Fischer (Copepoda: Crustacea). *Genetica* **79**, 207–214.

Stark, G.R., Debatisse, M., Giulotto, E. & Wahl, G.M. (1989) Recent progress in understanding mechanisms of mammalian DNA amplification. *Cell* **57**, 901–908.

Stingo, V., Capriglione, T., Rocco, L., Improta, R. & Morescalchi, A. (1989) Genome size and A-T rich DNA in selachians. *Genetica* **79**, 197–205.

Stoffler, D., Fahrenkrog, B. & Aebi, U. (1999) The nuclear pore complex: from molecular architecture to functional dynamics. *Current Opinion in Cell Biology* **11**, 391–401.

Storlazzi, A., Xu, L., Schwacha, A. & Kleckner, N. (1996) Synaptonemal complex (SC) component Zip1 plays a role in meiotic recombination independent of SC polymerization along the chromosomes. *Proceedings of the National Academy of Sciences of the USA* **93**, 9043–9048.

Strahl, B.D. & Allis, C.D. (2000) The language of covalent histone modifications. *Nature* **403**, 41–45.

Strauss, F. & Varshavsky, A. (1984) A protein binds to a satellite DNA repeat at three specific sites that would be brought into mutual proximity by DNA folding in the nucleosome. *Cell* **37**, 889–901.

Strissel, P.L., Espinosa, R., Rowley, J.D. & Swift, H. (1996) Scaffold attachment regions in centromere-associated DNA. *Chromosoma* **105**, 122–133.

Strouboulis, J. & Wolffe, A.P. (1996) Functional compartmentalization of the nucleus. *Journal of Cell Science* **109**, 1991–2000.

Struhl, K. (1998) Histone acetylation and transcriptional regulatory mechanisms. *Genes and Development* **12**, 599–606.

Su, T.T. & O'Farrell, P.H. (1998) Chromosome association of minichromosome maintenance proteins in *Drosophila* endoreplication cycles. *Journal of Cell Biology* **140**, 451–460.

Sudarsanam, P. & Winston, F. (2000) The Swi/Snf family: nucleosome complexes and transcriptional control. *Trends in Genetics* **16**, 345–351.

Suja, J.A., Antonio, C. & Rufas, J.S. (1992) Involvement of chromatid cohesiveness at the centromere and chromosome arms in meiotic chromosome segregation: a cytological approach. *Chromosoma* **101**, 493–501.

Sullivan, B.A. & Schwartz, S. (1995) Identification of centromeric antigens in dicentric Robertsonian translocations: CENP-C and CENP-E are necessary components of functional centromeres. *Human Molecular Genetics* **4**, 2189–2197.

Sullivan, D.T. (1995) DNA excision repair and transcription: implications for genome evolution. *Current Opinion in Genetics and Development* **5**, 786–791.

Sullivan, S.A., Aravind, L., Makalowska, I., Baxevanis, A.D. & Landsman, D. (2000) The Histone Database: a comprehensive WWW resource for histones and histone fold-containing proteins. *Nucleic Acids Research* **28**, 320–322.

Sumner, A.T. (1972) A simple technique for demonstrating centromeric heterochromatin. *Experimental Cell Research* **75**, 304–306.

Sumner, A.T. (1990) *Chromosome Banding.* Unwin Hyman, London.

Sumner, A.T. (1991) Scanning electron microscopy of mammalian chromosomes from prophase to interphase. *Chromosoma* **100**, 410–418.

Sumner, A.T. (1996) The distribution of topoisomerase II on mammalian chromosomes. *Chromosome Research* **4**, 5–14.

Sumner, A.T. (1998a) Induction of diplochromosomes in mammalian cells by inhibitors of topoisomerase II. *Chromosoma* **107**, 486–490.

Sumner, A.T. (1998b) The mitotic chromosome. *Advances in Genome Biology* **5A**, 211–261.

Sumner, A.T. (1998c) The structure of the centromeric region of CHO chromosomes. *Cell Biology International Reports* **22**, 127–130.

Sumner, A.T. & Buckland, R.A. (1976) Relative DNA contents of somatic nuclei of ox, sheep and goat. *Chromosoma* **57**, 171–175.

Sumner, A.T. & Leitch, A.R. (1999) Microscopy of chromosomes. In: *Light Microscopy in Biology: a Practical Approach* (ed. A.J. Lacey), 2nd edn, pp. 151–184. Oxford University Press, Oxford.

Sumner, A.T., Mitchell, A.R. & Ellis, P.M. (1998) A FISH study of chromosome fusion in the ICF syndrome: involvement of paracentric heterochromatin but not of the centromeres themselves. *Journal of Medical Genetics* **35**, 833–835.

Sun, T.-Q., Fenstermacher, D.A. & Vos, J.-M.H. (1994) Human artificial episomal chromosomes for cloning large DNA fragments in human cells. *Nature Genetics* **8**, 33–41.

Sun, X., Wahlstrom, J. & Karpen, G. (1997) Molecular structure of a functional *Drosophila* centromere. *Cell* **91**, 1007–1019.

Sunkel, C.E. & Coelho, P.A. (1995) The elusive centromere: sequence divergence and functional conservation. *Current Opinion in Genetics and Development* **5**, 756–767.

Surani, M.A. (1994) Genomic imprinting: control of gene expression by epigenetic inheritance. *Current Opinion in Cell Biology* **6**, 390–395.

Suraokar, M. & Bradley, A. (2000) Targeting sheep. *Nature* **405**, 1004–1005.

Sutani, T., Yuasa, T., Tomonaga, T., Dohmae, N., Takio, K. & Yanagida, M. (1999) Fission yeast condensin complex: essential roles of non-SMC subunits for condensation and Cdc2 phosphorylation of Cut3/SMC4. *Genes and Development* **13**, 2271–2283.

Sutherland, G.R. & Richards, R.I. (1995) Simple tandem DNA repeats and human genetic disease. *Proceedings of* the *National Academy of Sciences of the USA* **92**, 3636–3641.

Sutherland, G.R., Baker, E. & Richards, R.I. (1998) Fragile sites still breaking. *Trends in Genetics* **14**, 501–506.

Sybenga, J. (1999) What makes homologous chromosomes find each other in meiosis? A review and a hypothesis. *Chromosoma* **108**, 209–219.

Tahara, H., Tokutake, Y., Maeda, S. *et al.* (1997) Abnormal telomere dynamics of B-lymphoblastoid cell strains from Werner's syndrome patients transformed by Epstein–Barr virus. *Oncogene* **15**, 1911–1920.

Takisawa, H., Mimura, S. & Kubota, Y. (2001) Eukaryotic DNA replication: from pre-replication complex to initiation complex. *Current Opinion in Cell Biology* **12**, 690–696.

Talcott, B. & Moore, M.S. (1999) Getting across the nuclear pore complex. *Trends in Cell Biology* **9**, 312–318.

Tanaka, I., Akahori, Y., Gomi, K., Suzuki, T. & Ueda, K. (1999a) A novel histone variant localized in nucleoli of higher plant cells. *Chromosoma* **108**, 190–199.

Tanaka, T., Cosma, M.P., Wirth, K. & Nasmyth, K. (1999b) Identification of cohesin sites at centromeres and along chromosome arms. *Cell* **98**, 847–858.

Tanaka, M., Hennebold, J.D., Macfarlane, J. & Adashi, E.Y. (2001) A mammalian oocyte-specific linker histone gene *H1oo*: homology with the genes for the oocyte-specific cleavage stage histone (cs-H1) of sea urchin and the *B4/H1M* histone of the frog. *Development* **128**, 655–664.

Tassabehji, M., Metcalfe, K., Karmiloff-Smith, A. *et al.* (1999) Williams syndrome: use of chromosomal microdeletions as a tool to dissect cognitive and physical phenotypes. *American Journal of Human Genetics* **64**, 118–125.

Tausta, S.L., Turner, L.R., Buckley, L.K. & Klobutcher, L.A. (1991) High fidelity developmental excision of Tec1 transposons and internal eliminated sequences in *Euplotes crassus*. *Nucleic Acids Research* **19**, 3229–3236.

Telenius, H., Szeles, A., Keresö, J. *et al.* (1999) Stability of a functional murine satellite DNA-based artificial chromosome across mammalian species. *Chromosome Research* **7**, 3–7.

ten Hoopen, R., Manteuffel, R., Dolezel, J., Malysheva, L. & Schubert, I. (2000) Evolutionary conservation of kinetochore protein sequences in plants. *Chromosoma* **109**, 482–489.

Thiry, M. (1995) The interchromatin granules. *Histology and Histopathology* **10**, 1035–1045.

Thoma, F. (1999) Light and dark in chromatin repair: repair of UV-induced DNA lesions by photolyase and nucleotide excision repair. *EMBO Journal* **18**, 6586–6598.

Thurston, V.C., Zinkowski, R.P. & Binder, L.I. (1996) Tau

as a nucleolar protein in human nonneural cells in vitro and in vivo. *Chromosoma* **105**, 20–30.

Tian, X.C., Xu, J. & Yang, X. (2000) Normal telomere lengths found in cloned cattle. *Nature Genetics* **26**, 272–273.

Tiersch, T.R., Chandler, R.W., Wachtel, S.S. & Elias, S. (1989) Reference standards for flow cytometry and application in comparative studies of nuclear DNA content. *Cytometry* **10**, 706–710.

Tilghman, S.M. (1999) The sins of the fathers and mothers: genomic imprinting in mammalian development. *Cell* **96**, 185–193.

Tomizuka, K., Shinohara, T., Yoshida, H. *et al.* (2000) Double trans-chromosomic mice: maintenance of two individual human chromosome fragments containing Ig heavy and k loci and expression of fully human antibodies. *Proceedings of the National Academy of Sciences of the USA* **97**, 722–727.

Traut, W. & Marec, F. (1997) Sex chromosome differentiation in some species of Lepidoptera (Insecta). *Chromosome Research* **5**, 283–291.

Travers, A. (1999) The location of the linker histone on the nucleosome. *Trends in Biochemical Sciences* **24**, 4–7.

Trerè, D. (2000) AgNOR staining and quantification. *Micron* **31**, 127–131.

Tunnacliffe, A., Jones, C., Le Paslier, D. *et al.* (1999) Localization of Jacobsen syndrome breakpoints on a 40-Mb physical map of distal chromosome 11q. *Genome Research* **9**, 44–52.

Turner, B.M., Birley, A.J. & Lavender, J. (1992) Histone H4 isoforms acetylated at specific lysine residues define individual chromosomes and chromatin domains in *Drosophila* polytene nuclei. *Cell* **69**, 375–384.

Turner, J.A.M., Mahadevaiah, S.K., Benavente, R., Offenberg, H.H., Heyting, C. & Burgoyne, P.S. (2000) Analysis of male meiotic 'sex body' proteins during XY female meiosis provides new insights into their functions. *Chromosoma* **109**, 426–432.

Turner, J.M., Burgoyne, P.S. & Singh, P.B. (2001) M31 and macroH2A1.2 colocalise at the pseudoautosomal region during mouse meiosis. *Journal of Cell Science* **114**, 3367–3375.

Tweedie, S., Charlton, J., Clark, V. & Bird, A. (1997) Methylation of genomes and genes at the invertebrate–vertebrate boundary. *Molecular and Cellular Biology* **17**, 1469–1475.

Tyler, J.K. & Kadonaga, J.T. (1999) The 'dark side' of chromatin remodelling: repressive effects on transcription. *Cell* **99**, 443–446.

Ueda, K., Kinoshita, Y., Xu, Z.-J. *et al.* (2000) Unusual core histones specifically expressed in male gametic cells of *Lilium longiflorum*. *Chromosoma* **108**, 491–500.

Uemura, T., Ohkura, H., Adachi, Y., Morino, K., Shiozaki, K. & Yanagida, M. (1987) DNA topoisomerase II is required for condensation and separation of mitotic chromosomes in *S. pombe. Cell* **50**, 917–925.

Uhlmann, F. (2001) Chromosome cohesion and segregation in mitosis and meiosis. *Current Opinion in Cell Biology* **13**, 754–761.

Umbetova, G.H., Belyaeva, E.S., Baricheva, E.M. & Zhimulev, I.F. (1991) Cytogenetic and molecular aspects of position effect variegation in *Drosophila melanogaster*. IV. Underreplication of chromosomal material as a result of gene inactivation. *Chromosoma* **101**, 55–61.

Uozu, S., Ikehashi, H., Ohmido, N., Ohtsubo, H., Ohtsubo, E. & Fukui, K. (1997) Repetitive sequences: cause for variation in genome size and chromosome morphology in the genus *Oryza. Plant Molecular Biology* **35**, 791–799.

Vaccari, T., Beltrame, M., Ferrari, S. & Bianchi, M.E. (1998) *Hmg4*, a new member of the *Hmg1/2* gene family. *Genomics* **49**, 247–252.

Vafa, O. & Sullivan, K.F. (1997) Chromatin containing CENP-A and α-satellite DNA is a major component of the inner kinetochore plate. *Current Biology* **7**, 897–900.

van Geel, M., Eichler, E.E., Beck, A.F. *et al.* (2002) A cascade of complex subtelomeric duplications during the evolution of the hominoid and old world monkeys. *American Journal of Human Genetics* **70**, 269–278.

van Holde, K. & Zlatanova, J. (1996) What determines the folding of the chromatin fiber? *Proceedings of the National Academy of Sciences of the USA* **93**, 10548–10555.

van Holde, K. & Zlatanova, J. (1999) The nucleosome core particle: does it have structural and physiologic relevance? *Bioessays* **21**, 776–780.

Varley, J.M. & Morgan, G.T. (1978) Silver staining of the lampbrush chromosomes of *Triturus cristatus carnifex. Chromosoma* **67**, 233–244.

Varley, J.M., Macgregor, H.C., Nardi, I., Andrews, C. & Erba, H.P. (1980) Cytological evidence of transcription of highly repeated DNA sequences during the lampbrush stage in *Triturus cristatus carnifex. Chromosoma* **80**, 289–307.

Vaucheret, H. & Fagard, M. (2001) Transcriptional gene silencing in plants: targets, inducers and regulators. *Trends in Genetics* **17**, 29–35.

Vaziri, H., Schächter, F., Uchida, I. *et al.* (1993) Loss of telomeric DNA during aging of normal and trisomy 21 human lymphocytes. *American Journal of Human Genetics* **52**, 661–667.

Verheijen, R., van Venrooij, W. & Ramaekers, F. (1988) The nuclear matrix: structure and composition. *Journal of Cell Science* **90**, 11–36.

Vernet, G., Sala-Rovira, M., Maeder, M., Jacques, F. & Herzog, M. (1990) Basic nuclear proteins of the histoneless eukaryote *Gyphecodinium cohnii* (Pyrrho-phyta): two dimensional electrophoresis and DNA binding properties. *Biochimica Biophysica Acta* **1048**, 281–289.

Verreault, A. (2000) De novo nucleosome assembly: new pieces in an old puzzle. *Genes and Development* **14**, 1430–1438.

Verschure, P.J., van der Kraan, I., Manders, E.M.M. & van Driel, R. (1999) Spatial relationship between transcription sites and chromosome territories. *Journal of Cell Biology* **147**, 13–24.

Vig, B.K. (1987) Sequence of centromere separation: a possible role for repetitive DNA. *Mutagenesis* **2**, 155–159.

Viglianti, G.A. & Blumenfeld, M. (1986) Satellite DNA-correlated nucleosomal proteins in *Drosophila virilis. Biochemical Genetics* **24**, 79–92.

Vinogradov, A.E. (1995) Nucleotypic effect in homeotherms: body-mass-corrected basal metabolic rate of mammals is related to genome size. *Evolution* **49**, 1249–1259.

Vinogradov, A.E. (1997) Nucleotypic effect in homeotherms: body-mass independent resting metabolic rate of passerine birds is related to genome size. *Evolution* **51**, 220–225.

Vinogradov, A.E., Borkin, L.J., Günther, R. & Rosanov, J.M. (1990) Genome elimination in diploid and triploid *Rana esculenta* males: cytological evidence from DNA flow cytometry. *Genome* **33**, 619–627.

Visintin, R. & Amon, A. (2000) The nucleolus: the magician's hat for cell cycle tricks. *Current Opinion in Cell Biology* **12**, 372–377.

Vlad, M. & Macgregor, H.C. (1975) Chromomere number and its genetic significance in lampbrush chromosomes. *Chromosoma* **50**, 327–347.

Vlassova, I.E., Graphodatsky, A.S., Belyaeva, E.S. & Zhimulev, I.F. (1991) Constitutive heterochromatin in early embryogenesis of *Drosophila melanogaster. Molecular and General Genetics* **229**, 316–318.

Voet, T., Vermeesch, J., Carens, A. *et al.* (2001) Efficient male and female germline transmission of a human chromosomal vector in mice. *Genome Research* **11**, 124–136.

Vos, J.-M.H. (1997) The simplicity of complex MACs. *Nature Biotechnology* **15**, 1257–1259.

Vos, J.-M.H. (1999) Therapeutic mammalian artificial episomal chromosomes. *Current Opinion in Molecular Therapeutics* **1**, 204–215.

Vourc'h, C., Taruscio, D., Boyle, A.L. & Ward, D.C. (1993) Cell cycle-dependent distribution of telomeres, centromeres, and chromosome-specific sub-satellite domains in the interphase nucleus of mouse lymphocytes. *Experimental Cell Research* **205**, 142–151.

Vyskot, B., Araya, A., Veuskens, J., Negrutiu, I. & Mouras, A. (1993) DNA methylation of sex chromosomes in a dioecious plant, *Melandrium album*. *Molecular and General Genetics* **239**, 219–224.

Wachtler, F. & Stahl, A. (1993) The nucleolus: a structural and functional interpretation. *Micron* **24**, 473–505.

Waga, S. & Stillman, B. (1998) The DNA replication fork in eukaryotic cells. *Annual Review of Biochemistry* **67**, 721-751.

Wakayama, T., Shinkai, Y., Tamashiro, K.L. *et al.* (2000) Cloning of mice to six generations. *Nature* **407**, 318–319.

Wakefield, M.J., Keohane, A.M., Turner, B.M. & Graves, J.A. (1997) Histone underacetylation is an ancient component of mammalian X chromosome inactivation. *Proceedings of the National Academy of Sciences of the USA* **94**, 9665–9668.

Wakimoto, B.T. (1998) Beyond the nucleosome: epigenetic aspects of position-effect variegation in *Drosophila*. *Cell* **93**, 321–324.

Wallace, B.M.N., Searle, J.B. & Everett, C.A. (1992) Male meiosis and gametogenesis in wild house mice (*Mus musculus domesticus*) from a chromosomal hybrid zone; a comparison between 'simple' Robertsonian heterozygotes and homozygotes. *Cytogenetics and Cell Genetics* **61**, 211–220.

Wallrath, L.L. (1998) Unfolding the mysteries of heterochromatin. *Current Opinion in Genetics and Development* **8**, 147–153.

Wallrath, L.L. & Elgin, S.C.R. (1995) Position effect variegation in *Drosophila* is associated with an altered chromatin structure. *Genes and Development* **9**, 1263–1277.

Wallrath, L.L., Guntur, V.P., Rosman, L.E. & Elgin, S.C.R. (1996) DNA representation of variegating heterochromatic P-element inserts in diploid and polytene tissues of *Drosophila melanogaster*. *Chromosoma* **104**, 519–527.

Wang, J.C. (1996) DNA topoisomerases. *Annual Review of Biochemistry* **65**, 635–692.

Warburton, D. (1997) Human female meiosis: new insights into an error-prone process. *American Journal of Human Genetics* **61**, 1–4.

Warburton, P.E. & Cooke, H.J. (1997) Hamster chromosomes containing amplified human α-satellite DNA show delayed sister chromatid separation in the absence of de novo kinetochore formation. *Chromosoma* **106**, 149–159.

Waters, J.C. & Salmon, E.D. (1997) Pathways of spindle assembly. *Current Opinion in Cell Biology* **9**, 37–43.

Watson, J.D. & Crick, F.H.C. (1953) A structure for deoxyribonucleic acid. *Nature* **171**, 737–738.

Watson, J.M., Meyne, J. & Graves, J.A.M. (1996) Ordered tandem arrangement of chromosomes in the sperm heads of monotreme mammals. *Proceedings of the National Academy of Sciences of the USA* **93**, 10200–10205.

Weber, J.D., Taylor, L.J., Roussel, M.F., Sherr, C.J. & Bar-Sagi, D. (1999) Nucleolar Arf sequesters Mdm2 and activates p53. *Nature Cell Biology* **1**, 20–26.

Wei, X., Samarabandu, J., Bevdhar, R.S., Siegel, A.J., Acharya, R. & Berezney, R. (1998) Segregation of transcription and replication sites into higher order domains. *Science* **281**, 1502–1505.

Weiler, K.S. & Wakimoto, B.T. (1995) Heterochromatin and gene expression in *Drosophila*. *Annual Review of Genetics* **29**, 577–605.

Weiner, A.M. (2000) Do all SINEs lead to LINEs? *Nature Genetics* **24**, 332–333.

White, M.J.D. (1973) *Animal Cytology and Evolution*, 3rd edn. Cambridge University Press, Cambridge.

White, P.S., Sulman, E.P., Porter, C.J. & Matise, T.C. (1999) A comprehensive view of human chromosome 1. *Genome Research* **9**, 978–988.

Whitehead, C.M., Winkfein, R.J., Fritzler, M.J. & Rattner, J.B. (1997) ASE-1: a novel protein of the fibrillar centres of the nucleolus and nucleolus organizer region of mitotic chromosomes. *Chromosoma* **106**, 493–502.

Widom, J. (1989) Toward a unified model of chromatin folding. *Annual Review of Biophysics and Biophysical Chemistry* **18**, 365–395.

Wiens, G.R. & Sorger, P.K. (1998) Centromeric chromatin and epigenetic effects in kinetochore assembly. *Cell* **93**, 313–316.

Wieslander, L. (1994) The Balbiani ring multigene family: coding repetitive sequences and evolution of a tissue-specific cell function. *Progress in Nucleic Acid Research and Molecular Biology* **48**, 275–313.

Wigge, P.A. & Kilmartin, J.V. (2001) The Ndc80p complex from *Saccharomyces cerevisiae* contains conserved centromere components and has a function in chromosome segregation. *Journal of Cell Biology* **152**, 349–360.

Willard, H.F. (1996) Chromosome manipulation: a systematic approach toward understanding human chromosome structure and function. *Proceedings of the National Academy of Sciences of the USA* **93**, 6847–6850.

Willard, H.F. (1998) Centromeres: the missing link in the development of human artificial chromosomes. *Current Opinion in Genetics and Development* **8**, 219–225.

Willard, H.F. & Hendrich, B.D. (1999) Breaking the silence in Rett syndrome. *Nature Genetics* **23**, 127–128.

Williams, B.C., Murphy, T.D., Goldberg, M.L. & Karpen, G.H. (1998) Neocentromere activity of structurally acentric minichromosomes. *Nature Genetics* **18**, 30–37.

Wilson, K.L. (2000) The nuclear envelope, muscular dystrophy and gene expression. *Trends in Cell Biology* **10**, 125–129.

Winkler, G.S. & Hoeijmakers, J.H.J. (1998) From a DNA helicase to brittle hair. *Nature Genetics* **20**, 106–107.

Wolffe, A.P. (1995) Histone deviants. *Current Biology* **5**, 452–454.

Wolffe, A.P. (1998) *Chromatin, Structure and Function*, 3rd edn. Academic Press, London.

Wolffe, A.P. (1999) Architectural regulations and Hmg1. *Nature Genetics* **22**, 215–217.

Wolffe, A.P. & Pruss, D. (1996) Deviant nucleosomes: the functional specialization of chromatin. *Trends in Genetics* **12**, 58–62.

Wolffe, A.P. & Hayes, J.J. (1999) Chromatin disruption and modification. *Nucleic Acids Research* **27**, 711–720.

Wolffe, A.P., Khochbin, S. & Dimitrov, S. (1997) What do linker histones do in chromatin? *Bioessays* **19**, 249–255.

Wolffe, A.P., Jones, P.L. & Wade, P.A. (1999) DNA demethylation. *Proceedings of the National Academy of Sciences of the USA* **96**, 5894–5896.

Wolstenholme, J. & Angell, R.R. (2000) Maternal age and trisomy – a unifying mechanism of formation. *Chromosoma* **109**, 435–438.

Wong, A.K.C. & Rattner, J.B. (1988) Sequence organization and cytological localization of the minor satellite of mouse. *Nucleic Acids Research* **16**, 11645–11661.

Wong, A.K.C., Biddle, F.G. & Rattner, J.B. (1990) The chromosomal distribution of the major and minor satellite is not conserved in the genus *Mus*. *Chromosoma* **99**, 190–195.

Wood, K.W., Sakowicz, R., Goldstein, L.S.B. & Cleveland, D.W. (1997) CENP-E is a plus end-directed kinetochore motor required for metaphase chromosome alignment. *Cell* **91**, 357–366.

Woodcock, C.L. & Dimitrov, S. (2001) Higher-order structure of chromatin and chromosomes. *Current Opinion in Genetics and Development* **11**, 130–135.

Yamamoto, M. & Miklos, G.L.G. (1978) Genetic studies on heterochromatin in *Drosophila melanogaster* and their implications for the functions of satellite DNA. *Chromosoma* **66**, 71–98.

Yanagida, M. (1990) Higher-order chromosome structure in yeast. *Journal of Cell Science* **96**, 1–3.

Yang, F., O'Brien, P.C.M., Wienberg, J. & Ferguson-Smith, M.A. (1997) A reappraisal of the tandem fusion theory of karyotype evolution in the Indian muntjac using chromosome painting. *Chromosome Research* **5**, 109–117.

Yasuda, Y. & Maul, G.G. (1990) A nucleolar auto-antigen is part of a major chromosomal surface component. *Chromosoma* **99**, 152–160.

Yoder, J.A., Walsh, C.P. & Bestor, T.H. (1997) Cytosine methylation and the ecology of intragenomic parasites. *Trends in Genetics* **13**, 335–340.

Yokota, H., van den Engh, G., Hearst, J.E., Sachs, R.K. & Trask, B.J. (1995) Evidence for the organization of chromatin in megabase pair-sized loops arranged along a random walk path in the human G0/G1 interphase nucleus. *Journal of Cell Biology* **130**, 1239–1249.

Yu, Y. & Bradley, A. (2001) Engineering chromosomal rearrangements in mice. *Nature Reviews Genetics* **2**, 780–790.

Yucel, J.K., Marszalek, J.D., McIntosh, J.R., Goldstein, L.S.B., Cleveland, D.W. & Philp, A.V. (2000) CENP-meta, an essential kinetochore kinesin required for the maintenance of metaphase chromosome alignment in *Drosophila*. *Journal of Cell Biology* **150**, 1–11.

Yunis, J.J. (1981) Mid-prophase human chromosomes. The attainment of 2000 bands. *Human Genetics* **56**, 293–298.

Yunis, J.J., Kuo, M.T. & Saunders, G.F. (1977) Localization of sequences specifying messenger RNA to light-staining G-bands of human chromosomes. *Chromosoma* **61**, 335–344.

Zacharias, H. (1979) Underreplication of a polytene chromosome arm in the Chironomid *Prodiamesa olivacea*. *Chromosoma* **72**, 23–51.

Zacharias, H. (1993) Larger nuclei in the larval brain of *Drosophila nasutoides* often show underreplication, whereas metaphases provide a reliable DNA standard. *Genome* **36**, 294–301.

Zacharias, H. (2001) Key word: chromosome. *Chromosome Research* **9**, 345–355.

Zacharopoulou, A., Bourtzis, K. & Kerremans, Ph. (1991) A comparison of polytene chromosomes in salivary glands and orbital bristle trichogen cells in *Ceratitis capitata*. *Genome* **34**, 215–219.

Zakian, V.A. (1995) *Saccharomyces* telomeres: function, structure, and replication. In: *Telomeres*, pp. 107–137. Cold Spring Harbor Laboratory Press, New York.

Zakian, V.A. (1996) Structure, function, and replication of *Saccharomyces cerevisiae* telomeres. *Annual Review of Genetics* **30**, 141–172.

Zaragoza, M.V., Surti, U., Redline, R.W. Millie, E., Chakravarti, A. & Hassold, T.J. (2000) Parental origin and phenotype of triploidy in spontaneous abortions: predominance of diandry and association with the partial hydatidiform mole. *American Journal of Human Genetics* **66**, 1807–1820.

Zentgraf, H. & Franke, W.W. (1984) Differences of supranucleosomal organization in different kinds of chromatin: cell type-specific globular subunits containing different numbers of nucleosomes. *Journal of Cell Biology* **99**, 272–286.

Zetka, M.C. & Müller, F. (1996) Telomeres in nematodes. *Seminars in Cell and Developmental Biology* **7**, 59–64.

Zhang, P. (1999) The cell cycle and development: redundant roles of cell cycle regulators. *Current Opinion in Cell Biology* **11**, 655–662.

Zhang, X., Mar, V., Zhou, W., Harrington, L. & Robinson, M.O. (1999) Telomere shortening and apoptosis in telomerase-inhibited human tumor cells. *Genes and Development* **13**, 2388–2399.

Zhang, Y. & Reinberg, D. (2001) Transcription regulation by histone methylation: interplay between different covalent modifications of the core histone tails. *Genes and Development* **15**, 2343–2360.

Zhao, J., Hao, S. & Xing, M. (1991) The fine structure of the mitotic chromosome core (scaffold) of *Trilophidia annulata*. *Chromosoma* **100**, 323–329.

Zheng, B., Sage, M., Sheppeard, E.A., Jurecic, V. & Bradley, A. (2000) Engineering mouse chromosomes with Cre-loxP: range, efficiency, and somatic applications. *Molecular Cell Biology* **20**, 648–655.

Zhimulev, I.F. (1996) Morphology and structure of polytene chromosomes. *Advances in Genetics* **34**, 1–497.

Zhimulev, I.F. (1998) Polytene chromosomes, heterochromatin, and position effect variegation. *Advances in Genetics* **37**, 1–566.

Zhimulev, I.F. (1999) Genetic organization of polytene chromosomes. *Advances in Genetics* **39**, 1–589.

Zickler, D. & Kleckner, N. (1998) The leptotene–zygotene transition of meiosis. *Annual Review of Genetics* **32**, 619–697.

Zickler, D. & Kleckner, N. (1999) Meiotic chromosomes: integrating structure and function. *Annual Review of Genetics* **33**, 603–754.

Zimm, B.H. (1999) One chromosome: one DNA molecule. *Trends in Biochemical Sciences* **24**, 121–123.

Zink, D. & Cremer, T. (1998) Cell nucleus: chromosome dynamics in nuclei of living cells. *Current Biology* **8**, R321–R324.

Zink, D., Cremer, T., Saffrich, R. *et al.* (1998) Structure and dynamics of human interphase chromosome territories in vivo. *Human Genetics* **102**, 241–251.

Zinn, A.R., Page, D.C. & Fisher, E.M.C. (1993) Turner syndrome: the case of the missing sex chromosome. *Trends in Genetics* **9**, 90–93.

Zlatanova, J. & van Holde, K. (1998) Linker histones versus HMG1/2: a struggle for dominance? *Bioessays* **20**, 584–588.

Zou, H., McGarry, T.J., Bernal, T. & Kirschner, M.W. (1999) Identification of a vertebrate sister-chromatid separation inhibitor involved in transformation and tumorigenesis. *Science* **285**, 418–422.

Zumstein, L.A. & Lundblad, V. (1999) Telomeres: has cancer's Achilles' heel been exposed? *Nature Medicine* **5**, 1129–1130.

Zybina, E.V. & Zybina, T.G. (1996) Polytene chromosomes in mammalian cells. *International Review of Cytology* **165**, 53–119.

Index